FOCUS ON

Grades 5-8

MIDDLE SCHOOL

GEOLOGY

Teacher's Manual

Rebecca W. Keller, PhD

REAL SCIENCE 4 Kids

Illustrations: Rebecca W. Keller, PhD

Focus On Middle School Geology Teacher's Manual
ISBN 978-1-936114-87-0

Published by Gravitas Publications, Inc.
www.gravitaspublications.com

A Note From the Author

This curriculum is designed to give students both solid science information and hands-on experimentation. The middle school material is geared toward fifth through eighth grades, and much of the information in the text is very different from what is taught at this grade level in other textbooks. This is a *real* science textbook, so scientific terms are used throughout. It is not important at this time for students to master the terminology, but it *is* important that they be exposed to the real terms used to describe science.

For students, each chapter has two parts: a reading part in the *Focus On Middle School Geology Student Textbook* and an experimental part in the *Focus On Middle School Geology Laboratory Workbook*. In this teacher's manual, an estimate is given for the time needed to complete each chapter. It is not important that both the reading portion and the experimental portion be concluded in a single sitting. It may be better to have students do these on two separate days, depending on the interest level of the child and the energy level of the teacher. Also, questions not addressed in the *Teacher's Manual* may arise, and extra time may be required to investigate these questions before proceeding with the experimental section.

Each experiment is a *real* science experiment and not just a demonstration. They are designed to engage students in actual scientific investigation. The experiments are simple but are written the way real scientists actually perform experiments in the laboratory. With this foundation, it is my hope that students will eventually begin to think of their own experiments and test their own ideas scientifically.

Enjoy!

Rebecca W. Keller, PhD

How To Use This Manual

Each chapter in this *Focus On Middle School Geology Teacher's Manual* begins by providing additional information for the corresponding chapter in the *Focus On Middle School Geology Student Textbook*. This supplementary material is helpful when questions arise while students are reading the textbook. It is not necessary for students to learn this additional material since most of it is beyond the scope of this level. However, the teacher may find the information helpful when answering questions.

The second part of each chapter in the *Teacher's Manual* provides directions for the experiments in the *Laboratory Workbook* as well as answers to the questions asked in each experiment and review section. All of the experiments have been tested, but it is not unusual for an experiment to produce an unexpected outcome. Usually repeating an experiment helps both students and teacher see what might have occurred during the experimental process. Encourage the students to troubleshoot and investigate all possible outcomes. However, even repeating an experiment may not produce the expected outcome. **Do not worry if an experiment produces a different result.** Scientists don't always get the expected results when doing an experiment. The important thing is for students to learn about the scientific method and to make observations, think about what is taking place, and ask questions.

Getting Started

The experimentation process will be easiest if all the materials needed for the experiment are gathered together and made ready before beginning. It can be helpful to have a small shelf or cupboard or even a plastic bin dedicated to holding most of the necessary chemicals and equipment. The following Materials at a Glance chart lists all of the materials needed for each experiment. An additional chart lists the materials by type and quantity. A materials list is also provided at the beginning of each lesson.

Laboratory Safety

Most of these experiments use household items. Extra care should be taken while working with all materials in this series of experiments. The following are some general laboratory precautions that should be applied to the home laboratory:

▶ Never put things in your mouth without explicit instructions to do so. This means that food items should not be eaten unless tasting or eating is part of the experiment.

▶ Wear safety glasses while using glass objects or strong chemicals such as bleach.

▶ Wash hands before and after handling all chemicals.

▶ Use adult supervision while working with electricity, glassware, and knives, and while performing any step requiring a stove.

Materials at a Glance

Experiment 1	Experiment 2	Experiment 3	Experiment 4	Experiment 5

pencil
pen
colored pencils
small jar
trowel
binoculars
   (optional)
	pencil
	pen
	colored pencils
	compass
	small jar or
	   container with lid
	small items for
	   treasure
	trowel
		mineral samples: [2]
		   calcite
		   quartz
		   feldspar
		   hematite
		several rocks
		   collected by
		   students
		penny
		steel nail
		unglazed white
		   ceramic tile
		   (streak plate) [1]
		paper
		scissors
		marking pen
		tape
		vinegar or lemon
		   juice (optional)
		dropper bottle
		   (optional) [1]
			various materials
			   chosen by
			   students to make
			   model of Earth
				ingredients to make
				   brittle candy
				   (see recipe in
				   Chapter 5)
				1 jar smooth
				   peanut butter
				   (or substitute
				   whipped cream)
				118 ml (1/2 cup)
				   crushed graham
				   crackers
				—OR—
				materials chosen by
				   students to make a
				   model volcano

Experiment 6	Experiment 7	Experiment 8	Experiment 9	Experiment 10

2 liter (2 quart)
   plastic bottle with
   cap, squeezable
warm water
matches
	about 475 ml (2 cups)
	   each:
	   gravel [4]
	   sand [4]
	   dirt
	   powdered pottery
	     clay [3]
	water
	6-8 Styrofoam cups,
	   about 355 ml
	   (12 ounce) size
	pencil
	marking pen
	measuring cup,
	   graduated
	large bowl
	scissors
	stop watch or clock
	   with a second
	   hand (optional)
	tape (optional)
		pencil
		pen
		colored pencils
		compass
		small backpack
		water bottle
		notebook or extra
		   paper (optional)
			steel needle
			bar magnet
			slice of cork (such as
			   from a wine bottle)
			tape
			medium size bowl
			water
			1 additional bar
			   magnet (optional)
				pencil
				pen
				colored pencils
				imagination

Materials at a Glance
By Type

Equipment	Materials	Materials (cont.)

backpack, small
binoculars (optional)
bottle, dropper (optional) [1]
bottle, plastic with cap, 2 liter (2 quart), squeezable
bottle, water
bowl, large
bowl, medium
compass
jar, small
jar or container, small—with lid
magnet, bar, 1-2
measuring cup, graduated
nail, steel
needle, steel
penny
scissors
stop watch or clock with a second hand (optional)
tile, unglazed white ceramic (streak plate) [1]
trowel
	Focus On Middle School Astronomy Laboratory Workbook
	clay, powdered pottery, 475 ml (2 cups) [3]
	cork (such as wine bottle cork), one slice
	cups, 6-8 Styrofoam, about 355 ml (12 ounce) size
	dirt, 475 ml (2 cups)
	gravel, 475 ml (2 cups) [4]
	items, various small—for treasure
	matches
	mineral samples: [2]
	  calcite
	  feldspar
	  quartz
	  hematite
	notebook (optional)
	paper
	pen
	pen, marking
	pencil
	pencils, colored
	rocks collected by students, several
	sand, 475 ml (2 cups) [4]
	tape
	vinegar or lemon juice, small quantity (optional)
	water
	water, warm
		Experiment 4
		various materials chosen by students to make model of Earth
		
		Experiment 5
		ingredients to make brittle candy (see recipe in Chapter 5)
		1 jar smooth peanut butter (whipped cream can be substituted in the case of allergy to peanuts)
		118 ml (1/2 cup) crushed graham crackers
		
		—OR—
		
		materials chosen by students to make a model volcano

		Locations
		backyard and/or park library and/or internet neighborhood

[1] www.hometrainingtools.com — RM-TESTKIT, Mineral Test Kit (contains streak plate, dropper bottle, glass plate, & more)

[2] Obtain mineral samples from a local rock and mineral shop or www.hometrainingtools.com — RM-HARDNES, set of 9 minerals to go with Mohs scale of mineral hardness (hematite not included)

[3] Powdered clay — from art or hobby store. Can substitute wet clay if powdered clay can't be found.

[4] Gravel and sand — from your yard or a store that sells aquarium supplies

Note: Suppliers may discontinue items from time to time. For Materials List updates, go to www.gravitaspublications.com, click on the *Services* tab, and then the *Resources* link.

Contents

CHAPTER 1:   What Is Geology?   1

Experiment 1: Observing Your World   6
Review   12

CHAPTER 2:   Geologists' Toolbox   13

Experiment 2: Hidden Treasure   16
Review   21

CHAPTER 3:   Rocks, Minerals, and Soils   22

Experiment 3: Mineral Properties   28
Review   34

CHAPTER 4:   Earth's Layers   35

Experiment 4: Model Earth   40
Review   45

CHAPTER 5:   Earth's Dynamics   46

Experiment 5: Dynamic Earth   51
5A: Exploring Plate Tectonics   51
5B: Create a Model Volcano (alternate)   56
Review   59

CHAPTER 6:   The Atmosphere   60

Experiment 6: Exploring Cloud Formation   67
Review   70

CHAPTER 7:   The Hydrosphere   71

Experiment 7: Modeling an Aquifer   76
Review   81

CHAPTER 8:   The Biosphere   82

Experiment 8: My Biome   86
Review   90

CHAPTER 9:   The Magnetosphere   91

Experiment 9: Finding North   96
Review   99

CHAPTER 10: Earth as a System   100

Experiment 10: Solve One Problem   105
Review   112

Chapter 1: What Is Geology?

Overall Objectives   2

1.1   Introduction   2

1.2   What Is Geology?   2

1.3   Interpreting Geological Data   3

1.4   Why Study Earth?   3

1.5   What Do Geologists Study?   4

1.6   Geology and the Scientific Method   4

1.7   Summary   5

Experiment 1: Observing Your World   6

Review   12

Time Required

Text reading	30 minutes
Experimental	1 hour

Materials

pencil, pen, and colored pencils
small jar
trowel or spoon
binoculars (optional)

Overall Objectives

This chapter introduces the field of science called geology, which is the branch of science concerned with the study of Earth. In this chapter, students will learn the difference between historical and physical geology, the importance of interpreting geological data, what geologists do when they study Earth, and the different branches of physical geology.

1.1 Introduction

Most people take living on Earth for granted. Earth is our planetary home, but we don't often refer to Earth as the place where we live. Discuss with your students why people may not think about Earth as "home." For example:

> - *Telling someone where you live is a way to identify separate peoples, culture, and geography. Because everyone lives on Earth, calling Earth "home" is not a means of identification. It applies to everyone.*
>
> - *We don't easily observe other planets and have no daily experience of living on Earth compared to living on another planet—so we live in separate cities, states, provinces or countries, but not on separate planets.*
>
> - *The Earth is so large that we don't think about living on Earth as a planet in space.*

1.2 What Is Geology?

There are two main divisions of geology—physical geology and historical geology. Although this book will focus on physical geology, historical geology plays an important role in how we understand Earth and Earth's history.

Explore with the students the difference between historical and physical geology. For example:

> - *When a geologist takes a soil sample and does an experiment to determine the composition of the soil, is this physical or historical geology?* (physical)
>
> - *When a geologist evaluates a soil sample to determine the age of the rocks, minerals, and organic matter, is this physical or historical geology?* (both)
>
> - *When a geologist creates a narrative (story) about the origin of a soil sample, is this physical or historical geology?* (historical)

1.3 Interpreting Geological Data

Interpreting data is a basic task for scientists. Scientists, including geologists, collect data to understand the world around them. However, a collection of facts is not very useful in most areas of science. It's important for scientists to interpret those facts and try to understand what they mean.

Interpreting scientific facts is a human endeavor and, as such, is subject to human bias. Each human has a set of ideas, beliefs, values, and experiences that create a framework or lens through which to interpret everything. This framework or lens is called a worldview.

Because no two humans have identical ideas, beliefs, values, and experiences, no two humans have identical worldviews, and scientists are no exception. Our scientific understanding of the world around us is enriched by the differences in worldview between individual scientists and groups of scientists, but these differences in worldview also cause disagreements.

One major disagreement in the area of geology is the disagreement over the age of the Earth. This text will not address the age of the Earth and will not discuss different narratives about the historical age of the Earth. However, it is important for students to be aware that disagreements exist, and that scientists from different worldviews can and do interpret scientific data in different ways.

1.4 Why Study Earth?

Studying Earth is an important task for humans. In order to better live on and take care of our home planet, we should do all we can to understand it.

Have the students discuss how studying Earth can help us live better lives and take better care of Earth. For example:

- *How might knowing about rainfall help us grow food or establish cities?*
- *How might knowing about weather patterns help us prepare for hurricanes or snow storms?*
- *How might knowing about earthquakes help us build better buildings?*
- *How might studying the Earth's atmosphere help us protect delicate forests or the ocean?*

1.5 What Do Geologists Study?

This section discusses several different branches of geology. Geology is not just one discipline, but a group of subdisciplines that focus on different aspects of Earth.

These subdisciplines include:

- Geochemistry—the study of the chemical makeup of Earth's rocks, minerals, and soils.
- Structural geology—the exploration of how the Earth is arranged and how mountains, valleys, glaciers, etc. are formed.
- Resource geology—the research of natural resources such as gas, oil, and coal.
- Environmental geology—the branch of geology that explores how humans impact the Earth's environment.

Discuss with your students how these various branches of geology work individually and as a collective to help us understand Earth.

1.6 Geology and the Scientific Method

Geology is a science and utilizes the scientific method in the study of Earth. The scientific method is a way of organizing research in a series of steps that help scientists draw valid conclusions about their observations.

The scientific method has 5 major steps:

- *making observations*
- *formulating hypotheses*
- *performing experiments*
- *collecting results*
- *drawing conclusions*

One difficulty with geological research is the fact that many of Earth's features change slowly over long periods of time. For this reason, designing experiments can be challenging. Many things can happen to disrupt an established research program. For example, sudden changes like earthquakes can cause drastic changes very quickly and without warning, disrupting some long term experiments.

For this reason, many geological conclusions are a mixture of physical and historical geology. Geologists observe physical features and create historical narratives describing those features. This combination of physical observations and historical narratives forms what can be thought of as a scientific map.

A scientific map can be described as the scientists' best description of a particular area of study and is based on observations, data collected, and how well the data fit into a particular historical narrative. Scientific maps do change over time, and it's important that students understand that geology is a dynamic discipline where scientists disagree over what data mean and how to incorporate both physical and historical data into a valid description of Earth and Earth's processes.

1.7 Summary

Review the summary points of this chapter with the students.

- *Geology is the scientific discipline concerned with what Earth is made of, how it is put together, and how it changes over time.*

- *There are two broad categories of geology—physical geology, and historical geology.*

- *Physical geology is concerned with what Earth is made of, how it is put together, and its dynamic features. Historical geology is concerned with the history of Earth, how it came into being and how it has changed over time.*

- *There are several different branches of geology including geochemistry, structural geology, resource geology, and environmental geology.*

- *The scientific method is used by geologists to study Earth. The five steps of the scientific method include: making observations, formulating hypotheses, performing experiments, collecting results, and drawing conclusions.*

In this experiment students will explore their local world—the world around them.

Have the students read through the experiment and look through the experimental steps.

Discuss possible objectives for this experiment. Some suggestions are:

- *To learn more about the place where I live.*

- *To better understand my hometown, city, etc.*

- *To observe my surroundings and find out more about Earth though my observations.*

Discuss possible hypotheses for this experiment. Have the students discuss what they think they might learn as a result of observing where they live. Some suggested hypotheses are:

- *By making observations I will better understand where I live.*

- *By making observations I can understand how where I live changes over time,*

- *I will be able to assemble a list of features that describe where I live.*

In the boxes provided in the *Experiment* section, students can write about or draw what they see.

Experiment 1: Observing Your World   Date: _____

Objective
_____
_____

Hypothesis
_____
_____

Materials
▸ pencil, pen, and colored pencils
▸ small jar
▸ trowel
▸ binoculars (optional)

Experiment

❶ Step outside your front or back door and walk until your feet are on some type of ground (dirt, grass, or concrete).

❷ Observe where you are. Are you in a city? Are you in the country? Use the space below to draw or write what you see.

❸ Observe any geological features near you. Do you see mountains? Do you see lakes or rivers? Do you see the ocean? Do you see other geological features? Record what you are observing.

Have the students observe geological features that are nearby. If binoculars are available, have the students use the binoculars to see the details of geological features they observe.

In the box provided, they can write or draw what they see.

❹ Collect a small sample of dirt. (If you live in the city, walk to a park or some other place where you can collect a dirt sample.)

❺ Observe the dirt sample. Is it light in color? Dark? Does it contain small rocks? Large rocks? Does it have any organic matter (living things, such as grass or bugs)? Record what you observe.

Using the small jar and a trowel, have the students collect a sample of dirt. Encourage the students to examine the dirt carefully, noting color, texture, and the presence of organic matter.

Have the students observe any man-made structures nearby. These can include things such as homes, roads, buildings, parks, fences, and utility poles.

❻ Observe any man-made structures. Do you see buildings? Roads? Bridges? Other man-made structures? Record what you see.

Have the students explore any dynamic features near their home that may have changed how the area looks. For example, heavy rains or snowmelt runoff might have caused erosion.

❼ Observe any dynamic features in your area including the weather. Do you get earthquakes? Do you live near a volcano? Does it rain frequently, or do you get very little rain where you live? Do you have tornadoes, hurricanes, or severe snow storms?

❽ Think about the area in which you live. What is its history? How long has it looked the way it looks today? If you are in a city, how long has the city been there? What do you think it looked like before there were buildings, roads, or other structures? Write your observations below.

_____

_____

_____

_____

_____

_____

_____

_____

_____

_____

_____

_____

_____

_____

_____

_____

Have the students explore the history of their home. Have them research when their house was built, when their city was founded, or when the land surrounding them became farms or industrial parks.

Have the students use the chart to assemble all of the information they've collected.

Results

Assemble your data in the chart below.

Data Describing Where I Live

Geological Features	
Soil Type	
Man-made structures	
Dynamic Processes	
Weather	
History	

Conclusions

Imagine you met someone from the planet Alpha Centauri Bb. They have never been to Earth, and they ask you what Earth is like. Using the data you have collected, write a narrative (story) describing where you are from.

_____

_____

_____

_____

_____

_____

_____

_____

_____

_____

_____

_____

_____

_____

_____

_____

_____

Have the students use the data they have collected to write a narrative describing the area where they live. Have them think about how they would describe the area so that someone who has not seen it could imagine it.

Review

Define the following terms:

(Answers may vary.)

geology   *the study of Earth*

physical geology   *the study of the chemistry and physics of Earth*

historical geology   *the study of Earth in order to create a story about Earth's origins*

geochemistry   *the study of the chemistry of Earth*

structural geology   *the study of Earth's internal structure, form, and arrangement of rocks*

environmental geology   *the study of environmental changes caused by human activity*

resource geology   *the study of Earth's natural resources such as gas, oil, and coal*

worldview   *the way someone sees the world*

Chapter 2: Geologists' Toolbox

Overall Objectives 14

2.1 Introduction 14

2.2 Hand Tools 14

2.3 Electronic Tools 15

2.4 Other Tools 15

2.5 Summary 15

Experiment 2: Hidden Treasure 16

Review 21

Time Required

Text reading 30 minutes
Experimental 1 hour or more

Materials

pencil, pen, and colored pencils
compass
a small jar or container with a lid
small items to place in jar (treasure)
trowel

Overall Objectives

In this chapter students will explore some of the different tools geologists use to study Earth.

2.1 Introduction

As with any science, geology has been transformed as a result of technological advances. Modern tools allow geologists to study aspects of Earth's features and dynamics that were unavailable to early geologists.

Have the students look briefly at the entire chapter, noting various geological tools. Discuss with your students how these tools have changed our understanding of Earth. Ask open inquiry questions such as:

- *What advantages do modern geological tools have over older tools?*
- *What disadvantages might modern tools have?*
  (For example, the need for electricity or gas power, fragility, need for software updates, expense, etc.)
- *How do you think modern tools have shaped what we know about Earth?*

2.2 Hand Tools

Hand tools are easy to carry and easy to use, even if they are not necessarily technologically advanced. The rock hammer, map, and compass are tools that are easy to put in a backpack, are generally inexpensive, and are easy to use. Discuss with your students how hand tools help geologists study Earth. For example:

- *How might a rock hammer or crack hammer help a geologist study rocks, minerals, or fossils?*
- *How does a compass help geologists locate geological features?*
- *What types of maps are useful for hiking in mountain terrain? Why?*
- *What types of maps are useful for exploring different types of climates? Why?*
- *What advantage might a paper map have over an electronic map?*

2.3 Electronic Tools

Modern geologists use a variety of electronic tools to study Earth. Although electronic tools are dependent on a variety of other technologies working properly (e.g., batteries, satellites, motors), they can offer substantial advantages over older or non-technological tools. For example, a GPS can sometimes be more reliable than an out-of-date map, and a GPR can image below the surface of the Earth without disrupting the ground.

Have a discussion with the students about some potential advantages that electronic tools offer geologists. For example:

- *How might a GPS work better than a paper map for locating a geological feature?*
- *How would a GPR help locate groundwater without having to dig a hole?*
- *How might a seismograph help monitor earthquakes and aid evacuation plans?*

2.4 Other Tools

Other tools used by geologists include drills and rock and mineral kits. Discuss with your students how these tools help geologists study Earth. For example:

- *When might a geologist need to use a drill?*
- *How can a drill be used to explore the surface of the Earth?*
- *How can a rock or mineral kit be useful?*

2.5 Summary

Review the summary points of this chapter with the students.

- *Geologists use a variety of tools to study Earth.*
- *Rock hammers and crack hammers are useful for studying rocks and minerals.*
- *Different types of maps help geologists study different regions.*
- *Electronic tools help geologists explore aspects of Earth's features not available to them with conventional tools.*

In this experiment students will explore mapmaking by creating a map for finding a hidden treasure. Students will then test the accuracy of their map by having a friend try to find the treasure they bury.

Have the students read through the experiment and look through the experimental steps.

Discuss possible objectives for this experiment. Some suggestions are:

- *To explore mapmaking by creating a real map.*

- *To understand how maps are made and the difficulties that come up.*

- *To create and test a homemade map.*

Discuss possible hypotheses for this experiment. Some suggested hypotheses are:

- *I can create a map that is 80-100% accurate and I will know this by how quickly my friend finds the treasure.*

- *By creating a map and having it tested, I will learn about the difficulties of mapmaking.*

- *I can test the accuracy of my map by having a friend find my buried treasure.*

Have the students draw their map in the space provided. They will be using their footsteps as a tool to measure their map. Have them count their steps while walking heel-to-toe around a given space. This space can be square or rectangular or oddly shaped. Have them note on their map any obstacles they may encounter as they measure the space.

Experiment 2: Hidden Treasure   Date: _____

Objective
_____

_____

Hypothesis
_____

_____

Materials

- pencil, pen, and colored pencils
- compass
- a small jar or container with a lid
- small items to place in jar (treasure)
- trowel

Experiment

❶ Find some small objects to be your treasure and put them in the jar.

❷ Select an area near your home to make a map of. This area can be your front or back yard, a park, or other open space.

❸ Using each of your feet as a one-foot ruler, measure the outline of the area, walking heel-to-toe around it. Count your steps and notice if your path goes in a straight line or curves around objects.

❹ Make your map as accurate as possible, noting now many steps (feet) there are on each side of the area.

❺ Hold the compass and turn around until the needle lines up with the north (N) symbol. Note this direction (north) on your map, drawing an arrow and an "N." Now, note south (S), east (E), and west (W), again drawing arrows and using the letters.

❻ Draw the outline of your map in the following box.

Have the students use a compass to determine north, south, east, and west on their map.

Using a Compass

The needle on a compass will always point north (N). To read a compass, line the needle up with the north (N) label. If you wanted to travel north, you would head in this direction. To go south, you would travel in the opposite direction.

Once you have determined which way is north, you can find any other direction in which you want to go. For instance, if you want to go northeast, you would line the compass needle up with the north label and then travel in the direction of the northeast (NE) label on the compass.

For This Experiment

It may be helpful to the students to have them place their map on the ground in the same orientation as the landscape features. Have them hold the compass and stand next to the map. Direct them to turn until the compass needle points to north (N) and then draw an arrow and an N on their map to indicate which way is north. North will probably not be at the top of their map.

Explain that south is opposite north and have them mark south on their map with an arrow and an S. Next explain that when facing north, east will be to the right and west to the left. Have them mark east and west on their map.

You can also have the students use the compass to determine in which direction each side of their map is facing.

Have the students mark any details they observe and the distances between them.

❼ Add details to your map. Measure how far trees, shrubs, buildings, or other features are by walking heel-to-toe and counting your steps. For example, if you need to find out where a tree is located on your map, pick a starting point (say the corner of your map) and walk toward the object measuring the distance.

Example of some details

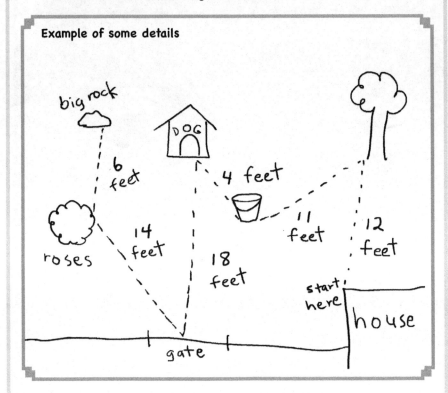

Have the students decide on a location where they will bury their treasure and mark this on their map. Or, if they are in a place where it is not possible to bury the treasure, they can find a place to hide it. They may want to make adjustments to their map to accommodate the treasure location.

Have the students give their map to a friend and have the friend see if they can find the buried treasure.

❽ Now pick a location in the area you have mapped and bury your treasure. Record the location of your treasure on your map.

❾ Give your map to a friend and see if they can find your buried treasure.

Results

❶ Observe how easy or difficult it is for your friend to find the buried treasure.

❷ Record the number of attempts it takes them to find your treasure. Also record how much help you have to give them so they can find the hidden treasure. Note any adjustments to your map.

❸ Record your data below:

Treasure Map Results

Number of Attempts	
Type of Help	
Adjustments to Map	

Have the students record the number of attempts their friend makes in order to locate the buried treasure.

Also have the students note any help they have to give to the friend so they can find the treasure. This help can be used to adjust the map.

For example:

- The treasure is 5 ft north and not 8 ft north.

- The treasure is northwest of the big tree and not north.

Have the students evaluate their map.

Ask questions to help with their evaluation. For example:

- *Was the map accurate enough so that the friend could find the buried treasure? Why or why not?*

- *What (if any) modifications to the map might you make?*

- *How might you make your map more accurate?*

- *Does the size of your feet compared with the size of your friend's feet make a difference?*

Conclusions

Based on your observations, evaluate the accuracy of your map.

_____

_____

_____

_____

_____

_____

_____

_____

_____

_____

_____

_____

_____

_____

_____

_____

Review

Answer the following questions:

(Answers may vary.)

▶ List three uses of a rock hammer or crack hammer.

   breaking rocks   *prying rocks apart*

   showing the size of a sample in a photo

▶ List three different types of maps and their uses.

   city map—shows where roads and landmarks are

   topographic map—shows if an area is flat or steep

   climate map—shows zones of temperature or precipitation

▶ What is a GPS?   *Global positioning system. Uses information from*

   satellites to show where you are and what time it is.

▶ What is a GPR?   *Ground penetrating radar. Uses radio waves that go*

   into the Earth to find things under the ground.

▶ What is the difference between a seismometer and a seismograph?

   A seismometer measures movements of the Earth's surface, and a

   seismograph draws them.

▶ What would be a good application of a core drill? _____

   To study the layers of Earth to see what minerals are there

▶ What is a rock and mineral test kit? _____

   It is used to find out what mineral is in a sample by testing for hardness,

   color, and how it reacts to acid.

Chapter 3: Rocks, Minerals, and Soils

Overall Objectives	23
3.1 Introduction	23
3.2 Minerals	23
3.3 Rocks	25
3.4 Testing Rocks and Minerals	25
3.5 Soils	26
3.6 Summary	27
Experiment 3: Mineral Properties	28
Review	34

Time Required

Text reading	30 minutes
Experimental	1 hour

Materials

known mineral samples:
    calcite   feldspar
    quartz   hematite
several rocks from your backyard or near your home
penny
nail
unglazed white ceramic tile (streak plate)
paper
scissors
marking pen
tape
vinegar or lemon juice (optional)
dropper bottle (optional)

Overall Objectives

In this chapter students will begin to explore the Earth by learning about the types of materials that compose it—minerals, rocks, and soils.

3.1 Introduction

In this section students are introduced to the concept of Earth having a crust, or outer layer, which is made of minerals, rocks, and soils. There is more discussion about the crust in Chapter 4.

Have the students think about the ground and some of the ways in which we use the ground—for example, when building houses and roads, growing plants, raising animals, or driving cars. Have the students think about different types of ground they know about. Use the following questions to guide the discussion. There are no right or wrong answers.

- *When you go outside, what type of ground do you find?*

- *Is the ground hard or soft? Rocky or sandy?*

- *When you think about what types of ground people use to put roads or buildings on, does the type of ground matter?*

- *What type of ground do you think is at the top of Mount Everest?*

- *What type of ground to you think is below the ocean?*

- *Can plants grow in any type of ground? Why or why not?*

3.2 Minerals

In this section students are introduced to minerals, the building blocks that make up the Earth's crust. The study of minerals is called mineralogy. Geologists define minerals as materials that: 1) are naturally formed, 2) are solid, 3) are inorganic (not containing carbon), and 4) have an organized internal structure. Only materials that meet these criteria are considered minerals. Synthetic diamonds and other useful materials produced by man-made chemical means are not considered to be minerals. Substances that have most of the characteristics of a mineral but don't have an organized internal structure are also not minerals but are called mineraloids. Examples of mineraloids include opals and obsidian (volcanic glass).

The majority of rock-forming minerals are silicates. A silicate is an oxygen-silicon compound with four oxygens bonded to one silicon in a tetrahedral shape. Silicon has a smaller atomic radius than oxygen which means that the silicon atom is smaller in size. It sits in the middle of the tetrahedron.

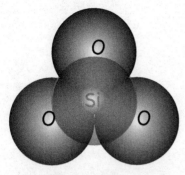

Space-filled model showing the tetrahedral shape of silicate. The four oxygens surround the silicon center.

Silicate is the building block of many minerals, and it joins to other silicate molecules, creating chains. These chains can be single, double, or sheets.

- Single chain silicate minerals include augite, jadeite, and rhodonite.

- Double chain silicate minerals include hornblende and riebeckite (asbestos).

- Sheets of silicates make up the mica group of minerals.

Minerals that do not have silicate as the main building block are called non-silicate minerals. Some examples of non-silicate minerals are dolomite, calcite, and halite.

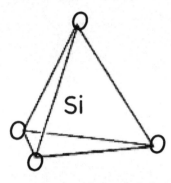

Line drawing of a tetrahedral shape. The oxygens occupy the corners of the tetrahedron and silicon sits in the middle.

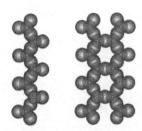

Single chain   Double chain

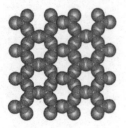

Sheet

Optional: Open Inquiry

These questions are meant for open discussion and further exploration, so answers are not provided. Allow students to use the local library or internet to research the ideas they find interesting.

Have the students compare silicon to carbon. Silicon is the building block of rocks and minerals—inorganic matter. Carbon is the building block of plants and animals—organic matter. Yet silicon sits just below carbon on the periodic table and has similar chemical properties. Both carbon and silicon have four valence electrons and typically form four bonds. Many carbon compounds, such as CO_2 and CH_4 have silicon derivatives. But carbon is the building block of life and silicon is not. Guide student discussion with the following questions.

- *Why do you think carbon is the building block of living things and silicon is the building block of non-living rocks and minerals?*

- *Do you think silicon-based life could exist? Why or why not?*

- *Do you think the "rules of chemistry" apply for the entire universe? Why or why not?*

3.3 Rocks

The three main types of rock are:

- igneous
- sedimentary
- metamorphic

Engage the students in a discussion about the three basic types of rock and how they are classified. Guide discussion with the following questions:

- *The earth's crust is made mostly of which type of rock?*
  igneous rock
- *How do igneous rocks form?*
  from molten magma that cools
- *What is the main chemical compound in magma?*
  silicate
- *What elements is the chemical compound silicate made of?*
  silicon and oxygen
- *What are sedimentary rocks?*
  rocks made from particles of igneous and other rocks and debris
- *How are sedimentary rocks formed?*
  by weathering (detrital sedimentary rocks) and chemical processes (chemical sedimentary rocks)
- *What is a metamorphic rock?*
  a rock that has undergone a transformation, or change
- *What are three processes that create metamorphosis for rocks?*
  heat, chemical reactions, and pressure

Help the students understand the basic differences between the three types of rock. Review the table and discuss how these types of rocks are similar or different. Discuss the types of textures of the three types of rocks. Discuss how it might be challenging to determine the difference between a layered sedimentary rock like shale and a foliated metamorphic rock like slate.

3.4 Testing Rocks and Minerals

In this section and in the experiment for this chapter, students are introduced to some of the tests that geologists perform in the field or laboratory to help identify minerals. In addition to the scratch and color tests, geologists also look at other identifiers, such as the texture of the sample, the shape of the crystals, whether or not it fluoresces, and if it is magnetic.

Tests can also be performed to see what characteristics a sample shows when it is broken. If the sample that is broken shows cleavage, it will break along a flat surface. If it fractures, it will have a conchoidal (shell-shaped) surface or an uneven surface. By putting together all the results of different tests, geologists are able to identify minerals.

Have the students look at the two Mohs scale of hardness charts in this section of the textbook. Guide discussion with questions such as the following:

- *Do you think you could use your fingernail to scratch test calcite? Why or why not?*
  No. Calcite is harder than a fingernail.

- *Which do you think is harder, calcite or quartz? Why?*
  Quartz. It has a higher number on the Mohs scale.

- *Will quartz scratch orthoclase?*
  Yes.

- *Could you do a streak test for the color of topaz? Why or why not?*
  No. Topaz is harder than the streak plate and would not leave a color streak.

3.5 Soils

The third material that makes up Earth is soil. The technical term for soil is regolith (from the Greek—*rhegos* means blanket, *lithos* means stone).

Soil covers most land surfaces and is the interface between life and rock. That is, living matter does not grow in rock, but in soil. Soil is composed of both rock particles and organic material, together with air and water. Although soils vary in composition, all soils contain rock particles, organic material, air, and water. Soils in which plants grow well typically have 25% air, 25% water, 5% organic matter, and 45% mineral matter.

The type of soil depends on the parent material—the rocks and minerals that break up into the smaller pieces that form the mineral matter of the soil.

Lead a discussion about different types of soil. Depending on where the student lives, the soil may or may not be very good for growing plants.

3.6 Summary

Review the summary points of this chapter with the students.

- *Earth's surface, or crust, is made of rocks, minerals, and soil.*

- *There are certain criteria that a material must meet in order to be considered a mineral:*

  1) *It must occur naturally.*
  2) *It must be a solid.*
  3) *It must be inorganic (not carbon-containing).*
  4) *It must have an ordered internal structure of atoms.*

- *There are eight elements that make up the majority of rock-forming minerals: oxygen, silicon, aluminum, iron, calcium, potassium, sodium, and magnesium.*

- *Rocks can be divided into three basic types: igneous, sedimentary, and metamorphic.*

- *Soil taxonomy is a system used to classify soils according to their properties.*

In this experiment students will learn how to test for different minerals using Mohs scale of mineral hardness and the streak test. Both of these tests are typical field tests geologists employ when trying to determine the mineral composition of various rocks.

Students will first experiment with using the field tests on known minerals and then try to determine if rocks from their yard or nearby contain any of the minerals they tested in the first part of the experiment.

Have the students read the entire experiment and then help them write an objective.

Some suggestions are:

- *To determine what minerals are in my backyard.*

- *To test rocks for the minerals calcite, quartz, feldspar, and hematite.*

- *To learn how to use field tests to determine which minerals are in rocks.*

Help the students write a hypothesis. The hypothesis should be specific and can be about their backyard rocks or the known minerals they test.

Some suggestions are:

- *Rocks in my backyard contain quartz.*

- *Rocks in my backyard do not contain quartz.*

- *Feldspar is harder than calcite.*

- *Quartz is softer than hematite.*

Experiment 3: MINERAL PROPERTIES   Date: _____

Objective
_____
_____

Hypothesis
_____
_____

Materials

- known mineral samples:
    calcite   feldspar
    quartz   hematite
- several rocks from your backyard or near your home
  (Have the students collect rocks that look different from each other.)
- penny
- nail
- unglazed white ceramic tile (streak plate)
- paper
- scissors
- marking pen
- tape

Experiment—Part 1

❶ The hardness of a mineral is determined by its resistance to being scratched. The Mohs scale of mineral hardness lists the relative hardness of ten common minerals as well as various objects that can be used in scratch testing minerals.

Test the hardness of the known mineral samples shown in the materials list above by using the objects listed in the Mohs hardness scale on the next page. Fill in the chart on the next page with your results.

Results

Mohs Scale of Mineral Hardness

Object	Hardness
fingernail	2.5
copper penny	3
steel nail	5.5
streak plate	6.5

(Answers will vary.)

Mineral Name	Hardness	Notes
calcite	*might be between 3 and 5*	*could not scratch with fingernail, could not scratch with penny, could scratch with nail, looks white and sort of shiny*
quartz	*more than 6.5*	*very hard, could not scratch, has a shiny surface with flat sides and a point at one end*
feldspar	*6*	*I think the streak plate scratched it but the steel nail did not; has an irregular surface*
hematite	*3.5-5*	*it scratched the penny and the nail scratched it, looks gray and lumpy*

Using the Mohs hardness scale provided in the *Laboratory Workbook*, have the students test each of the four known minerals for hardness.

The way to use the scale is as follows:

- *If a fingernail is able to scratch the mineral, the hardness is below 2.5.*

- *If a fingernail is not able to scratch the mineral, but a copper penny does, then the hardness is between 2.5 and 3.*

- *If a copper penny cannot scratch the mineral but a steel nail can, then the hardness is between 3 and 5.5.*

- *If a steel nail cannot scratch the mineral but an unglazed ceramic plate can (the mineral leaves a "streak"), then the hardness is between 5.5. and 6.5.*

- *If the mineral is not scratched by the ceramic plate (does not leave a streak) then the mineral is harder than 6.5.*

Have the students record other observations, such as size and color of the mineral samples. Also, have the students note any crystalline properties. It is important for students to record what they actually observe even if it's not what they think the "right" answer would be.

Typical hardness values are:

- Calcite: 3
- Quartz: 7
- Feldspar: 6
- Hematite: 5-6

Using the ceramic streak plate, have the students record the color streak left by the mineral. To determine the color streak, they will take each mineral and scrape it across the streak plate. The color left on the plate is the "color streak" of the mineral.

The color streak can be different from the color of the mineral sample as a whole. If this occurs, have the students note any differences.

You can also have the students try testing the different samples with vinegar or lemon juice (acids) to see if a chemical reaction occurs. A dropper bottle can be used to apply the acid. They might try using both vinegar and lemon juice to see if there is a difference.

Experiment—Part 2

❷ Do a streak test for each of the known minerals. Take each mineral sample and rub it firmly across the streak plate.

Results

Record the streak color of each mineral sample.

(Answers may vary.)

Mineral Name	Streak Color	Notes
calcite	white	it left a white powder on the streak plate
quartz	did not make a streak	it scratched the streak plate
feldspar	white	it left a white powder on the streak plate
hematite	reddish	the streak color is red but the rock is gray

Experiment—Part 3

❸ Using paper and tape, label the rocks you collected from your yard or near your house. Use a different number to identify each rock. Record the numbers in the following chart.

❹ Using the Mohs scale of mineral hardness, do a scratch test to find out the hardness of each unknown sample. Record your results.

❺ Do a streak test for each of the unknown samples and record your results.

Results

Rock #	Hardness	Streak Color	Notes

Using the hardness test and the streak test, have the students see if they can determine whether any of the known minerals they tested can be found in the rocks they have collected from their backyard or other local source.

They may or may not be able to determine whether the minerals they tested are in the rocks they've collected.

Based on the streak test and color test, have the students guess if any of the minerals are in their local rocks.

Other minerals can be tested if desired. You can find minerals at a local rock and mineral store or order them online from any rock and mineral source.

First, have the students discuss the different minerals they tested.

- *Was calcite harder than quartz?*

- *Was quartz harder than feldspar?*

- *Was feldspar harder than hematite?*

Next, have the students discuss the streak test.

- *Were any of the mineral samples a different color than their streak? Why or why not?*

Have them record their conclusions.

Conclusions—Part 1

By using the Mohs scale of mineral hardness, the streak color test, and your results in Part 1 and Part 2 of the experiment, were you able to determine the type of mineral or minerals that are in the rocks you collected near your home? Why or why not? Record your conclusions.

_____

_____

_____

_____

_____

_____

_____

_____

_____

_____

_____

_____

_____

_____

_____

_____

Conclusions—Part 2

The hardness test using the Mohs hardness scale and the streak test are "subjective" tests. That is, the outcome may vary depending on the type of ceramic plate used for the streak plate, the type of nail or penny, and individual interpretations of streak color.

Another way to test for the type of mineral is to use chemical analysis. An acid test uses hydrochloric acid to test for carbonate. Minerals containing calcium carbonate, such as calcite, will effervesce (bubble or foam) when acid is used to create a chemical reaction. Do you think this is a more objective type of test? Why or why not? Write your answers below.

_____

_____

_____

_____

_____

_____

_____

_____

_____

_____

_____

_____

_____

The second part of the *Conclusions* section asks the students to consider the subjective aspects of the hardness test and the streak color test. They are subjective in that one person might see a slightly different color or observe a different hardness for the same mineral than another person or objects used for testing might vary in actual hardness. Have the students think about how the subjectivity of the test might change the results.

An objective test is one where the observers' opinions may matter less, though this is not always the case. Have the students discuss whether or not chemical analysis might be more objective and less subjective than either the hardness test or streak test.

Review

Answer the following:

▶ In order to be considered a mineral, a material must:

_____occur naturally_____ , _____be a solid_____ ,
_____be inorganic_____ , and _have an ordered arrangement_
of atoms .

▶ The eight chemical elements that make up the majority of rock-forming minerals are:

oxygen

silicon

aluminum

iron

calcium

sodium

potassium

magnesium

▶ The three basic types of rocks are called:

igneous

sedimentary

metamorphic

▶ Some factors that affect the type of soil that is formed are:

_____parent material_____ , _____climate_____ ,
_____time_____ , and _____topography_____ .

Chapter 4: Earth's Layers

Overall Objectives   36

4.1   Introduction   36

4.2   Inside the Earth   36

4.3   The Crust   37

4.4   The Mantle   37

4.5   The Lithosphere   37

4.6   The Asthenosphere   38

4.7   The Mesosphere   39

4.8   The Core   39

4.9   Summary   39

Experiment 4: Model Earth   40

Review   45

Time Required

Text reading   30 minutes
Experimental   1 hour

Materials

Students are to choose materials to use
to make a model of Earth

Overall Objectives

In this chapter students will examine the layers of Earth. Students will learn the names of the different layers and their composition, overall dimensions, and other characteristics.

4.1 Introduction

So far, students have explored the science of geology, tools geologists use to study Earth, and Earth's basic chemical composition. This chapter begins to explore how Earth is put together. In subsequent chapters students will learn more about how the inner layers of Earth affect the outer layers and how Earth's different layers interact.

Earth is a "rock planet" which means that, unlike a gaseous planet, Earth is made of rocks and minerals. Have the students explore what they think the Earth looks like below the surface. Use questions such as:

- *How far below the surface of the Earth do you think rocks, minerals, and dirt extend?*

- *What else do you think is below the Earth's surface? Gas? Melted rocks? Water? Living things?*

- *What do you think is at the center of the Earth? Is the center hard? Soft? Both? Neither?*

4.2 Inside the Earth

In this section students will take a closer look at Earth's layers. Scientists have never been able to directly sample the material below the Earth's crust, but we know from other scientific experiments that the Earth is likely layered and that each of these layers is different.

There are three main layers called the crust, the mantle, and the core. The illustration in this section of the textbook shows the subdivisions within the layers. Help the students examine the graphic and map out the subdivisions. Have them notice that:

- *The crust has one layer and is the outermost layer of Earth.*

- *The mantle is made up of three subdivisions: the lithosphere, asthenosphere, and mesosphere.*

- *The core is made of both an inner core and an outer core.*

4.3 The Crust

The outermost layer of Earth is where we live and is called the crust. Help the students summarize the main points of this section. For example:

- *The crust is made of rocks, minerals, and soil.*
- *The crust is also called the outer shell of Earth.*
- *There are two types of crust—continental crust and oceanic crust.*
- *The continental crust is thicker than the oceanic crust.*

4.4 The Mantle

The mantle is just below the crust and includes three layers called the lithosphere, asthenosphere, and mesosphere. To date, scientists have not directly sampled the mantle. What we know about the lower layers is based on secondary evidence such as seismic waves.

Have the students discuss the challenges scientists face when attempting to drill through the crust to the mantle. For example:

- *Where would it be easier to reach the mantle—through the continental or the oceanic crust?* (the oceanic crust)
- *If you were going to drill through the oceanic crust what specialized equipment might you need?* (a boat, a drill, divers to place the drill and build a platform, etc.)
- *What kind of natural interference might you expect?* (hurricanes, severe storms, earthquakes, etc.)
- *Do you think drilling through the oceanic crust will cause any large-scale damage such as an earthquake? Why or why not?*

4.5 The Lithosphere

The lithosphere is the uppermost layer of the mantle. Scientists believe that the lithosphere is rigid and made of dense rock. Again, scientists have not directly sampled the lithosphere, but by using secondary evidence and inductive reasoning, they have drawn the conclusion that the lithosphere is rigid. Inductive reasoning takes a set of specific observations to draw a general conclusion. In the case of the lithosphere, we note specific observations about the behavior of the crust, such as the formation and location of mountains, volcanoes and earthquakes, and draw a general conclusion about the nature of the lithosphere.

Inductive Reasoning

The process of going from the specific to the general.

Deductive Reasoning:

The process of going from the general to the specific.

Inductive reasoning is a major part of scientific investigation. Although inductive reasoning is not always accurate, scientists can draw general conclusions from a limited data set.

Discuss with the students how scientists use inductive reasoning to draw conclusions. Contrast inductive reasoning with deductive reasoning and discuss the difference. Following are some examples of inductive and deductive reasoning.

- *I observe that the Sun rises in the east in New York, Paris, and Seattle; therefore, I can conclude that the Sun rises in the east everywhere on the planet.* (inductive reasoning)

- *I observe that water freezes when placed in our home freezer, the neighbor's freezer, and the restaurant freezer; therefore, I can conclude that water will freeze in any working freezer.* (inductive reasoning)

- *I observe that my cat is black; therefore, all cats are black.* (deductive reasoning).

- *I observe that my black cat likes to play with my dog; therefore, all black cats like to play with all dogs.* (deductive reasoning)

4.6 The Asthenosphere

The asthenosphere lies below the lithosphere and is believed to be soft and putty-like. Again, scientists have not directly observed the asthenosphere but have drawn conclusions by looking at secondary evidence. One type of evidence scientists use is seismic waves. Scientists can study how seismic waves travel through the inner Earth, and from this information they can create a model for the asthenosphere. Scientists can also make observations about how the lithospheric plates move, and from these observations they can infer that the asthenosphere is soft and malleable.

Discuss with the students how observing seismic waves can help scientists draw conclusions about the nature of the asthenosphere.

- *When a wave moves through a body of water and encounters a barrier, such as the side of the bathtub, is it possible to observe what happens to the wave?*

- *Do you think you could map the boundaries of a bathtub by observing only the waves and how they change? Why or why not?*

- *Do waves in the ocean or a lake behave differently when they hit a rocky shore than when they hit a sandy shore?*

4.7 The Mesosphere

The mesosphere is the area between the asthenosphere and the core. There is little known about the mesosphere, but scientists believe it is likely more solid than the asthenosphere due to increased pressure. It is also much thicker than either the asthenosphere or the lithosphere.

4.8 The Core

Scientists believe that Earth's core is composed mainly of iron and nickel. The core is divided into two layers—the liquid outer core, which contains convection currents, and the solid inner core. The inner and outer cores account for a significant portion of Earth's mass, with the inner core being the most dense.

Scientists also believe that the dynamics of the inner and outer core contribute to earthquakes and volcanic activity and are the cause of Earth's magnetic field. Earth's magnetic field will be discussed in Chapter 9.

Have the students discuss how the core might cause earthquakes and volcanoes. Use the following questions to guide the discussion.

- *How might the Earth's core and convection currents create earthquakes?*
- *How might the Earth's core and convection currents create volcanoes?*

4.9 Summary

Discuss the summary statements with the students.

- *The Earth is made of three main layers—the crust, the mantle, and the core.*
- *Life occurs on the outer layer, the crust.*
- *The mantle is subdivided into three layers: the lithosphere, the asthenosphere, and the mesosphere.*
- *The core has two layers—the inner core and the outer core.*
- *Geologists use seismic waves caused by seismic activity, such as earthquakes, to study the layers beneath the Earth's crust.*

In this experiment students will create a model of Earth using what they have learned in this chapter.

Students will be designing their own model and will decide what type of materials to use to create the model.

Have the students read through the experiment.

Discuss possible objectives for this experiment. Some suggestions are:

- *To explore building a model of Earth.*

- *To understand how to create accurate models of the Earth.*

- *To explore the benefits and limitations of building models.*

Discuss possible hypotheses for this experiment.

Some examples of hypotheses are:

- *I can create a model of Earth that accurately represents Earth's layers.*

- *I cannot create a model of Earth that accurately represents Earth's layers.*

- *I can create a model of Earth that teaches me something about Earth's layers.*

Experiment 4: Model Earth   Date: _____

Objective _____

_____

Hypothesis _____

_____

Materials

_____

_____

_____

_____

_____

_____

Experiment

In this experiment you will decide how to build an accurate model of Earth. Scientists use models to help them understand how things work. Creating accurate models is an important skill when doing science. The more accurately a model depicts reality, the more scientists can learn about how things work.

❶ On the following page, list what you know about Earth's layers. Record whether or not scientists believe the layer is soft or rigid, solid or liquid, rock or iron. Also record the depth of the layer and any other features the layer may exhibit.

❷ Use the student text, internet, or library to collect your information.

Features of Earth's Layers

Layer	Depth	Features
Crust		
Lithosphere		
Asthenosphere		
Mesosphere		
Inner Core		
Outer Core		

Have the students use the student textbook, the internet, or library to collect information about the layers. The more research they do, the more accurate the model they will be able to build and the more they will learn about Earth's layers.

In the chart provided have the students list the features and depths of the layers.

Have the students think about the characteristics of each layer and imagine materials they could use to create the layer so that the characteristics match.

Students can use clay, food items, Styrofoam, cloth, felt, dirt, or any other materials they think they could use to assemble a model of Earth.

Have the students choose the items from the chart of model materials that they will to use for their model of the Earth.

Help students think about what features they want to represent as they build their model.

For example:

- *They might want to build a model that represents the roundness of the Earth. In this case modeling clay of different colors could be used because it is easy to mold.*

- *They might want to represent the rigid lithosphere and the soft asthenosphere. In this case a pan with layered food items could be used (For example, peanut brittle (lithosphere) on top of peanut butter (asthenosphere).*

- *They might want to represent the hot liquid of the Earth's outer core surrounded by the firmer mantle. In this case it's best (and fun) to make a chocolate lava cake.*

CHOCOLATE LAVA CAKE

butter 113 grams (1/2 cup)
semi-sweet chocolate chips
    133 ml (1/2 cup + 1 Tbsp.)
2 whole eggs
2 egg yolks
powdered sugar
    192 ml (3/4 cup + 1 Tbsp.)
flour 94 ml (1/3 cup + 1 Tbsp.)

❸ Using the chart that you created on the previous page, go through the information you collected about each layer. Think about what material you could use to accurately represent each layer. Record the materials in the chart below.

Model Materials

Layer	Materials
Crust	
Lithosphere	
Asthenosphere	
Mesosphere	
Inner Core	
Outer Core	

❹ List the materials you plan to use for your model in the *Materials* section on the first page of this experiment.

❺ Assemble a model of Earth using these materials.

Results

❶ Observe how well your model represents Earth's geology. List your observations about how well your model depicts the overall architecture of Earth and the characteristics of each layer.

❷ Record your data below.

Model Results

Questions	Observations
Does your model have layers? Which ones? Describe each layer.	
Do the layers represent accurate depths in your model? How do you know?	
Do the layers represent accurate consistency? (For example, is the lithosphere rigid and the mesosphere soft?)	
Do the layers in the model accurately represent the change that occurs at the boundary between each of the layers? Why or why not?	

Microwave butter briefly until melted. Stir in chocolate chips until melted. Mix in whole eggs and yolks, then powdered sugar. Stir in flour. Pour into custard cups thoroughly greased with butter. Bake at 190° C (375° F) until edges are set and centers are still soft, about 10-13 minutes. Don't overbake. Makes about 4.

Students might like to poke a walnut or cherry into the cake to represent Earth's inner core. They might add Earth's crust by sprinkling nuts over the top of the cake.

Have the students determine how well their model represent Earth's geology.

The models are not going to be perfect, and in the process of model building some features will be more accurately represented than others. For example:

- *If students choose to layer food in a pan to represent layer textures, then the overall architecture will not be round.*

- *If students choose to create a round, layered Earth with modeling clay, they might use a rock or small Styrofoam ball for the solid inner core, a different color clay for each layer, and cloth for the crust. The clay layers will all be the same consistency.*

- *If they choose to use a Styrofoam ball, then there are no layers.*

- *If students choose to use a lava cake to represent the Earth's liquid center (outer core) and the mantle and then add a walnut or cherry for the solid inner core and nuts for the lithosphere, only the main layers will be represented. (But it's an extremely tasty model!)*

Have the students evaluate their models. Help them explore how challenging model building can be and the difficulties that arise when trying to build accurate models.

Conclusions

Based on your observations, evaluate the accuracy of your model.

_____

_____

_____

_____

_____

_____

_____

_____

_____

_____

_____

_____

_____

_____

_____

_____

_____

Review

Answer the following questions:

▶ List the three main layers of Earth.

_____ *crust* _____

_____ *mantle* _____

_____ *core* _____

▶ For each layer that has subdivisions, list the subdivision and the layer it belongs to. Describe the characteristics of each layer subdivision.

(Answers may vary.)

Subdivisions of Earth's Layers

Layer and Subdivision	Description
mantle lithosphere	rigid, divided into plates, holds the land masses of the crust
mantle asthenosphere	putty-like, made of melted rock, has convection currents, lithospheric plates float on it
mantle mesosphere	between the core and the asthenosphere, hotter and more solid than the asthenosphere
core outer core	liquid layer, has convection currents, made of iron and nickel, very dense, very hot
core inner core	solid, made of iron and nickel, very dense, very hot

Chapter 5: Earth's Dynamics

Overall Objectives   47

5.1   Introduction   47

5.2   Plate Tectonics   47

5.3   Mountains   48

5.4   Volcanoes   49

5.5   Earthquakes   49

5.6   Summary   50

Experiment 5: Dynamic Earth   51
        5A: Exploring Plate Tectonics   51
        5B: Create a Model Volcano (alternate)   56

Review   59

Time Required

Text reading   30 minutes
Experimental   1 hour

Materials—Experiment 5A

brittle candy (see p. 51 for ingredients)
1 jar smooth peanut butter
118 ml (1/2 cup) crushed graham crackers

Materials—Alternate Experiment 5B

students to determine

Overall Objectives

In this chapter students will explore Earth's dynamics. Students will be introduced to the theory of plate tectonics and will learn how mountains, volcanoes, and earthquakes form.

5.1 Introduction

Earth is a dynamic planet, meaning it is constantly changing. Features of Earth's surface can be changed quickly by natural phenomena, such as sudden earthquakes and volcanic activity. There are also changes that occur over very long periods of time, such as the formation of river valleys, the wearing away of land forms by glaciers, and the growth of mountains due to forces caused by the movement of tectonic plates.

Have the students discuss how Earth is constantly changing. Use questions such as:

- *How do you think mountains are formed?*
- *What do you think causes earthquakes?*
- *What happens to the surrounding area when a volcano erupts?*
- *What is a tsunami? How might a tsunami change the landscape?*

5.2 Plate Tectonics

Plate tectonics is a theory that has been proposed to explain how earthquakes, mountain ridges, and volcanoes occur. As mentioned in the previous chapters, scientists can't directly sample the layers below the Earth's crust. But scientists can put together pieces of secondary information and propose a theory to explain dynamic phenomena. So, scientists can draw conclusions based on data from observations such as the location, age, and activity of volcanoes; analysis of the composition and deformation of rock strata in mountains; and the location, type, and strength of earthquakes.

Have the students carefully examine the illustration of the cross section of Earth in this section of the student text. Ask questions to help them understand what the illustration is showing. For example:

- *Based on the illustration, what happens when two plates move towards each other and collide?*
  one plate can slide under the other plate, creating mountains
- *What happens when two plates move away from each other?*
  new crust is formed

- *What causes the movement of the plates?*
  convection in the mantle
- *Which part of the crust is thinnest?*
  oceanic crust

Next, have the students carefully examine the map of the world on the next page of the student text. This illustration shows cumulated data on volcanic and earthquake activity. Ask questions to help students understand what the illustration is showing.

- *What is the difference between divergent, convergent and transform plate boundaries?*
  (see map legend)
- *Do some volcanoes occur in the middle of plate boundaries? If so, what is the name of the areas where they occur?*
  Yes. These areas are called hotspots.
- *Which plate has the longest convergent boundary?*
  Pacific plate

5.3 Mountains

This section discusses three different types of mountains—folded mountains, fault-block mountains, and dome mountains. Have the students discuss these three types of mountains and how they are different from each other. For example:

- *Folded mountains occur when two plates push into each other.*
- *Fault-block mountains occur when blocks of land move up and down.*
- *Dome mountains occur when magma pushes land upwards but does not break the surface.*
- *Folded mountains are different from dome mountains because folded mountains are created when two plates push into each other and dome mountains are caused by magma under the ground. It is expected that folded mountains will be found at plate boundaries and dome mountains will be found away from plate boundaries.*

5.4 Volcanoes

Volcanoes occur when the pressure of magma pushing up through the mantle causes a break in the surface of the Earth. When this occurs, hot magma pushes out, creating new land masses and mountains. Magma that has emerged onto the surface of the Earth is called lava.

Help the students note that volcanoes can erupt suddenly and violently, like Mount St. Helens in Washington, or slowly over time, like the Mauna Loa volcano in Hawaii. The Hawaiian Islands are the tops of a chain of volcanoes whose bases are far below the surface of the ocean.

Have the students discuss the different types of volcanoes and how they form.

5.5 Earthquakes

According to the theory of plate tectonics, earthquakes can occur when two plates collide with, push on, or slide past each other, causing stresses to build up along the plate boundaries and also in the interior of the plates. Earthquakes occur along faults, or fractures, in the Earth's surface. The most active faults are at or near plate boundaries, such as the San Andreas Fault in California. But there are also some active areas within plate interiors, such as the New Madrid Seismic Zone located in the Mississippi Valley.

Discuss with the students the three different types of faults and how they cause different outcomes when an earthquake occurs (see the illustration in this section of the student text). Use questions such as:

- *What do you think happens to a street when a strike-slip fault earthquake occurs under it?*
- *What do you think happens when an earthquake of a normal fault type occurs under a street?*
- *Do you think you could tell the difference between a normal fault and a thrust fault by looking at the landscape after an earthquake has occurred? Why or why not?*

5.6 Summary

Discuss the summary statements with the students.

- *Earth is dynamic—it is constantly changing.*
- *Plate tectonics is a theory that helps explain Earth's dynamics.*
- *Three common types of mountains are folded mountains, fault-block mountains, and dome mountains.*
- *Three common types of volcanoes are cinder cone volcanoes, shield volcanoes, and composite volcanoes.*
- *Earthquakes are caused by stresses in the Earth and occur along fault lines.*

Experiment 5: Dynamic Earth

For this experiment you have the choice of performing either one of two experiments that explore Earth's dynamics. Read through both experiments first and then make your selection.

Experiment 5A: Exploring Plate Tectonics   Date: _____

Objective   _____

_____

Hypothesis   _____

_____

Materials

▸ brittle candy (recipe below)
▸ 1 jar smooth peanut butter
▸ 118 ml (1/2 cup) crushed graham crackers

Experiment

❶   Use the following instructions to make brittle candy.

Brittle Candy Recipe

Ingredients

237 ml (1 cup) white sugar

118 ml (1/2 cup) light corn syrup

1.25 ml (1/4 teaspoon) salt

59 ml (1/4 cup) water

28 grams (2 Tbsp) butter, softened

5 ml (1 teaspoon) baking soda

Equipment

2 liter (2 qt) saucepan

candy thermometer

cookie sheet, approx. 30x36 cm (12x14 inches)

spatulas

In this experiment students will explore Earth's dynamics. They are asked to choose one of two experiments—using food items to model plate tectonics or doing research to find directions for modeling a volcano and then making the model.

Have the students read through both experiments and choose the one they'd like to do. If time permits, students may perform both experiments.

Experiment 5A: Exploring Plate Tectonics

Students will build a model using food items to examine how the movement of Earth's tectonic plates causes earthquakes to occur and mountains to form.

Have the students read through the experiment and then discuss possible objectives for this experiment. Some suggestions are:

- *To explore how Earth's plates move on the putty-like asthenosphere.*

- *To experiment with a model of the crust, lithosphere, and asthenosphere to learn about plate tectonics.*

- *To explore the benefits and limitations of building models.*

Discuss possible hypotheses for this experiment. Some suggestions are:

- *I can create a model that will help me understand how plate tectonics works.*

- *My model will show me something I did not expect about plate tectonics.*

Students will make their own brittle candy that will be used to represent tectonic plates.

Help the students follow the recipe to make the brittle candy. It can be tricky to determine when the candy has cooked long enough. Make sure you check the candy often and use fresh cold water each time you test it.

Once the candy has cooled, have the students create a model of the three layers—crust, lithosphere, and asthenosphere.

If a student has allergies to peanuts, whipped cream can be used in place of peanut butter.

Brittle Candy Recipe (Continued)

Grease a large cookie sheet to put the candy on.

Measure sugar, corn syrup, salt, and water into a heavy 2 liter (2 quart) saucepan. Bring mixture to a boil over medium heat.

Stir until the sugar is dissolved.

Put the candy thermometer in the saucepan and continue cooking.

Stir frequently. The candy should be done when the temperature reaches 150° C (300° F). You can check whether it is done by dropping a small amount of the hot candy mixture into very cold water. If the candy separates into hard and brittle threads, it is done cooking.

Remove from heat and quickly stir in butter and baking soda.

Pour at once onto the greased cookie sheet and spread the mixture into a rectangle of about 30x36 centimeters (12x14 inches). You can use two buttered spatulas to spread the candy.

Allow to cool. Break the candy into large pieces.

❷ Spread a 1.25 cm (1/2 inch) thick layer of peanut butter on a plate.

❸ Mix 118 ml (1/2 cup) of crushed graham crackers with 59 ml (1/4 cup) of peanut butter.

❹ Take 2 large pieces of brittle candy and spread the peanut butter/graham cracker mixture on top of each of them.

❺ Place the two pieces of brittle candy (peanut butter/graham cracker side up) about 2.5-5 cm (1-2 inches) apart on top of the peanut butter.

❻ In the following chart write which layer each food item represents (the asthenosphere, the lithosphere, or the crust).

Earth's Layers Represented

Item	Earth's Layer
Peanut Butter	
Brittle Candy	
Graham Cracker/Peanut Butter Mixture	

❼ Gently holding the sides of the brittle candy pieces, move the candy around on top of the peanut butter and observe what happens.

Have the pieces bump into each other, scrape alongside each other, and move up or down with respect to each other.

Move the pieces quickly or slowly and observe the difference. Try to get one piece to slide under the other piece and observe what happens to the graham cracker topping.

In the box on the next page, write about or draw what you observe during this experiment. Think about what you have learned about plate tectonics and how your model relates to what you know.

Have the students move the brittle candy pieces on top of the peanut butter.

Have them bump the pieces together, scrape them alongside each other, or move the pieces up and down.

The students should pay special attention to the graham cracker/peanut butter topping. This layer represents the crust and is where the most visible changes will occur.

Have the students explore what happens when the brittle candy pieces are moved slowly, quickly, and collide or move past each other.

Explore with the students how well this works as a model of plate tectonics.

Have the students write their observations in the box provided.

Observations

Have the students review their observations and use them to record their answers to the questions in the chart provided.

Have the students discuss the ways in which their model reflects the theory of plate tectonics and how it does not.

For example:

- *The textures of the three foods may adequately represent the differences between the crust, lithosphere, and asthenosphere.*

- *The peanut butter may be pushed up to the surface where the "crust" is. This does not appear to happen in reality.*

- *The graham cracker/peanut butter crust may slide too easily off the brittle candy pieces which doesn't reflect reality.*

- *The graham cracker/peanut butter crust mounds up like mountains.*

Results

Record your observations in the following chart. Note what happened to the lower peanut butter layer, the brittle candy pieces, and the top graham cracker/peanut butter layer.

Model of Tectonic Plates

Questions	Observations
What happened when you moved the brittle candy pieces slowly on top of the peanut butter?	
What happened when you moved the brittle candy pieces quickly on top of the peanut butter?	
What happened when you collided two brittle candy pieces together?	
What happened when you moved the pieces up or down with respect to each other?	
What happened when you moved the two pieces side by side?	

Conclusions

Based on your observations, explain what you think happens when two of Earth's plates collide with each other, slide past each other, or move up and down with respect to each other. Do you think your model was a good representation of plate tectonics? Why or why not?

_____

_____

_____

_____

_____

_____

_____

_____

_____

_____

_____

_____

_____

_____

Have the students evaluate their results and their model.

Help the students explore how challenging model building can be and the difficulties that arise when trying to build accurate models.

Review

Answer the questions in the *Review* section on page 37.
(Page 59 of Teacher's Manual)

This is an optional alternate experiment in which students will do research to find directions for a model volcano to build. They will then write an experiment to go with this model.

Being able to construct models is a critical task in science. Knowing which models to use, how to use them, and what questions they could answer is important for doing science.

Have the students look on the internet or explore in the library to find various models for building a volcano. There are directions for baking soda models, clay models, and food models available.

Ask the students what question they would like to explore with their model. They may just want to see a violent explosion, and so they could make a model of a cone volcano to observe this, or they may want to explore the slow moving lava of a shield volcano.

Based on what they choose to explore, help them pick a model and design their own experiment. Help them list the steps of the experiment in a succinct manner.

Experiment 5B: Create A Model Volcano   Date: _____

Research various ways to create a model volcano. Choose a method and write your own experiment here.

Objective _____

_____

Hypothesis _____

_____

Materials

_____

_____

_____

_____

_____

Experiment

_____

_____

_____

_____

(Continue Experiment on next page.)

Results

Write about or draw what happened during your experiment.

Modeling a Volcano

Encourage the students to make careful observations and then write about or draw them.

Have the students evaluate their results and their model.

Help the students explore how challenging model building can be and the difficulties that arise when trying to build accurate models.

Have the students discuss the ways in which their model might reflect how a volcano works and the ways in which it is not accurate.

For example:

- *The lava from the model volcano was not hot.*

- *The model volcano erupted like a real volcano.*

- *In the model volcano the lava did not come from far below it.*

Conclusions

Based on your observations, evaluate the accuracy of your model. How was your model volcano like a real volcano and how was it different?

_____

_____

_____

_____

_____

_____

_____

_____

_____

_____

_____

_____

_____

_____

_____

_____

_____

_____

Review

Answer the following questions.

(Answers may vary.)

▶ Describe the theory of plate tectonics in your own words.

The lithosphere of Earth is rigid and made of separate plates that

float on the asthenosphere. The asthenosphere is putty-like and

has convection currents that move the plates around. Earthquakes,

volcanoes, and mountains are made as a result of plates moving

against each other.

▶ What are three types of mountains?

folded

fault-block

dome

▶ What are three types of volcanoes?

cinder cone

shield

composite

▶ What are three types of faults?

strike-slip

normal

thrust

Chapter 6: The Atmosphere

Overall Objectives   61

6.1   Introduction   61

6.2   Chemical Composition   61

6.3   Structure of the Atmosphere   63

6.4   Atmospheric Pressure   64

6.5   Gravity and the Atmosphere   65

6.6   The Greenhouse Effect   65

6.7   Summary   66

Experiment 6: Exploring Cloud Formation   67

Review   70

Time Required

Text reading   30 minutes
Experimental   1 hour

Materials

2 liter (2 quart) plastic bottle with cap
   (squeezable, like a soda bottle)
warm water
matches

Overall Objectives

This chapter will introduce students to Earth's atmosphere. Students will learn what it is, what it is composed of, and its physical structure, as well as the importance of Earth's atmosphere to the survival and existence of all organisms.

6.1 Introduction

The atmosphere is a thin layer of gases that encircles the Earth. The atmosphere on Earth is different from that of any other planet in our solar system and allows life as we know it to exist. Earth's atmosphere is responsible for supplying necessary chemical components, such as water and oxygen, that are required for life. It also keeps temperatures stable and mild.

Explore open inquiry with the following questions:

- *What do you think Earth would be like if it did not have an atmosphere?*
- *Could life survive on Earth if the atmosphere did not carry water? Why or why not?*
- *Do you think the air contains any chemical elements? Why or why not?*

6.2 Chemical Composition

The atmosphere is made up of about 78% nitrogen and 21% oxygen. The remaining 1% contains argon, carbon dioxide, ozone, and traces of other gases. This is called the dry atmosphere and does not include water vapor, the gaseous state of water. The amount of water vapor in the atmosphere varies from nearly zero to about 4% of volume.

Carbon dioxide

Although carbon dioxide makes up less than 1% of Earth's atmosphere, it is very important in maintaining life on Earth. It traps some of the infrared energy (heat) from the Sun, keeping Earth warm. However, if the amount of carbon dioxide in the atmosphere increases and it traps too much infrared energy, the Earth can begin to get warmer. Carbon dioxide is also essential to plants which use it as part of the process of photosynthesis and then release oxygen as a waste product. Plants are the major source of oxygen in the atmosphere.

There are both natural and man-made sources of carbon dioxide.

Ozone

Ozone is a molecule made of three oxygen atoms. Although ozone makes up less than .0001% of the atmosphere, it is essential for life on Earth. The ozone layer in the lower stratosphere prevents most of the harmful ultraviolet radiation that comes from the Sun from reaching the surface of Earth and also traps energy from the Sun, helping to keep Earth warm.

Water vapor and clouds

Water vapor is also essential for life. It is the most important gas for keeping Earth warm and is also part of Earth's water cycle. Water evaporates and enters the atmosphere as water vapor. The water vapor later condenses, turning back to the liquid state, forming clouds, and then falling to Earth as precipitation.

Since warm air can hold more water vapor than cold air, cooling the air to the point where it can no longer hold all the water vapor will cause the water vapor to condense and turn to the liquid state. The temperature at which condensation occurs is called the dewpoint. The dewpoint varies depending on the amount of water vapor in the air (humidity).

The most common way for air to be cooled is through lifting. Air moves into an area of the atmosphere that is at a lower pressure, causing the air to expand. Energy is required for this expansion, taking heat away from the air and cooling it. As the air cools, some of the water vapor will condense around dust particles in the air, forming water droplets.

The reverse happens as air sinks. As it encounters higher pressures at lower altitudes, the increased pressure squeezes the air, adding heat and allowing the air to once again hold more water vapor. This can cause clouds to evaporate as the liquid water turns to vapor.

in order for water droplets to fall as precipitation, they need to become larger and heavier. One process by which this happens is called the collision and coalescence process or the warm rain process. During this process, water droplets in clouds collide and stick together to form larger drops.

Since warm air can hold more water vapor than cold air and heat makes water evaporate, the highest levels of water vapor in the atmosphere are over the oceans in the equatorial region. The lowest levels of water vapor are found over the dry deserts and at the poles where the air is cold and the water is tied up as ice.

Oxygen

Oxygen is the most important atmospheric gas for animal life. Animals breathe it in and use it for various chemical processes in the body.

Have the students discuss the chemical composition of Earth's atmosphere. Guide inquiry with questions such as:

- *Do you think Earth would stay warm without the atmosphere? Why or why not?*

- *Why do you think the amount of carbon dioxide in the atmosphere is important?*

- *What do you think would happen if there were no water vapor in the atmosphere?*

- *Why do you think oxygen is an important gas in the atmosphere?*

6.3 Structure of the Atmosphere

Using criteria such as differences in temperature, chemical composition, movement, and density of the gases, scientists have divided the atmosphere into five layers: troposphere, stratosphere, mesosphere, thermosphere, and exosphere. The upper boundary of each layer is called a pause, the area where the greatest change in temperature and other characteristics occurs. The pauses are called: tropopause, stratopause, mesopause, and thermopause. Since the exosphere extends into space, it is not generally considered to have a pause.

Troposphere

The troposphere is the layer of the atmosphere closest to the surface of the Earth and is the layer we live in and where weather occurs. It extends to a height of between 6-20 km (4-12 miles) Most of the mass of Earth's atmosphere exists within the troposphere, and the atmosphere becomes less and less dense as the altitude increases.

Stratosphere

The layer above the troposphere is the stratosphere which extends to about 50 km (31 miles) above Earth's surface. The stratosphere is dry and contains fewer gas molecules; therefore, it is more stable than the troposphere. Airplanes fly in the lower stratosphere to avoid turbulence caused by weather in the troposphere.

The ozone layer is found in the lower stratosphere. The ozone layer is important to life because ozone absorbs harmful ultraviolet radiation coming from the Sun, which both protects living things from severe burns and warms the Earth.

Mesosphere

Above the stratosphere is the mesosphere with an altitude of about 50 km (31 miles) to about 85 km (53 miles). The top of the mesosphere is the coldest part of the atmosphere with temperatures of approximately -90° C (-130° F). Most meteors burn up in the mesosphere.

Thermosphere

The thermosphere is above the mesosphere and reaches an altitude ranging from about 500-1000 km (310-620 miles). The variations in altitude result from differences in the heat coming from the Sun due to changes in the amount of energy released by the Sun or differences between day and night.

Exosphere

The exosphere is the outer layer of the atmosphere and is sometimes considered to be part of space. Atoms and molecules escape into space freely from the exosphere.

Have the students discuss the various layers of the atmosphere. Guide inquiry with questions such as the following.

- *In what ways do you think the troposphere is important to life on Earth?*

- *How do you think the exosphere is different from the other layers of the atmosphere?*

- *If you were the pilot of a big airplane, which layer of the atmosphere would you want to fly in? Why?*

- *How do you think scientists decide how thick each layer of the atmosphere is?*

6.4 Atmospheric Pressure

Atmospheric pressure (air pressure) is the force exerted by the weight of the molecules in the Earth's atmosphere. It is defined as the force per unit area exerted against a surface by the weight of the air above that surface. Barometers are instruments that measure atmospheric pressure either in millibars or inches of mercury.

As altitude increases, atmospheric pressure decreases because the atmosphere is less dense—there are fewer gas molecules per volume. With less atmospheric pressure and fewer molecules per volume, the remaining molecules will expand into the available space, causing a cooling effect.

- *What do you think the atmosphere would be like at the top of Mount Everest?*

- *Why do you think atmospheric pressure is important to life?*

6.5 Gravity and the Atmosphere

Earth's atmosphere is held in place by the force of gravity acting on the atoms and molecules of the gases. The escape velocity is the minimum speed needed for an object to escape from the gravitational field of a planet, star, or moon, and Earth's escape velocity prevents atmospheric gases from escaping into space. The gravitational field of the Moon is much weaker than that of Earth; therefore, the escape velocity of the Moon is low and is not enough to hold a significant atmosphere.

Guide inquiry with questions such as:

- *What do you think would happen to the atmosphere if Earth's gravity were weaker? Why?*
- *Do you think a jet plane could fly fast enough to reach escape velocity? Why or why not?*

6.6 The Greenhouse Effect

A greenhouse traps energy from the Sun, thus staying warm inside even in cold weather. The greenhouse effect in the atmosphere works in a similar manner. During the day energy from the Sun warms the surface of the Earth. At night the Earth cools and heat is radiated into the atmosphere where some of it is trapped by gases in the atmosphere and some escapes back into space. The trapping of heat by gases in the atmosphere keeps the Earth warm and at about the same yearly average temperatures. However, if the greenhouse effect becomes too strong, the Earth will begin to warm up. The gases involved in the greenhouse effect are called greenhouse gases and include carbon dioxide, water vapor, and methane, among others.

The way the atmosphere works and the ways in which Earth's spheres interact are very complicated and not yet well understood. Although scientists are able to measure a warming trend of the Earth, they are not able to definitively state its cause. Human activities have increased the levels of carbon dioxide in the atmosphere, and some scientists think this is the cause of the global warming trend. Other scientists think that Earth is in a natural warming cycle, and still others think that there could be a number of contributing factors.

Have the students discuss the greenhouse effect. Use questions such as:

- *In what ways is Earth's atmosphere like a greenhouse?*
- *How do you think the greenhouse effect is beneficial to life on Earth? How do you think it might be harmful?*

6.7 Summary

- *The atmosphere is a thin layer of gases that surrounds the Earth.*
- *Earth's atmosphere is divided into five layers: troposphere, stratosphere, mesosphere, thermosphere, and exosphere.*
- *Nitrogen is the most abundant gas in the atmosphere and oxygen the next most abundant.*
- *Atmospheric pressure is the force exerted by the weight of air molecules in Earth's atmosphere.*
- *The greenhouse effect is the trapping of energy from the Sun by the atmosphere, keeping Earth warm.*

Experiment 6: Exploring Cloud Formation   Date: _____

Objective _____

_____

Hypothesis _____

_____

Materials

▸ 2 liter (2 quart) plastic bottle with cap
▸ warm water
▸ matches

Experiment

❶ Pour warm water into the plastic bottle until it is about 1/4 full. Put the cap on the bottle.

❷ Light a match and remove the cap from the bottle. Drop the match in the bottle and quickly replace the cap.

❸ Squeeze the plastic bottle hard near the bottom and release. Notice what happens to the air in the bottle as you do this.

❹ Record your observations in the chart in the *Results* section.

❺ Repeat this experiment, filling the plastic bottle 1/2 full, 2/3 full, and then almost full. After each experiment, empty the bottle and start with fresh warm water. Record your observations each time.

In this experiment students will explore how clouds are formed.

Have the students read the entire experiment and then write an objective. Some examples of objectives are:

- *To make clouds in a bottle.*

- *To understand how clouds are made.*

- *To see what happens to air and water in a bottle if a lit match is thrown in.*

Have the students write a hypothesis. Some examples are:

- *The air in the bottle will make a cloud.*

- *The air in the bottle will change as the bottle is squeezed and released.*

- *The lit match will go out when it hits the water.*

Step ❶: Putting the cap on the bottle will cause water vapor to form.

Step ❷: Students may need help with this step since the bottle cap must be removed and replaced quickly in order to keep water vapor from escaping.

The smoke from the match will act as particulate matter for the water vapor to condense on.

Step ❸: Squeezing the bottle will increase the air pressure inside. Releasing the bottle will cause the air pressure to decrease, lowering the air pressure and thus cooling the air as it expands. This should cause clouds to appear. If clouds don't appear, try using warmer water.

Have the students record their observations of each part of the experiment. Have them write down what they see and also whether they think any of the variable might have changed. For example:

- *Was the water the same temperature for each part of the experiment?*

- *Was the cap on or off for a longer time?*

- *Was the bottle harder or easier to squeeze?*

- *Was there more or less smoke?*

Results

Cloud Formation Observations

Water Level in Bottle	Observations
1/4 Full	
1/2 Full	
2/3 Full	
Almost Full	

Conclusions

What do you think is happening in the water bottle as the ratio of water vapor, smoke (particles), and air pressure changes?

_____

_____

_____

_____

_____

_____

_____

_____

_____

_____

_____

_____

_____

_____

_____

_____

_____

Have the students write valid conclusions based on the data they have collected. Explain that the ratio means the quantity of the different factors relative to each other.

To answer questions the students may have, refer to section *6.2 Chemical Composition, Water vapor and clouds* in this book which discusses cloud formation and precipitation.

Review

Answer the following:

(Answers may vary.)

▶ The most plentiful gas in the atmosphere is _____*nitrogen*_____ at about ___*78*___%.

▶ The five layers of the atmosphere are:

troposphere   *thermosphere*

stratosphere   *exosphere*

mesosphere

▶ Why is it important for carbon dioxide to be in the atmosphere?

It helps keep Earth warm, and plants need to use it to make food during

photosynthesis.

▶ What is atmospheric pressure?

the weight of the air molecules pushing on things

▶ In your own words, explain the greenhouse effect.

A greenhouse traps heat that comes from the Sun so it can stay warm

inside. The atmosphere is similar. Sunlight warms the Earth. At night

the Earth cools off, and some of the heat escapes back into space and

some of it is trapped by gases in the atmosphere. This keeps Earth warm.

Chapter 7: The Hydrosphere

Overall Objectives72

7.1Introduction72

7.2The Hydrologic Cycle72

7.3Oceans73

7.4Surface Water74

7.5Groundwater74

7.6Human Interaction and Pollution75

7.7Summary75

Experiment 7: Modeling an Aquifer76

Review81

Time Required

Text reading30 minutes
Experimental1 hour

Materials

about 475 ml (2 cups) each:pencil
    gravelmarking pen
    sandlarge measuring cup
    dirtlarge bowl
    powdered pottery clayscissors
waterstop watch or clock with a
6 Styrofoam cups, about 355 ml   second hand (optional)
    (12 ounce) size

Overall Objectives

In this chapter students will be introduced to the hydrosphere—all the water that surrounds Earth—on its surface, under the ground, and in the atmosphere. They will learn about ocean water and freshwater, how water cycles through Earth's systems, and why it is important to keep Earth's water clean.

7.1 Introduction

All of Earth's water considered as a whole is called the hydrosphere. Water is required for the existence of life on Earth and is found all the way from up in the atmosphere down into the crust of the Earth.

Guide students' inquiry with questions such as:

- *What do you think the hydrosphere is?*
- *Where do you think most of Earth's water is found?*
- *Could you live without water? Why or why not?*
- *Do you think water stays in one place or does it travel long distances? Why?*

7.2 The Hydrologic Cycle

Earth has many cycles that move elements and molecules through Earth's various systems. The hydrologic cycle describes how water is cycled around the planet. It is interesting to note that the same water is used over and over, so the water that exists now is the same water that was on Earth at the time there were dinosaurs.

During the process of evaporation, water from the oceans and land surfaces turns into its gaseous state and enters the atmosphere as water vapor. Some of the water vapor in the air condenses and forms clouds which are carried around the planet by wind currents and can travel very long distances before the water falls to Earth as rain or snow. Most of the evaporated water comes from the oceans since they contain the vast majority of Earth's water. The land masses receive water from the ocean when winds blow clouds over the land. Much of the water that falls on land is returned to the ocean by rivers.

Groundwater is water that has infiltrated the soil from precipitation or from the bottom of lakes, rivers, and other bodies of water. Groundwater may flow underground and into lakes and rivers, thus returning groundwater to the oceans.

Guide a discussion about the hydrologic cycle by asking questions such as the following:

- *Do you think water goes from the land to the oceans? If so, how?*
- *Do you think water goes from the oceans to the land? If so, how?*
- *What different forms does water come in?*
- *In what ways do you think the different forms that water comes in affect life on Earth?*

7.3 Oceans

The oceans contain 97% of Earth's water and affect its temperatures. Oceans absorb and then radiate heat energy from the Sun, providing a warming effect, and evaporation of the ocean water contributes a cooling effect. Winds carry warmer or cooler air from the oceans over the land masses, and ocean currents carry water around the globe, both of which cause a warming or cooling effect.

The ocean currents that travel all the way around the world are called the global conveyor belt and are driven by the temperature and salinity of the water. The global conveyor belt is thought of as starting in the Norwegian Sea where the warm water of the Gulf Stream ocean current comes in contact with the cold air in the atmosphere. As some of the heat from the ocean is transferred to the air above it, the water cools. Since cold water is denser than warm water, it sinks, and warmer water moves in to take its place. The warmer water pushes the cold water south, beginning the conveyor belt effect. Eventually, the colder waters warm up, and the cycle continues.

Have the students discuss oceans. Guide inquiry by asking questions such as:

- *Do you think ocean water moves from place to place? Why or why not?*
- *Do you think ocean water is the same temperature from top to bottom? Why or why not?*
- *How do the oceans get salty?*
- *What causes tides?*
- *Are the oceans important to life on Earth? Why or why not?*

7.4 Surface Water

It is surprising to find out how little of Earth's water exists as freshwater on the surface of the land masses. Although it is essential for life, surface water makes up only .3% of all of Earth's water. Surface water travels through the hydrologic cycle by means of precipitation, evaporation into the atmosphere, runoff into the oceans, and infiltration into the ground.

Glaciers hold about 2% of Earth's total quantity of water in the form of ice. This water becomes available as the glaciers melt. Depending on the location of the glacier, the meltwater will either flow onto the land or into an ocean.

Discuss surface water with the students. Ask questions such as:

- *Where do you think you would find surface water?*
- *Does surface water make up most of the water on Earth? Why or why not?*
- *What do you think might happen if all of the ice in glaciers melted? Where would it go? What affects might it have on the Earth?*

7.5 Groundwater

Groundwater is found below the surface of the Earth. Precipitation and water that is held in lakes and other bodies of water infiltrate the soil and rocks in the ground. The porosity and permeability of the soil and rocks affect the flow of groundwater. Porosity refers to the size and quantity of void spaces in a material, and permeability refers to how quickly (or slowly) the water flows through the material. Rock that is fractured can also hold and transmit water.

An aquifer is made of porous materials in the ground that hold enough water to supply a well or a naturally occurring spring. Aquifers can be surrounded by materials that are not porous or permeable and thus be contained in a certain area, or they can stretch over long distances, carrying the water to other areas.

Have the students discuss groundwater. Further the discussion with questions such as:

- *Do you think groundwater is important to life? Why or why not?*
- *Do you think water can travel under the ground? Why or why not?*
- *If you get water from a well, where does it come from?*

7.6 Human Interaction and Pollution

Since the same water is recycled over and over, it is important that we keep our water sources clean. There are many ways that people pollute water. Garbage, human waste, and industrial by-products are put into lakes, rivers, and the oceans. Pesticides, fertilizers, and animal waste from farms can infiltrate the ground and pollute groundwater. Acidic gases resulting from the burning of fossil fuels are emitted into the air where they mix with water vapor and later fall to Earth as acid rain.

Have the students discuss water pollution. To guide the discussion, ask questions such as:

- *What do you think might happen if we did not have clean water?*
- *What human activities do you think could result in the pollution of water?*
- *What things do you think people could do to reduce water pollution?*

7.7 Summary

Review the main points of this chapter with the students.

- *Hydrosphere is the term for all the water that surrounds the Earth. It includes water vapor in the atmosphere and the water on and in the ground and in the oceans.*
- *The hydrologic cycle is the process by which water continuously moves around the Earth.*
- *The oceans contain about 97% of all the Earth's water.*
- *Surface water is found on the land masses in lakes, marshes, swamps, streams, and rivers.*
- *Groundwater is found in the soil and rocks below Earth's surface.*
- *Human activities contribute to water pollution.*

In this experiment students will test different materials for permeability. Permeability is a term that is new to the students and is being introduced in this experiment.

Have the students read through the experiment and write an objective. Some sample objectives are:

- *To see how fast water runs through different materials.*

- *To make a model aquifer and see if water will go through it.*

- *To try to find out what makes a good aquifer.*

Discuss possible hypotheses with the students. Some examples are:

- *The water will go through each material at a different speed.*

- *I can't make a good aquifer because it will be too small.*

- *Layering different materials in an aquifer will change how effective it is.*

Have the students decide how many holes to put in the bottom of each cup and how big the holes should be.

If you are unable to obtain powdered pottery clay, wet pottery can be substituted.

Experiment 7: Modeling an Aquifer   Date: _____

Objective _____

_____

Hypothesis _____

_____

Materials

▸ gravel, about 475 ml (2 cups)

▸ sand, about 475 ml (2 cups)

▸ dirt, about 475 ml (2 cups)

▸ powdered pottery clay, about 475 ml (2 cups)

▸ water

▸ 6 Styrofoam cups, about 355 ml (12 ounce) size

▸ pencil

▸ marking pen

▸ measuring cup

▸ large bowl

▸ scissors

▸ stop watch or clock with a second hand (optional)

Experiment—Part A

❶ Use the pencil to poke holes in the bottom of each cup. Put the same number of holes in each cup and try to make all the holes about the same size.

❷ Put 237 ml (1 cup) of each material (sand, gravel, dirt, and clay) in its own cup. Label each cup with the name of the material it contains.

❸ One at a time, hold each cup over a large bowl and pour about 118 ml (4 ounces) of water into the cup. Note how long it takes for the water to drain through each material. The speed at which water travels through a porous material is called the material's permeability. You can use a timer with a second hand or visually note how quickly or slowly the water runs through each material compared to the others.

❹ In the *Results* section, note how long it takes the water to go through each material and write down any other observations.

Help the students note how long it takes the water to run through each material. They can visually note the speeds at which the materials let water through, but it might be easier for them to use a stopwatch or clock with a second hand to measure the permeability rate. Have them record their results in the chart provided.

Have the students discuss whether they think there is a relationship between the porosity and the permeability of these materials.

Results

Permeability Test

Substance	Relative Permeability & Notes

Have the students discuss what characteristics they think might be present in an effective aquifer and how they might represent these characteristics in the aquifers they build. Have them think about whether different types of aquifers might be good for different uses.

To guide the discussion, ask questions such as:

- *Would you want the aquifer's water to be contained in one area? Why or why not?*

- *Would you want the water to flow through the aquifer quickly? Slowly? Why or why not?*

- *Would you want the water in the aquifer to be able to travel for a long distance? Why or why not?*

- *Do you think the aquifer would have layers of different materials or be made of all the same material? Why or why not?*

If students would like to make a longer aquifer, one or more additional Styrofoam cups can be taped to the first one.

Experiment—Part B

❶ Aquifers hold and carry water under the ground. Look at the results of your tests in *Part A* and think about how you could layer the materials you tested to make an effective aquifer.

❷ Take one of the remaining Styrofoam cups that has holes in the bottom. Using scissors, cut it in half lengthwise (cut down each side and across the bottom) so you have two equal parts. Cut the second cup in half in the same manner.

❸ There are different types of aquifers, including those that hold the water in one area, let it flow short distances, or carry it long distances. Think about how you might use the materials you tested to create different types of aquifers.

❹ Use fresh samples of the materials for this part of the experiment. Take one half of a Styrofoam cup and in it put two or more layers of the materials you tested. This is your aquifer. If you would like to make a longer aquifer, you can tape cup halves together. In the chart in the *Results* section, list the order of the layers in each aquifer you make.

❺ Hold your aquifer at a slight angle over a large bowl. Pour water on the higher end of your aquifer and note what happens. Write your observations in the *Results* section.

❻ Repeat Steps ❹-❺ to make several more aquifers.

Results

Record the layers in each aquifer and your observations about water flow.

Aquifers	
Aquifer Layers (In order from top to bottom)	**Observations**

Have the students write valid conclusions based on the data they have collected.

Conclusions

By doing this experiment, what did you learn about permeability? Do you think permeability and porosity are related? Why or why not? What did you learn about aquifers? In what ways do you think your model aquifers were similar or dissimilar to real aquifers?

_____

_____

_____

_____

_____

_____

_____

_____

_____

_____

_____

_____

_____

_____

_____

Review

Answer the following:

(Answers may vary.)

▶ What is a cycle? _____ *a series of events that repeats* _____

▶ What is the hydrologic cycle? _____ *The way water moves continuously*

around the Earth. The Sun heats the water in the oceans and on land which

makes it evaporate so it goes into the atmosphere. It turns into clouds and

returns to Earth as rain and snow. Rivers take the water back to the ocean.

▶ The names of the five oceans are:

Pacific Ocean

Atlantic Ocean

Indian Ocean

Southern Ocean

Arctic Ocean

▶ What percentage of Earth's water is held in the oceans? _____ *97%* _____

▶ Where is surface water found? _____ *on top of the land in rivers, streams,*

lakes, swamps, marshes, etc.

▶ What is groundwater? _____ *water found in rocks and soil under the*

ground

Chapter 8: The Biosphere

Overall Objectives   83

8.1   Introduction   83

8.2   Cycles in the Biosphere   83

8.3   Ecosystems   85

8.4   Summary   85

Experiment 8: My Biome   86

Review   90

Time Required

Text reading   30 minutes
Experimental   1-2 hours

Materials

pencil, pen, and colored pencils
compass
small backpack
water bottle
notebook or extra paper (optional)

Overall Objectives

In this chapter students are introduced to the biosphere and the interaction of life with various "spheres" in the Earth system. Students will learn about some of the ways in which elements are cycled through life forms and through Earth systems. The concept of ecosystems will be introduced.

8.1 Introduction

Biosphere refers to the part of Earth that supports life. This includes parts of the crust, the hydrosphere, and the lower atmosphere. Plants and animals not only live on Earth but interact with it. They affect the surface of the Earth by providing organic matter to enrich soils and make them able to support plant life. Plants can affect landforms by fracturing rocks with their roots and by preventing erosion with their root structures. Animals can affect landforms by making trails that can lead to erosion, by building dams and burrows in the ground, and by stirring up the soil and moving stones with their hooves.

Guide inquiry with questions such as the following:

- *Do you think the hydrosphere is important to the biosphere? Why or why not?*

- *Do you think the atmosphere is part of the biosphere? Why or why not?*

- *In what ways do you think plants and animals might affect the other spheres of Earth?*

- *Do you think animals could exist if there were no soil? Why or why not?*

8.2 Cycles in the Biosphere

In this chapter students explore some of the cycles in the biosphere that make life possible by recycling elements through the Earth's systems. As you review each cycle with the students, it would be helpful to have them look at the accompanying illustration in the student text to help them understand how the cycle works.

The Oxygen Cycle

Plants and animals are the primary means by which oxygen is cycled. During photosynthesis plants take in sunlight, carbon dioxide, and water to make food for themselves in the form of sugars. They release oxygen as a waste product. Animals breathe in the oxygen and then give off carbon dioxide as a waste product which is then used by plants, continuing the cycle. In this way oxygen atoms are reused.

The Carbon Cycle

Carbon is an element that is needed by all living things and makes up about 18% of the human body. During photosynthesis plants use the carbon from carbon dioxide to make molecules needed for them to live. Animals obtain carbon by eating plants or other animals that have eaten plants. When plants and animals die, decomposition returns carbon to the atmosphere as carbon dioxide where it is again available to plants.

The Nitrogen Cycle

Although nitrogen makes up about 78% of Earth's atmosphere and is essential to life, nitrogen is not usable by plants in the form in which it exists in the atmosphere. Nitrogen is carried from the atmosphere by precipitation and enters the soil as rainwater or snowmelt infiltrates it. Bacteria in the soil use a process called nitrogen fixation to convert the nitrogen to a form usable by plants, and the nitrogen is then be taken up by the roots. Nitrogen becomes available to animals when the animals eat the plants or other animals that have eaten plants. Nitrogen reenters the atmosphere when plants, animals, and animal manure decompose.

The Energy Cycle

In the oxygen, nitrogen, and carbon cycles, elements that exist on Earth are cycled through plants and animals, and the same atoms are used over and over. In contrast, in the energy cycle heat and light energy from the Sun is continuously coming to Earth where it can be absorbed and radiated or reflected back into space. Plants use energy from the Sun during photosynthesis, and animals indirectly obtain energy from the Sun by eating plants or animals that have eaten plants.

Have the students review the illustration in this section of the student text and help them observe how incoming energy from the Sun is distributed.

Guide inquiry about cycles in the biosphere by asking questions such as:

- *Do you think plants are important to the air we breathe? Why or why not?*

- *Do you think animals are part of the energy cycle? If so, how?*

- *If there were no bacteria in the soil, do you think life could exist? Why or why not?*

- *Does the energy cycle work in the same way that the oxygen, carbon, and nitrogen cycles do? Explain how it works.*

8.3 Ecosystems

Environment refers to the set of conditions surrounding an organism in the region where it lives. Ecologists study living things in their environments. An ecosystem is a specific area that contains a community of living things existing under similar conditions. An ecosystem can be any size—from very small to very large. On the other hand, a biome is always a very large region and is an ecosystem that is defined by the climate, soils, and plant life that exist within it.

Scientists disagree on how many biomes there are, but most seem to agree that there are at least five main biomes: aquatic, desert, forest, grasslands, and tundra. Each of these can then be subdivided according to climate, soils, plant life, and other characteristics.

8.4 Summary

Discuss the summary statements with the students.

- *The biosphere is the part of Earth that supports life.*
- *The biosphere includes the lower part of the atmosphere, the part of the crust where life exists, and the oceans.*
- *Some important cycles in the biosphere are the oxygen cycle, the carbon cycle, the nitrogen cycle, and the energy cycle.*
- *Ecology is the study of the relationship of living things to the world around them.*
- *An ecosystem is a specific area that contains a community of living things existing under similar conditions. An ecosystem can be any size.*
- *A biome is a large ecosystem that is defined by the climate, soils, and plant life that exist within it.*

In this experiment students will observe the environment in which they live.

Have the students read the entire experiment and then write an objective. Some examples are:

- *To observe the environment where I live.*

- *To look at things in nature and then figure out what kind of biome I live in.*

- *To observe plants and animals in my area and record my observations.*

Have the students write a hypothesis. Some examples are:

- *I will be able to make enough observations of plants and animals and other things to tell what biome I live in.*

- *By observing the environment outdoors, I can tell what biome I live in.*

- *I live in a city so I won't be able to tell what biome I'm in.*

Have the students think about what route they would like to take on their hike. If they live in a city, help them decide where they could hike to see the most plants and animals.

Encourage the students to look all around them—to observe living things both big and small, up high and down low. Have them take their time and observe carefully.

Students may wish to take along a notebook or extra paper so they can take more notes and/or make drawings.

Experiment 8: My Biome   Date: _____

Objective
_____

_____

Hypothesis
_____

_____

Materials

▸ pencil, pen, and colored pencils
▸ compass
▸ small backpack
▸ water bottle
▸ notebook (optional)

Experiment

❶ Pack a small backpack with your water bottle, pencil, *Laboratory Workbook*, and compass—and maybe an extra notebook.

❷ Take a one to two hour hike in your surrounding environment.

❸ Record what you see. Pay attention to everything around you. Carefully observe the plant and animal life, where they live, and what they are doing. Use the compass to see which direction you're going in.

❹ Describe the types of plants in your surroundings and the types of animals (both large and small) and observe their interactions. Are ants crawling on plants? Are there dogs playing in the grass? Are there cats looking for birds or insects to catch? Are birds flying?

❺ Observe the weather and how the weather may affect the type of plant or animal life in your surroundings. Is there snow on the ground and do the plants look lifeless? Is it warm and sunny and are there insects out foraging for food? Is it raining? Or windy? Are the birds sitting quietly waiting for the storm to pass?

Observations of My Biome

More Observations of My Biome

Have the students record their observations in the boxes provided. They may also use additional paper.

Have the students select the observations they think are the most important in describing their biome.

Results

Summarize your observations below.

Summary of My Biome Observations

Types of Plants	
Types of Animals	
Interactions	
Weather	

Conclusions

Use the chart on page 80 of the student text to determine the type of biome you live in. Record the name of your biome, describe it, and list any unique features you discovered.

_____

_____

_____

_____

_____

_____

_____

_____

_____

_____

_____

_____

_____

_____

_____

_____

Have the students review the biome chart in the student text, decide which type of biome they live in, and write about it.

Review

Answer the following questions:

(Answers may vary.)

▸ What is the biosphere? _____

_____the part of the Earth where living things exist_____

_____

▸ Name four cycles that occur in the biosphere.

_____oxygen cycle_____

_____carbon cycle_____

_____nitrogen cycle_____

_____energy cycle_____

▸ In your own words, describe one of the cycles in the biosphere.

_____In the carbon cycle, plants use carbon dioxide to make food and to get_____

_____carbon to make molecules. Animals get carbon by eating plants or other_____

_____animals that eat plants. The plants and animals die and carbon dioxide_____

_____goes back into the air._____

_____

▸ What is an ecosystem? _____

_____a specific area that has a community of similar living things living under_____

_____similar conditions_____

Chapter 9: The Magnetosphere

Overall Objectives   92

9.1   Introduction   92

9.2   Magnets   92

9.3   Earth's Magnetic Field   92

9.4   What Is the Magnetosphere?   93

9.5   Magnetic North and South Poles   94

9.6   Magnetic Field Reversals   94

9.7   Summary   95

Experiment 9: Finding North   96

Review   99

Time Required

Text reading	30 minutes
Experimental	1 hour

Materials

steel needle
bar magnet
slice of cork
tape
medium size bowl
water
1 additional bar magnet (optional)

Overall Objectives

In this chapter students will learn about magnets and magnetism. They will gain an understanding of Earth's magnetic field and how it interacts with the Sun to form the magnetosphere.

9.1 Introduction

Earth's magnetic field can be used for navigation with a compass, and it causes the northern and southern lights.

9.2 Magnets

It may help the students' understanding of magnets and magnetism to have two bar magnets available for them to experiment with. Have them observe how like poles repel and unlike poles attract and how magnets will stick to some metallic objects and not others.

A magnet is an object that produces a force called magnetism which causes certain metallic objects to either attract or repel each other. The area around a magnet that is affected by the force of magnetism is called the magnetic field.

A lodestone is a form of the mineral magnetite and has naturally occurring magnetic properties. The force of magnetism was discovered by observing lodestones and how they affect some metallic objects as well as how they are affected by the geomagnetic field. Materials that are magnetic or can be influenced by a magnetic field are called ferromagnetic, with iron and nickel being the most common.

A magnet has two poles which are called the north pole and the south pole, and these have opposite charges (positive and negative). Since like charges repel and unlike charges attract, when the same poles of two magnets are near each other, they will repel, and when a north and a south pole are near each other, they will attract. (See the *Focus On Middle School Physics Student Textbook* for a more complete discussion of magnets.)

Guide inquiry with questions such as the following:

- *What are some ways that magnets might be useful?*
- *Do you think magnets occur in nature? Why or why not?*
- *If you had two magnets, do you think you could make two of the same type of pole stick together? Why or why not?*
- *What advantage could it be that two opposite poles on a magnet attract each other?*

9.3 Earth's Magnetic Field

Like a bar magnet, Earth has a north pole and a south pole and thus is said to have a dipole magnetic field. *Di* is from Greek and means two. The needle of a compass points north because the magnetized metallic needle is attracted to Earth's North Magnetic Pole. Earth's magnetic field is referred to as the geomagnetic field and comes out from the areas at the poles and surrounds the Earth.

Scientists think that Earth's magnetic field is electromagnetic. An electromagnet can be built by running an electric current through a wire that is coiled around an iron rod. With this type of magnet, the magnetic field can be strengthened by moving the rod up and down within the coil.

The geomagnetic field is thought to be generated deep within Earth in the liquid outer core where electrical forces are created due to convection currents caused by variations in the temperature, pressure, and composition of molten iron and nickel. In addition to convection currents, the spin of the Earth causes a spiralling, or coiling, motion in the molten materials. This is called the Coriolis force which also aligns the spirals (coils) of molten materials into a north/south orientation. The resulting electrical forces work as an electromagnet, creating the geomagnetic field.

Guide a discussion about Earth's magnetic field using questions such as:

- *Do you think Earth's geomagnetic field has any features that are similar to a bar magnet? If so, what are they?*
- *How can you tell that Earth has a geomagnetic field?*
- *How do you think the geomagnetic field is made?*

9.4 What Is the Magnetosphere?

The magnetosphere is the area in space that contains Earth's geomagnetic field. The magnetosphere is created by the interaction of the geomagnetic field with solar wind, the rapidly moving gases emitted by the Sun. Solar winds travel at a rate of about 400 km/second (1 million miles per hour), although the speed varies. The area where the solar wind collides with the geomagnetic field is called the bow [bau] shock, and this collision causes a flattening effect on the part of the geomagnetic field that is facing the Sun. The magnetosphere deflects most of the of the solar wind gases which then flow around the outside of the magnetosphere, pulling it out behind Earth in a long tail called the magnetotail. The magnetosphere protects Earth from receiving too much harmful radiation from the Sun, thus making Earth habitable.

Solar storms are blasts of hot gases from the Sun. Auroras occur when the charged particles from solar storms enter Earth's atmosphere at the poles, interacting with charged particles in the atmosphere. Solar flares are huge explosions on the surface of the Sun that can release as much energy as a billion megatons of TNT. Solar flares don't affect Earth as much as do coronal mass ejections (CMEs) which are violent ejections of charged particles from the Sun. CMEs can cause geomagnetic storms in the magnetosphere that can disrupt electrical power grids on Earth and communications satellites in orbit. The most spectacular auroras are caused by CMEs.

Have the students discuss the magnetosphere. Guide inquiry by asking questions such as:

- *How is the magnetosphere is formed?*
- *Would the magnetosphere change if there were no solar wind? If so, how?*
- *What would happen to life on Earth if there were no magnetosphere?*
- *Does anything happen on Earth that shows that there has been a big solar storm? If so, describe what happens.*

9.5 Magnetic North and South Poles

Earth's geomagnetic field is slightly tilted with respect to the Geographic North and South Poles which are located at the ends of Earth's axis of rotation. When a compass points to north it is pointing to the Magnetic North Pole rather than the Geographic North Pole. The magnetic poles wander over time, and this is thought to be caused by changes in the flow of the molten metals in the outer core. In the 1600s the Magnetic North Pole was traveling south. Then, in the mid 1800s it gradually turned and began heading back toward the Geographic North Pole. Currently, the north pole is moving at a rate of about 64 km (40 miles) per year, and in the early 20th century the rate was about 16 km (10) miles per year.

Guide student discussion with questions such as:

- *Could you find the Geographic North Pole with a compass? Why or why not?*
- *Over time, will a compass needle always point to the same location on Earth? Why or why not?*
- *Why do you think the Magnetic North Pole gradually changes its position?*

9.6 Magnetic Field Reversals

By taking core samples of ancient rocks and studying the magnetically induced alignment of metallic elements, paleomagnetists have been able to determine where the magnetic poles were located at different times in Earth's history. They can also see that Earth's magnetic field has reversed many times, with north becoming south and south becoming north. From fossil records, scientists have concluded that magnetic field reversals do not affect life on Earth—they do not leave Earth temporarily without a magnetic field, cause mass extinctions, or affect the rotation of the Earth.

Have the students discuss magnetic field reversals and paleomagnetism. Ask questions such as:

- *Does the Magnetic North Pole ever make a big shift in position? Why or why not?*
- *How can scientists study the geomagnetic field and its changes over time?*

9.7 Summary

Review the summary statements with the students.

- *Magnetism is the force that causes certain metallic objects to attract or repel each other.*
- *The magnetic field is the area around a magnet that is affected by magnetism.*
- *Electrical forces created by the movement of molten metals in Earth's outer core create the geomagnetic field.*
- *The magnetosphere is created by the interaction of Earth's geomagnetic field with gases from the Sun.*
- *Solar wind is the steady stream of hot gases that comes from the Sun.*
- *Paleomagnetism is the study of ancient rocks to learn the history of Earth's magnetic field.*

In this experiment students will make a compass and determine how to use it to find magnetic north.

Before the experiment begins, take a cork (such as from a wine bottle) and cut off a piece big enough to float upright in water and keep the needle taped to the top of it dry.

Have the students read the entire experiment and then write an objective. Some possible objectives are:

- *To make a compass that works.*

- *To find out how to find north.*

- *To see if a needle can become magnetized and used to find north.*

Have the students write a hypothesis. For example:

- *A compass can be made with a needle and used to find north.*

- *A bar magnet will make a needle into a magnet.*

- *The direction of north can be found with a homemade compass.*

Have the students test the needle to make sure it's magnetized before taping it to the cork.

When the cork stops turning on the water, the needle should be pointing in a north/south direction. However, this does not determine which end of the needle is pointing north.

Experiment 9: Finding North   Date: _____

Objective _____
_____

Hypothesis _____
_____

Materials

▸ steel needle

▸ bar magnet

▸ slice of cork

▸ tape

▸ medium size bowl

▸ water

Experiment

❶ Take the bar magnet and slowly stroke the needle against it for about 45 seconds. This will magnetize the needle.

❷ Find an object that the magnet will stick to. Now see if the needle will stick to that object. If the needle doesn't stick, rub it against the magnet for a while more, and then test it again.

❸ Center the magnetized needle on the top surface of the piece of cork and tape it in place.

❹ Pour water into the bowl until it is almost full, and carefully place the cork in the center of the bowl so it is floating, needle-side up. The needle should not be touching the side of the bowl.

❺ Observe what happens. Can you tell which way is north? Why or why not?

❻ In the following box, note your observations.

Observations

---

❼ In the box in the *Results* section, draw a simple map of the room you are in. Indicate the walls, doorway, and any windows.

❽ Place your map on the table next to your cork and needle compass. Turn the map so that it is in the same orientation as the room. (Match the direction of the walls on your map to the walls in the room.)

In the middle of the map, draw a line that goes in the same direction as the needle of your compass.

❾ Think about the approximate locations of sunrise and sunset in relation to the room you're in. Knowing that the sun rises in the east and sets in the west, mark the approximate locations of east (E) and west (W) on your map.

❿ Can you now tell which end of the needle is pointing north? If so, mark it on your map with the symbol N.

Results

Map to the Magnetic North Pole

Have the students write about or draw their observations in the box provided.

In order to find out which end of the needle is pointing north, students are asked to make a map. In the *Results* section, have them draw a simple map of the walls of the room where the experiment is being performed. They should indicate the doorway(s) and windows to help orient the map. They may also indicate furniture or other features if they wish to do so.

Have the students place their map on the table next to the bowl of water containing the compass. Help them turn the map so that it is in the same orientation as the walls of the room. Now have them draw a line in the middle of the map that matches the orientation of the needle. This is the north/south direction.

Help the students think about the approximate location of sunrise and sunset in relation to the room they're in. Once they have noted the approximate location of east and west on their map, they will be able to tell which end of the needle is pointing north and can note this on their map. If they have forgotten which direction north is compared to east and west, have them look back at their map in *Experiment 2: Hidden Treasure.*

Since the locations of east and west are only an approximation on this map, they will not be at right angles (90°) from north.

Have the students review their observations and write valid conclusions.

Conclusions

Based on your observations, what conclusions can you make about the ease or difficulty of making a compass and finding north. What other conclusions can you draw from your observations?

_____

_____

_____

_____

_____

_____

_____

_____

_____

_____

_____

_____

_____

_____

Review

Answer the following questions:

(Answers may vary.)

▶ Another name for Earth's magnetic field is:

geomagnetic field

▶ In your own words describe how the magnetosphere is created.

Solar winds from the Sun collide with Earth's magnetic field, flattening

it. Then the solar wind slides around the sides of the magnetic field,

pulling it out into the magnetotail.

▶ Are the Magnetic North Pole and the Geographic North Pole the same? Why or why not?

No. The Geographic North Pole is at the end of Earth's axis of rotation

and the Magnetic North Pole is at the end of Earth's geomagnetic field.

They are in different places.

▶ When the Magnetic North Pole turns into the Magnetic South Pole, this event is called:

geomagnetic reversal

▶ Paleomagnetism is:

the study of ancient rocks to learn about Earth's geomagnetic field

Chapter 10: Earth as a System

Overall Objectives   101

10.1   Introduction   101

10.2   Interdependence   101

10.3   Natural Events   102

10.4   Human Activities   102

10.5   Earth System Science   103

10.6   Summary   104

Experiment 10: Solve One Problem   105

Review   112

Time Required

Text reading   30 minutes
Experimental   1 hour or more

Materials

pencil, pen, and colored pencils
imagination

Overall Objectives

In this chapter students will begin to look at the ways in which the different spheres of Earth interact with each other.

10.1 Introduction

Geologists study each of Earth's spheres individually. The way each sphere works to renew itself and stay in balance is extremely complicated. To add to this complexity, the various spheres interact with each other while working together to form the Earth system as a whole.

Guide inquiry with questions such as the following:

- *Do you think an erupting volcano would interact with more than one of Earth's spheres? How and which ones?*
- *Which spheres are involved when it rains? How are they affected?*
- *How would an earthquake affect different spheres and which ones would it affect?*
- *Would a hurricane affect any spheres besides the atmosphere? How?*

10.2 Interdependence

Earth's various spheres interact with and depend on each other. A tropical rainforest is given as an example of interdependence.

The plant life of the biosphere in the rainforest interacts with the atmosphere during photosynthesis when plants exchange carbon dioxide and oxygen. The atmosphere carries water from the hydrosphere, and it falls as rain, interacting with the lithosphere and the biosphere as the rainfall pools and seeps into the ground. The atmosphere also provides nitrogen to the soil in the lithosphere. Bacteria in the biosphere fix the nitrogen so plants, also in the biosphere, can take the nitrogen in through their roots along with water from the hydrosphere. The atmosphere and magnetosphere allow the proper amount of energy from the Sun to fall on the rainforest where the plants can use this energy during photosynthesis.

A disruption of the functioning of any one of these spheres could affect the entire rainforest. For example, too little or too much rain, or too little or too much energy from the Sun could kill trees and other plants. Too little vegetation would deprive animals of food and shelter.

Guide inquiry with questions such as the following:

> • *What are some additional ways different spheres might interact in a tropical rainforest?*
>
> • *Think about the area in which you live. How do the different spheres interact to make the area the way it is?*
>
> • *How could a change in one sphere affect other spheres in the area where you live? How would this make your area different?*

10.3 Natural Events

Natural events are outside human control. Volcanic eruptions, earthquakes, hurricanes, and tsunamis can cause widespread devastation and can affect many of Earth's spheres.

The huge eruption of Mount Tambora in Indonesia in 1815 is estimated to have killed about 10,000 people with flows of hot rock, gas, and ash. An even greater problem occurred as a result of the volcano spewing a large quantity of sulfates into the atmosphere, causing a cooling effect on the climate of the northeastern United States, Canada, and western Europe, as well as Indonesia. People sometimes refer to 1816 as "Eighteen Hundred and Froze to Death" or "The Year Without a Summer" because there were freezing temperatures even in the middle of the summer, killing crops and causing great hardship. It is estimated that over 100,000 more people in Indonesia died as a result of the climatic effects of the volcano and perhaps hundreds of thousands in Europe and North America.

Guide inquiry with questions such as the following:

> • *What other natural events can you think of that can affect Earth's spheres? Which spheres would be affected and in what ways?* (tornadoes, blizzards, high winds, forest fires, etc.)
>
> • *Do you think scientists will someday be able to predict natural events ahead of time? Why or why not?*

10.4 Human Activities

Human activities can affect the various Earth spheres. For example, in the vicinity of a mine the lithosphere, biosphere, atmosphere, and hydrosphere can be disrupted. The burning of fossil fuel in machinery can lead to the formation of acid rain which can affect the atmosphere, hydrosphere, and biosphere. Plants are dug up, the flow of ground and surface water could be changed, and animal habitats may be destroyed.

Guide inquiry with the following questions:

- *What are some human activities that you think might affect various spheres in a negative way? Which spheres would be affected?*

- *What are some human activities that might affect various spheres in a positive way? Which spheres would be affected?*

10.5 Earth System Science

Scientists in many different branches of science have long studied each of Earth's spheres separately. Earth system science is a relatively new scientific discipline that gathers data from these studies of individual spheres and puts the data together to begin to understand how Earth operates as an integrated system. These scientists are interested in finding out more about how the different spheres interact, Earth changes that are occurring and what they mean, the ways human activities affect Earth's system, and ways to protect and conserve resources.

To conduct a study of a geological event, an earth system scientist would ask which spheres caused the event, which spheres were affected as a result of the event, and how did the changes in a sphere that resulted from the event then affect other spheres.

Some Earth system scientists are looking for new approaches to help find solutions to problems such as pollution and overuse of resources. These scientists focus on sustainability—finding innovative ways to use resources in ways that do not deplete them or damage Earth. Sustainable forms of energy production that don't create pollution include solar panels, windmills, and hydroelectric dams. There are new wastewater treatment facilities that use the interaction of different spheres to purify water by circulating it through a wetland-like environment rather than using chemicals. There are scientists who are studying new uses for plants, such as mushrooms, for medicines and to help clean the environment.

Space research is also a part of Earth system science. With advances in technology, scientists can learn more about the magnetosphere and how it works, how the Sun affects the Earth and the rest of the solar system, and how the universe outside our solar system interacts with Earth.

Guide student discussion about Earth system science with questions such as:

- *Why do you think Earth system science is an important area of study?*

- *What questions do you think an Earth system scientist would ask when studying a volcanic eruption?*

- *Do you think it is important for scientists to study technology such as solar panels and windmills? Why or why not?*

- *If you were an Earth system scientist, what would you want to study?*

10.6 Summary

Review the summary statements with the students.

- *The different spheres of Earth are interdependent, interacting with and relying on each other in order to function properly.*

- *A change in conditions in one sphere can cause changes to other spheres.*

- *Human activities can lead to changes in Earth's spheres.*

- *Earth system science is a field of study in which information from the study of Earth's individual spheres is combined in order to look at Earth as a whole.*

- *Many Earth system scientists are interested in sustainability— finding out how to use resources in ways that do not deplete them or damage the Earth system.*

Experiment 10: Solve One Problem   Date: _____

Objective   _____

_____

Hypothesis   _____

_____

Materials

▶ pencil, pen, and colored pencils

▶ imagination

Experiment

❶ Spend a day observing your neighborhood. Go outside and watch how the people in your neighborhood interact with their surroundings. Observe how people's interactions affect other people's lives, plant life, and other animal life.

❷ Make a list of all the problems you discover. Include any details you notice. For example, maybe your neighbor is elderly and can't get her trash to the curb, so it piles up, spills over, and then the dogs eat it and scatter it across the street and get sick. Or maybe your neighbors walk their dog but never pick up their dog's waste and, unhappily, other people step in it, tracking it into the local store where it contaminates a cantaloupe that has fallen on the floor on top of it.

❸ Record how many different spheres these problems influence. For example, trash in the yard interacts with: Earth's crust (trash in the dirt), the biosphere (animals eat the trash), the atmosphere (the trash gives off an odor), and possibly the hydrosphere (if it rains and the water becomes contaminated).

In this experiment students will explore concepts of Earth system science.

Students are asked to observe their neighborhood and identify any problems they notice. They are then asked to think of solutions to these problems.

Have the students read through the entire experiment and then write an objective. Following are some examples:

- *To find solutions to problems in the neighborhood.*

- *To see what problems the neighbors have.*

- *To find out how I can help my neighbors.*

Next, have the students write a hypothesis. For example:

- *I can get my friends together to help a neighbor with a problem.*

- *I can make problem solving into a non-profit business so I can help lots of neighbors.*

- *I can identify several problems in the neighborhood.*

Have the students spend some time walking through their neighborhood and noticing what people are doing and how they interact with each other, with animals and plants, etc. Have them take their *Laboratory Workbook* or a notebook with them so they can write down their observations, including details.

In this chart students will note the problems they have observed and any details about the problems. Then they will note which of Earth's spheres are affected by each problem.

Observations of Problems

Problem	Spheres Influenced

❹ Examine the problems and imagine a solution to each problem that you could actually carry out. For example, if your neighbor can't get the trash to the curb in time for the trash pick up, then maybe you could volunteer to move the trash for her. List your solutions below.

Solutions to Problems

Problem	Solution

An important part of Earth system science is looking at human activities, seeing how they affect various spheres, and then finding solutions to problems that are created by these human activities.

Encourage the students to use their imagination in thinking of solutions to the problems they've noticed. Let them record their ideas even if they seem impractical. The important thing is for the students to collect data and begin to analyze them in a problem solving manner.

Here and on the next page again let the students use their imagination even if their expanded solutions seem impractical.

❺ Now that you have solved the problems, explore how you might expand your solutions to solve the problems for a larger number of people. For example, perhaps you can recruit 5 friends to volunteer so 6 elderly people will have help moving their trash to the curb every week. List how you could expand your solutions to serve more people.

Expanded Solutions

Problem	Expansion

6. Now that you have found a way to expand your solutions to serve a greater number of people, explore how you could turn one of your ideas into a business that would allow you to generate resources and jobs for others while helping to solve more problems. For example, you could start a non-profit company and get donations from the city government to help recruit more volunteers to help more elderly with their trash. Describe below how your idea could become a business or non-profit organization.

Have the students answer the questions using the data they've collected in this experiment.

Summarize your results below.

Results

Questions	Ideas
What was the problem?	
Which of Earth's spheres did the problem influence?	
What was the solution?	
How could you expand the solution to serve more people?	
How could you turn your solution into a business or non-profit?	
When will you start implementing your solution in real life?	

Conclusions

It is said that *knowledge is power*. Describe below how understanding the science of the Earth—how it is put together and how it works—can help give you the power to solve real-life problems.

_____

_____

_____

_____

_____

_____

_____

_____

_____

_____

_____

_____

_____

_____

_____

_____

_____

_____

_____

Have the students discuss what they have learned about the interaction of Earth's spheres and Earth system science. Then have them review the charts they made during this experiment and use the data in describing how a knowledge of Earth system science can help them solve problems.

Review

Answer the following questions:

(Answers will vary.)

▶ Describe how the Earth's spheres are interdependent.

Earth's spheres interact and rely on each other. In a tropical rainforest the rain comes from the atmosphere and falls on the ground where it goes into the soil of the lithosphere and the plants in the biosphere can use it to make food. The atmosphere and magnetosphere let in the right amount of energy from the Sun so plants in the biosphere can do photosynthesis.

▶ Describe how human activities can cause changes in Earth's spheres.

For example, a big mine puts a big hole in the ground and this hole changes the lithosphere and can change the water flow in the hydrosphere. Plants are dug up and animals move out, changing the biosphere. Machines can pollute the air making acid rain which affects the atmosphere, hydrosphere, and biosphere.

▶ How do Earth system scientists study Earth?

They take data that has come from the study of individual spheres and put the data together to learn about how all the spheres of Earth interact with each other. They look at the ways that Earth is changing to see if they can find out what this means. They look at how resources can be used in a sustainable way so they are not used up and don't harm the Earth.

Made in the USA
Columbia, SC
04 September 2018

We support ASE

MW00838190

FUNDAMENTALS OF

Automotive Maintenance
and Light Repair

SECOND EDITION

STUDENT WORKBOOK

JONES & BARTLETT
LEARNING

World Headquarters
Jones & Bartlett Learning
5 Wall Street
Burlington, MA 01803
978-443-5000
info@jblearning.com
www.jblearning.com

Jones & Bartlett Learning books and products are available through most bookstores and online booksellers. To contact Jones & Bartlett Learning directly, call 800-832-0034, fax 978-443-8000, or visit our website, www.jblearning.com.

17782-4

Production Credits

General Manager: Kimberly Brophy
VP, Product Development: Christine Emerton
Product Owner: Kevin Murphy
Product Development Manager: Amanda Brandt
Development Editor: Carly Mahoney
Project Specialist: Brooke Haley
Director of Marketing: Brian Rooney
VP, Manufacturing and Inventory Control: Therese Connell
Composition: S4Carlisle Publishing Services
Project Management: S4Carlisle Publishing Services
Cover Design: Scott Moden
Text Design: Scott Moden
Senior Media Development Editor: Shannon Sheehan
Rights Specialist: Maria Leon Maimone
Cover Image: Front cover: © E+/Getty Images. Back cover: © blue jean images RF/Getty Images
Printing and Binding: CJK Group Inc.
Cover Printing: CJK Group Inc.

6048

Printed in the United States of America
23 22 21 20   10 9 8 7 6 5 4 3 2

Contents

CHAPTER 1 Careers in Automotive Technology . . . 1

CHAPTER 2 Introduction to Automotive Technology 6

CHAPTER 3 Introduction to Automotive Safety . . . 15

CHAPTER 4 Personal Safety 23

CHAPTER 5 Vehicle Service Information and Diagnostic Process 27

CHAPTER 6 Hand and Measuring Tools 36

CHAPTER 7 Power Tools & Equipment 45

CHAPTER 8 Fasteners and Thread Repair 49

CHAPTER 9 Vehicle Protection and Jack and Lift Safety 54

CHAPTER 10 Vehicle Maintenance Inspection . . . 61

CHAPTER 11 Communication and Employability Skills 67

CHAPTER 12 Motive Power Theory—SI Engines 73

CHAPTER 13 Engine Mechanical Testing 91

CHAPTER 14 Lubrication System Theory 97

CHAPTER 15 Servicing the Lubrication System 110

CHAPTER 16 Cooling System Theory 115

CHAPTER 17 Servicing the Cooling System 129

CHAPTER 18 Automatic Transmission Fundamentals 136

CHAPTER 19 Maintaining the Automatic Transmission/Transaxle 147

CHAPTER 20 Hybrid and Continuously Variable Transmissions 152

CHAPTER 21 Manual Transmission/Transaxle Principles 160

CHAPTER 22 The Clutch System 172

CHAPTER 23 Driveshafts, Axles, and Final Drives 190

CHAPTER 24 Wheels and Tires Theory 198

CHAPTER 25 Servicing Wheels and Tires 207

CHAPTER 26 Steering Systems Theory 212

CHAPTER 27 Servicing Steering Systems 225

CHAPTER 28 Suspension Systems Theory 230

CHAPTER 29 Servicing Suspension Systems 246

CHAPTER 30 Wheel Alignment 252

CHAPTER 31 Principles of Braking 257

CHAPTER 32 Hydraulic and Power Brakes Theory 267

CHAPTER 33 Servicing Hydraulic Systems and Power Brakes 281

CHAPTER 34 Disc Brake Systems Theory 285

CHAPTER 35 Servicing Disc Brakes 290

CHAPTER 36 Drum Brake Systems Theory 299

CHAPTER 37 Servicing Drum Brakes 309

CHAPTER 38 Wheel Bearings 319

CHAPTER 39 Electronic Brake Control 327

CHAPTER 40 Principles of Electrical Systems 338

CHAPTER 41 Electrical Components and Repair 345

CHAPTER 42 Meter Usage and Circuit Diagnosis 350

CHAPTER 43 Battery Systems 355

CHAPTER 44 Starting and Charging Systems . . . 362

CHAPTER 45 Lighting Systems 375

CHAPTER 46 Body Electrical System 385

CHAPTER 47 Principles of Heating and
Air-Conditioning Systems. 393

CHAPTER 48 Ignition Systems 402

CHAPTER 49 Gasoline Fuel Systems 420

CHAPTER 50 Engine Management System 428

CHAPTER 51 On-Board Diagnostics. 433

CHAPTER 52 Induction and Exhaust. 439

CHAPTER 53 Emission Control 451

CHAPTER 54 Alternative Fuel Systems. 460

Careers in Automotive Technology

At the start of each chapter, you will find the Learning Objectives from the textbook. These are your objectives as you make your way through the exercises in this workbook and the chapter in your textbook. The following activities have been designed to help you refresh your knowledge of the material in this chapter.

Learning Objectives

After reading this chapter, you will be able to:

- LO 1-01 Outline the history of the automobile.
- LO 1-02 Describe the careers in the automotive service sector.
- LO 1-03 Describe each type of repair facility.
- LO 1-04 Describe the importance of automotive industry certification and ongoing training.

Matching

Match the following terms with the correct description or example.

- **A.** Automotive Service Excellence (ASE)
- **B.** Brake technician
- **C.** Drivability technician
- **D.** Heavy line technician
- **E.** Lube technician
- **F.** Service consultant/advisor
- **G.** Shop foreman

_____ 1. A technician who undertakes major engine, transmission, and differential overhaul and repair.

_____ 2. A technician who carries out scheduled maintenance activities.

_____ 3. A supervisor who oversees the work of technicians and staff.

_____ 4. An independent, nonprofit organization dedicated to the improvement of vehicle repair through the testing and certification of automotive professionals.

_____ 5. A technician who services vehicle brake systems.

_____ 6. A technician who diagnoses and identifies mechanical and electrical faults that affect vehicle performance and emissions.

_____ 7. A customer service worker who works with both customers and technicians.

Multiple Choice

Read each item carefully, and then select the best response.

_____ 1. Who is generally acknowledged to have invented the modern automobile around 1885?
- **A.** Henry Ford
- **B.** Karl Benz
- **C.** Armand Peugeot
- **D.** Charles Rolls

_____ 2. Who is a repair shop's first point of contact for customers seeking vehicle repairs?
- **A.** Shop foreman
- **B.** Service consultant/advisor
- **C.** Drivability technician
- **D.** Service manager

_____ **3.** What type of technician might specialize in particular vehicle systems, such as engines, transmissions, or final drives?
- **A.** Shop foreman
- **B.** Light line technician
- **C.** Drivability technician
- **D.** Heavy line technician

_____ **4.** What type of technician diagnoses and repairs faults; replaces or overhauls brake systems; and tests the components of disc, drum, and power brake systems used on all types of vehicles?
- **A.** Light line technician
- **B.** Heavy line technician
- **C.** Brake technician
- **D.** Drivability technician

_____ **5.** What type of technician works with computer-controlled systems to service, identify, and repair faults on electronically controlled vehicle systems such as fuel injection, ignition, antilock braking, cruise control, and automatic transmissions?
- **A.** Electrical technician
- **B.** Heavy line technician
- **C.** Shop foreman
- **D.** Service manager

_____ **6.** What type of technician regularly uses meters, oscilloscopes, circuit wiring diagrams, and solder equipment?
- **A.** Light line technicians
- **B.** Transmission specialists
- **C.** Electrical technicians
- **D.** Heavy line technicians

_____ **7.** Which of the following is a key skill of a service manager?
- **A.** Test-driving vehicles
- **B.** Changing oil and filters
- **C.** Creating positive work environments
- **D.** Having a good understanding of electrical theory

_____ **8.** What types of shops are usually independent and focus on one type of service, such as transmission service, electrical system repair, or emission system diagnosis?
- **A.** Dealerships
- **B.** Specialty shops
- **C.** Franchises
- **D.** Fleet shops

_____ **9.** Which of the following programs receives access to new vehicle technology as well as manufacturer service information to help prepare students for working on today's vehicles and technology?
- **A.** National Automotive Technicians Education Foundation
- **B.** Automotive Service Excellence
- **C.** Automotive Youth Educational Systems
- **D.** Advanced Engine Performance certification

_____ **10.** Technicians who handle refrigerants or work on AC systems are required to have which of the following?
- **A.** Environmental Protection Agency Section 609 certification
- **B.** ASE A7 Heating, Ventilation and Air Conditioning certification
- **C.** NATEF Refrigerant certification
- **D.** AYES R134a certification

_____ **11.** Today's vehicles are assembled on high-volume production lines, with robots used for many of the assembly processes, including welding seams.
- **A.** True
- **B.** False

_____ **12.** Heavy line technicians diagnose and replace the mechanical and electrical components of motor vehicles, such as gaskets, belts, hoses, timing belts, water pumps, radiators, alternators, and starters.
   A. True
   B. False

_____ **13.** Drivability technicians perform wheel alignments and wheel balancing, and they diagnose and replace faulty steering system components.
   A. True
   B. False

_____ **14.** In larger shops, roles may be assigned to separate electrical and drivability technicians, whereas in smaller shops, one technician could perform both roles.
   A. True
   B. False

_____ **15.** Electrical technicians test and replace faulty charging system components, starter motors, and related items such as batteries.
   A. True
   B. False

_____ **16.** Light line technicians use electronic test equipment, scan tools, pressure transducers, exhaust gas analyzers, lab scopes, meters, and circuit wiring diagrams to locate electrical, fuel, and emission systems faults.
   A. True
   B. False

_____ **17.** A service manager's job is to oversee technicians' work in order to ensure that customers receive quality repair work.
   A. True
   B. False

_____ **18.** Because dealership technicians are working on the latest vehicles, they are right at the cutting edge of technology.
   A. True
   B. False

_____ **19.** The automotive service industry in the United States is generally not subject to licensure requirements.
   A. True
   B. False

_____ **20.** The Automotive Youth Educational Systems (AYES) is an independent, nonprofit organization dedicated to the improvement of vehicle repair through the testing and certification of automotive professionals.
   A. True
   B. False

_____ **21.** Henry Ford applied two concepts that helped make the Model T affordable for the masses, _____ and the assembly line.
   A. Interchangeability
   B. Certification
   C. Electric vehicles
   D. More frequent maintenance intervals

_____ **22.** A(n) _____ changes oil and filters and carries out lubrication, fluid inspection, fluid service, and tire rotations.
   A. Lot attendant
   B. Lube technician
   C. Transmission specialist
   D. Electrical technician

_____ **23.** A(n) _____ diagnoses, repairs, and services steering system components and suspension systems on all types of vehicles.
   A. Service consultant
   B. Drivability technician
   C. Chassis technician
   D. Electrical technician

_____ **24.** A(n) _____ diagnoses, replaces, maintains, identifies faults with, and repairs electrical wiring and computer-based equipment in vehicles.
  A. Electrical technician
  B. Heavy line technician
  C. Chassis technician
  D. Light line technician

_____ **25.** Electrical technicians use meters, oscilloscopes, test instruments, and circuit wiring diagrams to diagnose _____.
  A. Vehicle emission failures
  B. Electrical faults
  C. Damaged steering and suspension systems
  D. Leaky brake systems

_____ **26.** A _____ works with computer-controlled engine management systems to service, identify, and repair faults on electronically controlled vehicle systems such as fuel injection, ignition, and automatic transmissions.
  A. Lube technician
  B. Service manager
  C. Chassis and brake technician
  D. Drivability technician

_____ **27.** A(n) _____ primarily works on the other components of the drivetrain, including the drive shafts and differentials.
  A. Transmission specialist
  B. Electrical technician
  C. Heavy line technician
  D. Lube technician

_____ **28.** A _____ oversees the work of all types of technicians and staff, and communicates with customers and external suppliers.
  A. Service consultant
  B. Shop foreman
  C. Transmission specialist
  D. Lot attendant

_____ **29.** _____ are affiliated with a specific vehicle manufacturer.
  A. Fleet shops
  B. Independent shops
  C. Dealerships
  D. Specialty shops

_____ **30.** To earn _____, technicians are required to pass one or more ASE certification tests and have two years of qualifying work experience as a technician.
  A. ASE certification
  B. AYES certification
  C. EPA Section 609 certification
  D. ASA certification

_____ **31.** Tech A says that newer vehicles require less maintenance than older vehicles. Tech B says that service intervals for an older vehicle can be extended if new oils are used. Who is correct?
  A. Tech A
  B. Tech B
  C. Both Tech A and Tech B
  D. Neither Tech A nor Tech B

_____ **32.** Tech A says that Henry Ford is credited with the invention of the automobile. Tech B says that Carl Benz is credited with the invention of the automobile. Who is correct?
  A. Tech A
  B. Tech B
  C. Both Tech A and Tech B
  D. Neither Tech A nor Tech B

_____ **33.** Tech A says that the production of vehicles today requires a mix of robotic and human assembly to be profitable. Tech B says that most parts on a car are preassembled before they reach the assembly line for higher assembly numbers per day. Who is correct?
   A. Tech A
   B. Tech B
   C. Both Tech A and Tech B
   D. Neither Tech A nor Tech B

_____ **34.** Tech A says that the automotive industry is highly technical and only a certain few people will find jobs. Tech B says the automotive industry is wide open with job opportunities for almost every level of skill. Who is correct?
   A. Tech A
   B. Tech B
   C. Both Tech A and Tech B
   D. Neither Tech A nor Tech B

_____ **35.** Tech A says that a technician can specialize in different areas based on his or her interest and ability. Tech B says that when a technician specializes in a certain area, he or she will only work on certain vehicle models. Who is correct?
   A. Tech A
   B. Tech B
   C. Both Tech A and Tech B
   D. Neither Tech A nor Tech B

_____ **36.** Tech A says that the foreman is the frontline contact for customer relations. Tech B says that the service consultant is the frontline contact for customer relations. Who is correct?
   A. Tech A
   B. Tech B
   C. Both Tech A and Tech B
   D. Neither Tech A nor Tech B

_____ **37.** Tech A says that dealership technicians generally have access to manufacturers' training to help prepare them as technicians. Tech B says that an independent shop works on a wide variety of equipment that requires a broad skill level in technicians. Who is correct?
   A. Tech A
   B. Tech B
   C. Both Tech A and Tech B
   D. Neither Tech A nor Tech B

_____ **38.** Tech A says that AYES certifies technicians. Tech B says that ASE certifications can help get you a job. Who is correct?
   A. Tech A
   B. Tech B
   C. Both Tech A and Tech B
   D. Neither Tech A nor Tech B

_____ **39.** Tech A says that the maintenance requirements of a vehicle have not changed since the creation of the automobile. Tech B says that manufacturers are predicting 25,000-mile (40,000-km) service intervals. Who is correct?
   A. Tech A
   B. Tech B
   C. Both Tech A and Tech B
   D. Neither Tech A nor Tech B

_____ **40.** Tech A says that a technician can progress to different jobs within the industry. Tech B says that careers in the automotive industry include new car assembly lines. Who is correct?
   A. Tech A
   B. Tech B
   C. Both Tech A and Tech B
   D. Neither Tech A nor Tech B

Introduction to Automotive Technology

At the start of each chapter, you will find the Learning Objectives from the textbook. These are your objectives as you make your way through the exercises in this workbook and the chapter in your textbook. The following activities have been designed to help you refresh your knowledge of the material in this chapter.

Learning Objectives

After reading this chapter, you will be able to:

- LO 2-01 Identify vehicle body types and their characteristics.
- LO 2-02 List the functions of common vehicle systems and describe general vehicle operation.
- LO 2-03 Describe drivetrain layouts and their major components.
- LO 2-04 Describe torque and identify engine configurations.

Matching

Match the following terms with the correct description or example.

A. Differential gear set
B. Four-wheel drive
C. Horizontally opposed engine
D. In-line engine
E. Longitudinal

F. Piston engine
G. Rotary engine
H. Torque converter
I. Unibody design
J. VR engine

_____ **1.** An engine in which the cylinders are arranged side by side in a single row.

_____ **2.** An internal combustion engine that uses cylindrical pistons moving back and forth in a cylinder to extract mechanical energy from chemical energy.

_____ **3.** A term used to describe an engine configuration that uses a single bank of cylinders staggered at a shallow 15-degree V.

_____ **4.** An engine that uses a triangular rotor turning in a housing instead of conventional pistons.

_____ **5.** A term used to describe the front-to-back engine orientation when mounted in the engine compartment.

_____ **6.** The arrangement of gears between two axles that allows each axle to spin at its own speed when the vehicle is going around a corner.

_____ **7.** A drivetrain layout in which the engine drive has either two wheels or four wheels, depending on which mode is selected by the driver.

_____ **8.** A vehicle design that does not use a rigid frame to support the body. The body panels are designed to provide the strength for the vehicle.

_____ **9.** A device that is turned by the crankshaft and transmits torque to the input shaft of an automatic transmission.

_____ **10.** An engine with two banks of cylinders, 180 degrees apart, on opposite sides of the crankshaft. It is also called a flat engine or a boxer engine.

Multiple Choice

Read each item carefully, and then select the best response.

_____ **1.** Which vehicle design has an enclosed body with a maximum of four doors and a trunk located in the rear of the vehicle accessible from a trunk lid?
   A. Coupe
   B. Sedan
   C. Hatchback
   D. Station wagon

_____ **2.** Which type of vehicle acts like both a full-size van and a pickup in that it has a heavier-duty chassis so it can carry heavier loads?
   A. Sedan
   B. Hatchback
   C. Sport utility vehicle
   D. Minivan

_____ **3.** What type of chassis design was first used in aircraft and then spread to automobiles?
   A. Unibody
   B. Body-on-frame
   C. Dual shell
   D. Steel ladder

_____ **4.** In which type of drivetrain layout are all four wheels driven by the engine all of the time?
   A. Front-wheel drive
   B. Rear-wheel drive
   C. All-wheel drive
   D. Four-wheel drive

_____ **5.** All of the following criteria are used to define the drivetrain layout, EXCEPT _____.
   A. Engine position
   B. Transmission type
   C. Engine orientation
   D. Type of drive

_____ **6.** Which of the following engine designs is the most powerful compared with its overall dimensions, but more complicated and expensive than the other engines?
   A. V8
   B. Flat 6
   C. W12
   D. In-line 4

_____ **7.** Which type of engine uses a single bank of cylinders, staggered at a shallow 15-degree V within the bank?
   A. Horizontally opposed
   B. W
   C. V
   D. VR

_____ **8.** Which type of axle uses the engine's torque to turn the wheels (drive the vehicle) and at the same time support the weight of the vehicle?
   A. Live axle
   B. Dead axle
   C. Transaxle
   D. Solid axle

_____ **9.** The twisting force applied to a shaft is known as what?
   A. Horsepower
   B. Torque
   C. Work
   D. Load

_____ **10.** Which designation is NOT used when measuring torque?
  A. Foot-pound
  B. Inch-pound
  C. Newton meter
  D. Inch meter

_____ **11.** Reducing the number of doors to the passenger compartment makes the vehicle structure more rigid.
  A. True
  B. False

_____ **12.** A station wagon has an extended roof that goes all the way to the rear of the vehicle. It is similar to a van but not as tall.
  A. True
  B. False

_____ **13.** Body-on-frame is the term used when a vehicle body is mounted on a rigid frame or chassis.
  A. True
  B. False

_____ **14.** Mechanical energy can be converted into chemical energy in two primary ways: through the operation of an internal combustion engine or through the operation of an electric motor.
  A. True
  B. False

_____ **15.** The suspension system makes the connection between the steering wheel and the road wheels so the driver can point the vehicle in the intended direction of travel.
  A. True
  B. False

_____ **16.** A drivetrain is classified by type, cylinder arrangement, number of cylinders/rotors, and total engine displacement in cubic inches or liters.
  A. True
  B. False

_____ **17.** Multi-cylinder internal combustion automotive engines are produced in four common configurations.
  A. True
  B. False

_____ **18.** V engines have two banks of cylinders sitting side by side in a V arrangement sharing a common crankshaft.
  A. True
  B. False

_____ **19.** Horizontally opposed engines are very powerful for their size, but they do not use conventional pistons that slide back and forth inside a straight cylinder.
  A. True
  B. False

_____ **20.** The automatic transmission uses a torque converter instead of a clutch.
  A. True
  B. False

_____ **21.** Part-time 4WD means the vehicle is usually driven in two-wheel drive and switched to full-time when needed by engaging the transfer case.
  A. True
  B. False

_____ **22.** A transfer case locks the drive shafts together and directs torque through them to both axles.
  A. True
  B. False

_____ **23.** All transfer cases use a viscous coupling to split the drive between the front and rear wheels.
  A. True
  B. False

_____ **24.** A _____ has only two doors.
A. Sedan
B. Coupe
C. Hatchback
D. Station wagon

_____ **25.** A _____ is available in three-door and five-door designs.
A. Coupe
B. Minivan
C. Sedan
D. Hatchback

_____ **26.** A _____ is an automobile that can convert from having an enclosed top to having an open top by means of a roof that can be removed, retracted, or folded away.
A. Convertible
B. Sport utility vehicle
C. Minivan
D. Coupe

_____ **27.** A pickup, or _____, carries and tows cargo.
A. Sedan
B. Hatchback
C. Coupe
D. Truck

_____ **28.** A _____ can easily be used to carry out functions that would otherwise require several different vehicles.
A. Sedan
B. Coupe
C. Convertible
D. Sport utility vehicle

_____ **29.** The _____ is the underlying supporting structure for vehicles—similar to the skeleton of a human—on which additional components are mounted.
A. Hatch
B. Chassis
C. Transmission
D. Unibody

_____ **30.** The _____ design is constructed of a large number of steel sheet metal panels that are precisely formed in presses and spot-welded together into a structural unit.
A. Chassis
B. Body-on-frame
C. Unibody
D. Powertrain

_____ **31.** Stored _____ energy is converted to mechanical energy to propel a vehicle down the road.
A. Chemical
B. Hydraulic
C. Manual
D. Industrial

_____ **32.** As the pistons move up and down, they rotate the crankshaft, turning the _____ or flex plate, which is bolted to the engine crankshaft.
A. Camshaft
B. Clutch cable
C. Flywheel
D. Pressure plate

_____ **33.** The _____ system evens out the road shocks caused by irregular road surfaces.
A. Powertrain
B. Charging
C. Safety
D. Suspension

_____ **34.** Manufacturers mount engines in one of two orientations, _____ or _____, depending on which design best fits the vehicle and the rest of the drivetrain.
  A. Longitudinal, Transverse
  B. Front, Rear
  C. Rear, Transverse
  D. Front, Mid

_____ **35.** In a piston engine, the way engine cylinders are arranged is called the engine _____.
  A. Drive
  B. Configuration
  C. Position
  D. Layout

_____ **36.** _____ engines are sometimes referred to as "flat" engines and are commonly found in 4- and 6-cylinder configurations.
  A. Rotary
  B. In-line
  C. Horizontally opposed
  D. Piston

_____ **37.** Axles come in two configurations: _____ axle and _____ axle.
  A. Live, Dead
  B. Extra, Lazy
  C. Front-wheel, All-wheel
  D. Traction, Torque

_____ **38.** A vehicle with a manual transmission uses a _____ to engage and disengage the engine from the transmission.
  A. Torque converter
  B. Clutch
  C. Driveshaft
  D. Transaxle

_____ **39.** What type of chassis design is shown in the image?

  A. Steel ladder-frame
  B. Unibody
  C. Platform
  D. Skeleton

_____ **40.** What type of chassis design is shown in the image?

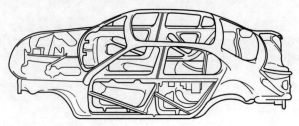

  A. Steel ladder-frame
  B. Unibody
  C. Platform
  D. Skeleton

_____ **41.** Why is the engine tilted in this illustration?

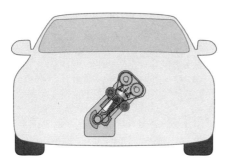

    A. To allow for a larger engine to be installed
    B. To increase the fuel economy
    C. To lower the engine and hood height
    D. To increase ease of maintenance

_____ **42.** What type of rear axle configuration is shown in the image?

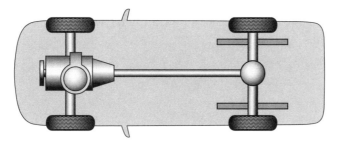

    A. Live axle
    B. Transaxle
    C. Transverse axle
    D. Dead axle

_____ **43.** What type of axle configuration is shown in the image?

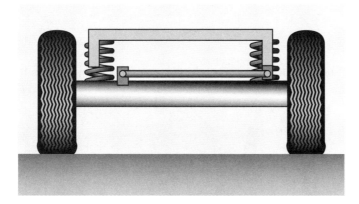

    A. Live axle
    B. Transaxle
    C. Transverse axle
    D. Dead axle

_____ **44.** The arrow in the image is pointing to what part of the engine in the manual transmission setup?

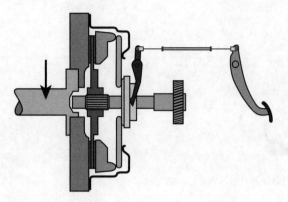

   A. Clutch plate
   B. Pressure plate
   C. Crankshaft
   D. Flywheel

_____ **45.** The arrow in the image is pointing to what part of the engine in the manual transmission setup?

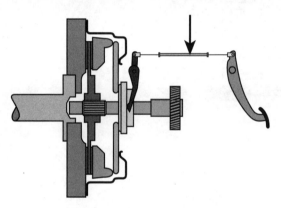

   A. Clutch cable
   B. Clutch pedal
   C. Clutch cover
   D. Diaphragm spring

_____ **46.** The arrow in the image is pointing to what part of the engine in the manual transmission setup?

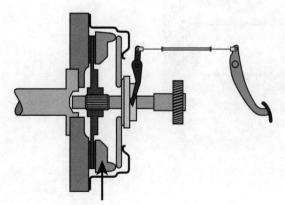

   A. Clutch pedal
   B. Crankshaft
   C. Diaphragm spring
   D. Pressure plate

_____ **47.** The arrow in the image is pointing to what part of the engine in the manual transmission setup?

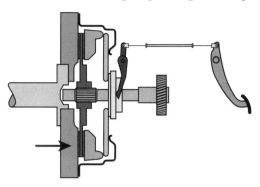

- **A.** Clutch fork
- **B.** Flywheel
- **C.** Pressure plate
- **D.** Clutch pedal

_____ **48.** The arrow in the image is pointing to what part of the final drive assembly?

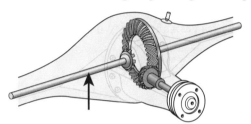

- **A.** Rear axle
- **B.** Ring gear
- **C.** Pinion gear
- **D.** Clutch fork

_____ **49.** The arrow in the image is pointing to what part of the final drive assembly?

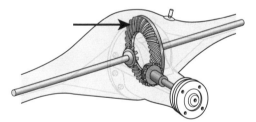

- **A.** Pinion gear
- **B.** Crankpin
- **C.** Pressure plate
- **D.** Ring gear

_____ **50.** The arrow in the image is pointing to what part of the final drive assembly?

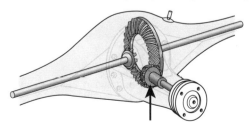

- **A.** Pinion gear
- **B.** Ring gear
- **C.** Crankpin
- **D.** Rear axle

_____ **51.** Tech A says that most vehicles today are built with a ladder-frame. Tech B says that most vehicles today do not have a frame. Who is correct?
  A. Tech A
  B. Tech B
  C. Both Tech A and Tech B
  D. Neither Tech A nor Tech B

_____ **52.** Tech A says that gasoline is energy in chemical form. Tech B says that chemical energy is converted to mechanical energy in the combustion process of an engine. Who is correct?
  A. Tech A
  B. Tech B
  C. Both Tech A and Tech B
  D. Neither Tech A nor Tech B

_____ **53.** Tech A says that a live axle is an axle that cannot be steered. Tech B says that a dead axle supports the weight of the vehicle but doesn't power it. Who is correct?
  A. Tech A
  B. Tech B
  C. Both Tech A and Tech B
  D. Neither Tech A nor Tech B

_____ **54.** Tech A says that the gear ratio in the transmission helps the vehicle gain speed. Tech B says that the gear ratio of the final drive can be selected with a gearshift lever. Who is correct?
  A. Tech A
  B. Tech B
  C. Both Tech A and Tech B
  D. Neither Tech A nor Tech B

_____ **55.** Tech A says that pushing on the brake pedal stops the vehicle by converting chemical energy into thermal energy. Tech B says that brake pedal pressure is transmitted to the brakes hydraulically. Who is correct?
  A. Tech A
  B. Tech B
  C. Both Tech A and Tech B
  D. Neither Tech A nor Tech B

_____ **56.** Tech A says that one advantage of a mid-engine design is better weight distribution. Tech B says that the rear-engine design is commonly used in four-wheel drive vehicles. Who is correct?
  A. Tech A
  B. Tech B
  C. Both Tech A and Tech B
  D. Neither Tech A nor Tech B

_____ **57.** Tech A says that the purpose of the battery is to charge the vehicle. Tech B says that in-line engines are generally easier to work on than V engines. Who is correct?
  A. Tech A
  B. Tech B
  C. Both Tech A and Tech B
  D. Neither Tech A nor Tech B

_____ **58.** Tech A says that the natural angle of a V6 is 90 degrees. Tech B says that the natural angle of a V10 is 60 degrees. Who is correct?
  A. Tech A
  B. Tech B
  C. Both Tech A and Tech B
  D. Neither Tech A nor Tech B

_____ **59.** Tech A says that the definition of torque is how far the crankshaft twists. Tech B says that torque can relate to tightness of bolts. Who is correct?
  A. Tech A
  B. Tech B
  C. Both Tech A and Tech B
  D. Neither Tech A nor Tech B

Introduction to Automotive Safety

At the start of each chapter, you will find the Learning Objectives from the textbook. These are your objectives as you make your way through the exercises in this workbook and the chapter in your textbook. The following activities have been designed to help you refresh your knowledge of the material in this chapter.

Learning Objectives

After reading this chapter, you will be able to:

- LO 3-01 Adhere to workplace safety guidelines.
- LO 3-02 Locate shop safety features and equipment.
- LO 3-03 Prevent fires and operate fire safety equipment.
- LO 3-04 Research material data using safety data sheets (SDS).
- LO 3-05 Work near hazardous on-vehicle systems.

Matching

Match the following terms with the correct description or example.

A. Environmental Protection Agency

B. Hazardous environment

C. Safety data sheet

D. Occupational Safety and Health Administration

E. Personal protective equipment

F. Threshold limit value

_____ **1.** Equipment designed to protect the technician, such as safety boots, gloves, clothing, protective eyewear, and hearing protection.

_____ **2.** A sheet that provides information about handling, use, and storage of a material that may be hazardous.

_____ **3.** The maximum allowable concentration of a given material in the surrounding air.

_____ **4.** Government agency created to provide national leadership in occupational safety and health.

_____ **5.** A place where hazards exist.

_____ **6.** Federal government agency that deals with issues related to environmental safety.

Multiple Choice

Read each item carefully, and then select the best response.

_____ **1.** An unsafe work environment includes which of the following?

  A. A well-organized shop layout

  B. Good supervision

  C. Safety training

  D. Taking shortcuts

_____ **2.** Which federal government agency deals with issues related to environmental safety?

  A. SDS

  B. EPA

  C. OSHA

  D. NIOSH

_____ **3.** Which of the following is a document that describes the steps required to safely use a piece of equipment, such as the vehicle hoist?

  A. Regulation

  B. Policy

  C. Procedure

  D. Compliance

_____ **4.** A _____ can be used to identify hazards and risks within the work environment.
- **A.** Risk analysis
- **B.** Procedure
- **C.** Vulnerability analysis
- **D.** Threat assessment

_____ **5.** All of the following are components of safety signs, EXCEPT _____.
- **A.** Signal word
- **B.** Background color
- **C.** Warning light
- **D.** Pictorial message

_____ **6.** Which signal word indicates a potentially hazardous situation, which, if not avoided, may result in minor or moderate injury?
- **A.** Danger
- **B.** Hazard
- **C.** Warning
- **D.** Caution

_____ **7.** A _____ droplight is one that is designed in such a way that the electrical parts can never come into contact with the outer casing of the device.
- **A.** Grounded
- **B.** Ground fault
- **C.** Double-insulated
- **D.** Cordless

_____ **8.** What class of fires involves flammable liquids or gaseous fuels?
- **A.** Class A
- **B.** Class B
- **C.** Class C
- **D.** Class D

_____ **9.** To operate a fire extinguisher, you should follow which of the following acronyms?
- **A.** PULL
- **B.** PASS
- **C.** PAGE
- **D.** APES

_____ **10.** Which of the following is the concentration of hazardous material in the air you breathe that your shop must not exceed?
- **A.** The threshold limit value
- **B.** Ten percent
- **C.** It varies according to shop standards.
- **D.** Fifteen percent

_____ **11.** Good _____ is about always making sure the shop and your work surroundings are neat and kept in good order.
- **A.** Ventilation
- **B.** Housekeeping
- **C.** Accounting
- **D.** Leadership

_____ **12.** In high-exposure situations, such as vehicles running in the shop, a mechanical means of _____ is required.
- **A.** Ventilation
- **B.** Cleaning
- **C.** Protection
- **D.** Welding

_____ **13.** You can reduce the risk of injury from lifting anything by _____.
- **A.** Breaking down the load into smaller quantities
- **B.** Lifting without assistance
- **C.** Using your back to lift
- **D.** Counting to 10 before lifting

_____ **14.** Which of the following is an example of a mechanical means of ventilation?
   A. Open door
   B. Fume hood
   C. Open window
   D. Closed door

_____ **15.** SDS refers to items of safety equipment like safety footwear, gloves, clothing, protective eyewear, and hearing protection.
   A. True
   B. False

_____ **16.** There is the possibility of an accident occurring whenever work is undertaken.
   A. True
   B. False

_____ **17.** Shop policies and procedures ensure that the shop operates according to OSHA and EPA laws and regulations.
   A. True
   B. False

_____ **18.** An OSHA document for the shop that describes how the shop complies with legislation is an example of a procedure.
   A. True
   B. False

_____ **19.** There are three standard signal words—danger, warning, and caution.
   A. True
   B. False

_____ **20.** Signal words on a warning sign allow the safety message to be conveyed to people who are illiterate or who do not speak the local language.
   A. True
   B. False

_____ **21.** Whenever a vehicle's engine is running, toxic gases are emitted from its exhaust.
   A. True
   B. False

_____ **22.** An engine fitted with a catalytic converter can be run safely indoors.
   A. True
   B. False

_____ **23.** Always keep circuit breaker and electrical panel covers open to provide easy access.
   A. True
   B. False

_____ **24.** The danger of a gasoline fire is always present in an automotive shop.
   A. True
   B. False

_____ **25.** Fire extinguishers are marked with pictograms depicting the types of fires that the extinguisher is approved to fight.
   A. True
   B. False

_____ **26.** Eye wash stations are used to flush the eye with potassium chloride in the event that you get foreign liquid or particles in your eye.
   A. True
   B. False

_____ **27.** In the United States, it is required that workplaces have an SDS for every chemical that is on site.
   A. True
   B. False

_____ **28.** Always use compressed air to blow brake and clutch dust from components and parts before brake service.
   A. True
   B. False

_____ **29.** Used oil and fluids will often contain dangerous chemicals and impurities and need to be safely recycled or disposed of in an environmentally friendly way.
   A. True
   B. False

_____ **30.** _____ are a safe way of escaping danger and gathering in a safe place where everyone can be accounted for in the event of an emergency.
   A. Safety inspections
   B. Signs
   C. Evacuation routes
   D. Machinery guards

_____ **31.** OSHA stands for the _____.
   A. Occupational Safety and Health Administration
   B. Operational Standards and Hazards Association
   C. Oil Spill Hazard Agency
   D. Operations and Shop Health Administration

_____ **32.** EPA stands for the _____.
   A. Environmental Protection Agency
   B. Environmental Policy Administration
   C. Evacuation Procedures Agency
   D. Emissions and Pollutants Agency

_____ **33.** Whose responsibility is it to know and follow the rules?
   A. The federal government's
   B. The shop supervisor's
   C. Everyone's
   D. The service consultant's

_____ **34.** The signal word _____ indicates a potentially hazardous situation, which, if not avoided, could result in death or serious injury.
   A. Danger
   B. Warning
   C. Caution
   D. Hazard

_____ **35.** _____ are used to separate walkways and pedestrian traffic from work areas.
   A. Machinery guards
   B. Painted lines
   C. Doors and gates
   D. Handrails

_____ **36.** _____ is extremely dangerous, as it is odorless and colorless and can build up to toxic levels very quickly in confined spaces.
   A. Carbon dioxide
   B. Carbon monoxide
   C. Hydrocarbon
   D. Nitrogen

_____ **37.** All circuit breakers and fuses should be clearly _____ so you know which circuits and functions they control.
   A. Labeled
   B. Painted
   C. Covered
   D. Engraved

_____ **38.** Three elements must be present at the same time for a fire to occur: _____, _____, and _____.
   A. Fuel, smoke, oxygen
   B. Fuel, oxygen, heat
   C. Wood, wind, heat
   D. Gas, flame, smoke

_____ **39.** Class B fire extinguishers are designed for use on flammable liquids or gaseous fuels and are marked with a _____.
  A. Green triangle
  B. Red square
  C. Blue circle
  D. Yellow pentagram

_____ **40.** _____ are designed to smother a small fire and are very useful in putting out a fire on a person.
  A. Safety data sheets
  B. Fuel retrievers
  C. Fuel vapor covers
  D. Fire blankets

_____ **41.** SDS stands for _____.
  A. Storage and Disposal System
  B. Safety Data Sheet
  C. Safety Direction Standard
  D. Spill Disposal System

_____ **42.** _____ filters can trap very small particles and prevent them from being redistributed into the surrounding air.
  A. HEPA
  B. SDS
  C. HCS
  D. PPE

_____ **43.** Coming into frequent or prolonged contact with used _____ can cause dermatitis and other skin disorders, including some forms of cancer.
  A. Face masks
  B. Coolant
  C. Brake cooler
  D. Engine oil

_____ **44.** Shop _____ are valuable ways of identifying unsafe equipment, materials, or activities so they can be corrected to prevent accidents or injuries.
  A. Safety inspections
  B. Wash stations
  C. Spill kits
  D. Inventory procedures

_____ **45.** What step of fire extinguisher operation is shown in the image?

  A. Pull out the pin.
  B. Squeeze the handle.
  C. Squeeze the pin.
  D. Pinch the pin.

_____ **46.** What step of fire extinguisher operation is shown in the image?

A. Aim the pin.
B. Aim the nozzle.
C. Squeeze the nozzle.
D. Release the handle.

_____ **47.** What step of fire extinguisher operation is shown in the image?

A. Aim the nozzle.
B. Squeeze the pin.
C. Squeeze the nozzle.
D. Squeeze the handle.

_____ **48.** What step of fire extinguisher operation is shown in the image?

   A. Squeeze the nozzle.
   B. Release the handle.
   C. Sweep the nozzle.
   D. Release the pin.

_____ **49.** Tech A says that exposure to solvents may have long-term effects. Tech B says that accidents are almost always avoidable. Who is correct?
   A. Tech A
   B. Tech B
   C. Both Tech A and Tech B
   D. Neither Tech A nor Tech B

_____ **50.** Tech A says that after an accident you should take measures to avoid it in the future. Tech B says that it is OK to block an exit for a shop. Who is correct?
   A. Tech A
   B. Tech B
   C. Both Tech A and Tech B
   D. Neither Tech A nor Tech B

_____ **51.** Tech A says that both OSHA and the EPA can inspect facilities for violations. Tech B says that a shop safety rule does not have to be reviewed once put in place. Who is correct?
   A. Tech A
   B. Tech B
   C. Both Tech A and Tech B
   D. Neither Tech A nor Tech B

_____ **52.** Tech A says that all hazards can be removed from a shop. Tech B says that it is a good practice to disconnect an air gun while inspecting it. Who is correct?
   A. Tech A
   B. Tech B
   C. Both Tech A and Tech B
   D. Neither Tech A nor Tech B

_____ **53.** Tech A says that both caution and danger are signal words that indicate a potentially hazardous situation. Tech B says that an exhaust extraction hose is not needed if the vehicle is only going to run for a few minutes. Who is correct?
   A. Tech A
   B. Tech B
   C. Both Tech A and Tech B
   D. Neither Tech A nor Tech B

_____ **54.** Tech A says that if you are unsure of what personal protective equipment (PPE) to use to perform a job, you should just use what is nearby. Tech B says that air tools are less likely to shock you than electrically powered tools. Who is correct?
- **A.** Tech A
- **B.** Tech B
- **C.** Both Tech A and Tech B
- **D.** Neither Tech A nor Tech B

_____ **55.** Tech A says that a safety data sheet (SDS) contains information on procedures to repair a vehicle. Tech B says that you only need an SDS if your safety may be in danger. Who is correct?
- **A.** Tech A
- **B.** Tech B
- **C.** Both Tech A and Tech B
- **D.** Neither Tech A nor Tech B

_____ **56.** Tech A says that firefighting equipment includes safety glasses. Tech B says that a class A fire extinguisher can be used to fight an electrical fire only. Who is correct?
- **A.** Tech A
- **B.** Tech B
- **C.** Both Tech A and Tech B
- **D.** Neither Tech A nor Tech B

_____ **57.** Tech A says that a good way to clean dust off brakes is with compressed air. Tech B says that asbestos may be in current auto parts. Who is correct?
- **A.** Tech A
- **B.** Tech B
- **C.** Both Tech A and Tech B
- **D.** Neither Tech A nor Tech B

_____ **58.** Tech A says that when cleaning brake and clutch components, the wash station should be placed directly under the component. Tech B says that you should follow state and local regulations when disposing of used oil. Who is correct?
- **A.** Tech A
- **B.** Tech B
- **C.** Both Tech A and Tech B
- **D.** Neither Tech A nor Tech B

Personal Safety

At the start of each chapter, you will find the Learning Objectives from the textbook. These are your objectives as you make your way through the exercises in this workbook and the chapter in your textbook. The following activities have been designed to help you refresh your knowledge of the material in this chapter.

Learning Objectives

After reading this chapter, you will be able to:

- LO 4-01 Dress for the workplace.
- LO 4-02 Comply with hand protection guidelines.
- LO 4-03 Comply with protective headgear guidelines.

Matching

Match the following terms with the correct description or example.

A. Barrier cream

B. Ear protection

C. Gas welding goggles

D. Heat buildup

E. Headgear

F. Respirator

G. Welding helmet

_____ **1.** Protective gear worn when the sound levels exceed 85 decibels, when working around operating machinery for any period of time, or when the equipment you are using produces loud noise.

_____ **2.** Protective gear designed for gas welding; they provide protection against foreign particles entering the eye and are tinted to reduce the glare of the welding flame.

_____ **3.** A dangerous condition that occurs when a glove can no longer absorb or reflect heat and heat is transferred to the inside of the glove.

_____ **4.** A cream that looks and feels like a moisturizing cream but has a specific formula to provide extra protection from chemicals and oils.

_____ **5.** Protective gear designed for arc welding; it provides protection against foreign particles entering the eye, and the lens is tinted to reduce glare of the welding arc.

_____ **6.** Protective gear that includes items like hairnets, caps, or hard hats.

_____ **7.** Protective gear used to protect the wearer from inhaling harmful dusts or gases.

Multiple Choice

Read each item carefully, and then select the best response.

_____ **1.** Proper footwear provides protection against which of the following?
   A. High-intensity light
   B. Cuts
   C. Carbon monoxide
   D. Electricity

_____ **2.** What type of gloves will protect your hands from burns when welding or handling hot components?
   A. Light-duty
   B. Leather
   C. Chemical
   D. General-purpose cloth

_____ 3. What type of protectant prevents chemicals from being absorbed into your skin and should be applied to your hands before you begin work?
   A. Sunscreen
   B. Moisturizer
   C. Barrier cream
   D. Sanitizer

_____ 4. What type of gloves should be used to protect your hands from exposure to greases and oils?
   A. Chemical
   B. Light-duty
   C. Leather
   D. General-purpose cloth

_____ 5. When using a _____, always make sure the cartridge is the correct type for the contaminant in the atmosphere.
   A. Dust mask
   B. Welding helmet
   C. Respirator
   D. Face shield

_____ 6. The most common type of eye protection is a pair of safety glasses, which must be marked with _____ on the lens and frame.
   A. Z87
   B. PPE
   C. UV
   D. Polarized

_____ 7. What type of protection can be worn instead of a welding mask when using or assisting a person using an oxyacetylene welder?
   A. Respirator
   B. Full face shield
   C. Safety glasses
   D. Gas welding goggles

_____ 8. Personal protective equipment (PPE) includes clothing, shoes, safety glasses, hearing protection, masks, and respirators.
   A. True
   B. False

_____ 9. Leather gloves should always be worn when using solvents and cleaners.
   A. True
   B. False

_____ 10. Avoid picking up very hot metal with leather gloves because it causes the leather to harden, making it less flexible during use.
   A. True
   B. False

_____ 11. Light-duty gloves are designed for use in cold temperatures, particularly during winter, so that cold tools do not stick to your skin.
   A. True
   B. False

_____ 12. If a proper barrier cream is not available, a standard moisturizer is a suitable replacement.
   A. True
   B. False

_____ 13. Ear protection that covers the entire outer ear usually has higher noise-reduction ratings than in-the-ear type.
   A. True
   B. False

_____ 14. A disposable dust mask should not be used if chemicals, such as paint solvents, are present in the atmosphere.
   A. True
   B. False

_____ **15.** When grinding, you should wear a pair of safety glasses underneath your face shield for added protection.
  A. True
  B. False

_____ **16.** Tinted safety glasses are not designed to be worn outside in bright sunlight conditions.
  A. True
  B. False

_____ **17.** Ultraviolet radiation can burn your skin like a sunburn.
  A. True
  B. False

_____ **18.** Safety goggles provide much the same eye protection as safety glasses, but with added protection against harmful chemicals that may splash up behind the lenses of glasses.
  A. True
  B. False

_____ **19.** Always remove watches, rings, and jewelry before starting work.
  A. True
  B. False

_____ **20.** It is a good idea to keep a spare set of _____ in the workshop in case a toxic or corrosive fluid is spilled on the ones you are wearing.
  A. Gloves
  B. Work clothes
  C. Safety glasses
  D. Hearing protection

_____ **21.** Your _____ can protect you from bumping your head on a vehicle when the vehicle is raised on a hoist.
  A. Face shield
  B. Respirator
  C. Hard hat
  D. Eye protection

_____ **22.** Never use _____ such as gasoline or kerosene to clean your hands.
  A. Barrier creams
  B. Solutes
  C. Cleaners
  D. Solvents

_____ **23.** There are two types of breathing devices: disposable _____ and _____.
  A. Dust masks, respirators
  B. Face shields, helmets
  C. Dust masks, face shields
  D. Respirators, helmets

_____ **24.** A _____ has removable cartridges that can be changed according to the type of contaminant being filtered.
  A. Respirator
  B. Dust mask
  C. Face shield
  D. Helmet

_____ **25.** The light from a welding arc is very bright and contains high levels of _____ radiation.
  A. Atomic
  B. Ultraviolet
  C. Alpha
  D. Beta

_____ **26.** It is necessary to use _____ when using solvents and cleaners, epoxies, and resins, or when working on a battery.
  A. A respirator
  B. Coveralls
  C. A face shield
  D. Leather gloves

_____ **27.** Tech A says that PPE does not include clothing. Tech B says that the PPE used should be based on the task you are performing. Who is correct?
   A. Tech A
   B. Tech B
   C. Both Tech A and Tech B
   D. Neither Tech A nor Tech B

_____ **28.** Tech A says that protective clothing that is not in good condition should be replaced. Tech B says that safety glasses are always adequate to protect your eyes regardless of the activity. Who is correct?
   A. Tech A
   B. Tech B
   C. Both Tech A and Tech B
   D. Neither Tech A nor Tech B

_____ **29.** Tech A says that appropriate work clothes include loose-fitting clothing. Tech B says that you should always wear cuffed pants when working in a shop. Who is correct?
   A. Tech A
   B. Tech B
   C. Both Tech A and Tech B
   D. Neither Tech A nor Tech B

_____ **30.** Tech A says that proper footwear may include both leather- and steel-toed shoes. Tech B says that leather-soled shoes provide slip resistance. Who is correct?
   A. Tech A
   B. Tech B
   C. Both Tech A and Tech B
   D. Neither Tech A nor Tech B

_____ **31.** Tech A says that a hat can help keep your hair clean when working on a vehicle. Tech B says that chemical gloves may be used when working with solvent. Who is correct?
   A. Tech A
   B. Tech B
   C. Both Tech A and Tech B
   D. Neither Tech A nor Tech B

_____ **32.** Tech A says that you should wear gloves only when it is absolutely necessary. Tech B says that leather gloves are used to pick up very hot pieces of metal. Who is correct?
   A. Tech A
   B. Tech B
   C. Both Tech A and Tech B
   D. Neither Tech A nor Tech B

_____ **33.** Tech A says that barrier creams are used to make cleaning your hands easier. Tech B says that hearing protection needs to be worn only by people operating loud equipment. Who is correct?
   A. Tech A
   B. Tech B
   C. Both Tech A and Tech B
   D. Neither Tech A nor Tech B

_____ **34.** Tech A says that dust masks should be used when painting. Tech B says that a respirator should be used when the TLV for a chemical is exceeded. Who is correct?
   A. Tech A
   B. Tech B
   C. Both Tech A and Tech B
   D. Neither Tech A nor Tech B

_____ **35.** Tech A says that tinted safety glasses can be worn when working outside. Tech B says that welding can cause a sunburn. Who is correct?
   A. Tech A
   B. Tech B
   C. Both Tech A and Tech B
   D. Neither Tech A nor Tech B

Vehicle Service Information and Diagnostic Process

At the start of each chapter, you will find the Learning Objectives from the textbook. These are your objectives as you make your way through the exercises in this workbook and the chapter in your textbook. The following activities have been designed to help you refresh your knowledge of the material in this chapter.

Learning Objectives

After reading this chapter, you will be able to:

- LO 5-01 Utilize information systems.
- LO 5-02 Identify vehicle information.
- LO 5-03 Complete a repair order.
- LO 5-04 Explain Strategy-Based Diagnosis and the three Cs.

Matching

Match the following terms with the correct description or example.

A. Belt routing label
B. Lemon law buyback
C. Refrigerant label
D. Service campaign and recall
E. Shop or service manual

F. Technical Service Bulletin (TSB)
G. Vehicle Emission Control Information (VECI) label
H. Title history
I. Vehicle Identification Number (VIN)
J. Vehicle Safety Certification (VSC) label

_____ 1. A detailed account of a vehicle's past.

_____ 2. A label that lists the type and total capacity of refrigerant that is installed in the air conditioning system.

_____ 3. Information issued by manufacturers to alert technicians to unexpected problems or changes to repair procedures.

_____ 4. A label that lists a diagram of the serpentine belt routing for the engine accessories.

_____ 5. A consumer protection law used in some states to identify a new vehicle that has undergone several unsuccessful attempts to repair the same fault.

_____ 6. A unique serial number that is assigned to each vehicle produced.

_____ 7. A label certifying that the vehicle meets the Federal Motor Vehicle Safety, Bumper, and Theft Prevention Standards in effect at the time of manufacture.

_____ 8. A corrective measure conducted by manufacturers when a safety issue is discovered with a particular vehicle.

_____ 9. Manufacturer's or after-market information on the repair and service of vehicles.

_____ 10. A label used by technicians to identify engine and emission control information for the vehicle.

Match the following steps with the correct sequence for using a shop manual.

A. Step 1   **D.** Step 4

B. Step 2   **E.** Step 5

C. Step 3

_____ **1.** Locate the correct section that will contain the information you need. The first page of the shop manual is usually a table of contents.

_____ **2.** Decide what information you need to know about the job and about the vehicle. Make sure you know the make, model, and year of manufacture of the vehicle, and the type and size of the engine. You should also have the vehicle identification number handy.

_____ **3.** Locate the vehicle specifications by consulting the specifications page in the proper section.

_____ **4.** Find the appropriate shop manual for the make, model, and year of the vehicle you are working on.

_____ **5.** Locate the service procedures in the proper section.

Match the following steps with the correct sequence for using a service information program.

A. Step 1   **E.** Step 5

B. Step 2   **F.** Step 6

C. Step 3   **G.** Step 7

D. Step 4

_____ **1.** The search engine will provide a list of possible matches for you to select from. If the initial search does not produce what you are looking for, try changing the search criteria. Keep searching until you find the information.

_____ **2.** Enter the vehicle identification information into the system in the appropriate places: year, make, model, engine, and, possibly, VIN.

_____ **3.** Log in to the application using the appropriate username and password.

_____ **4.** Once the general details for the item are displayed, gather the specific information on the specifications or repairs. You may need more than one piece of information.

_____ **5.** Search for the information you require to perform the service or repair.

_____ **6.** If necessary, start the computer and select the service information program.

_____ **7.** Print out or write down the information needed. Put this on a clipboard, and take it with you to perform the service or repair.

Match the following steps with the correct sequence for using a labor guide.

A. Step 1   **E.** Step 5

B. Step 2   **F.** Step 6

C. Step 3   **G.** Step 7

D. Step 4   **H.** Step 8

_____ **1.** Check for any "combination" time that would need to be added to the base job when a related job is also being completed.

_____ **2.** Find the labor operation either by working your way through the menu tree or by typing a keyword into the search bar.

_____ **3.** Calculate the total time and multiply it by the shop's hourly labor rate. You now have the correct figure to estimate the charge for the particular service.

_____ **4.** Once you locate the labor operation, there are usually two columns that list the time. The first one is "warranty time." The second time listed is the "customer pay time."

_____ **5.** Decide what specific labor operations you need to locate. Make sure you know the year, make, model, engine, and any other pertinent details of the vehicle.

_____ **6.** Enter the vehicle information into the system.

_____ **7.** Check for any "additional" time. This is extra time needed to deal with situations that occur on a relatively common basis, such as vehicle-installed options that are not common to all vehicles, like wheel locks.

_____ **8.** Log in to the labor estimating system.

Match the following steps with the correct sequence for locating parts information on the computer.

A. Step 1   **D.** Step 4
B. Step 2   **E.** Step 5
C. Step 3

_____ **1.** Search for the parts you require to conduct the service or repair.

_____ **2.** Gather information on the identified parts, including part numbers, location, availability, and cost.

_____ **3.** Enter the year, make, model, and engine and/or VIN information into the system in the appropriate places.

_____ **4.** Log in to the application using the appropriate username and password.

_____ **5.** The search engine will provide a list of possible matches for you to select from. If the initial search does not produce what you are looking for, try changing the search criteria. Keep searching until you find the information.

Match the following steps with the correct sequence for using the three Cs to document a repair.

A. Step 1   **E.** Step 5
B. Step 2   **F.** Step 6
C. Step 3   **G.** Step 7
D. Step 4   **H.** Step 8

_____ **1.** Retest the vehicle to be sure the fault has been corrected.

_____ **2.** Review the information you collect from the tests. To review effectively, you need to understand how the systems work and interact.

_____ **3.** Identify and document the correction required, including work activities and parts required or used. Make repairs or replace parts to complete the repair.

_____ **4.** Identify and document the root cause of the concern. Research shop manuals and conduct tests to identify the cause of the concern. This may require a number of tests across multiple systems.

_____ **5.** Using the three Cs, document the repair process required for a repair order.

_____ **6.** Fill in the required repair order with details of the work conducted.

_____ **7.** Identify and document the concern. This should be on the repair order. Obtain as much information as possible, as this will help you to understand the concern. Identify what the concern is and which vehicle systems are involved.

_____ **8.** Always work safely and use the proper tools and correct personal protective equipment (PPE).

Multiple Choice

Read each item carefully, and then select the best response.

_____ **1.** What type of information publication comes in two types—factory and after-market?
A. Owner's manual
B. Service manual
C. Technical service bulletin
D. Recall procedure guide

_____ **2.** Who pays for the costs associated with a mandatory recall?
A. Consumer
B. Repair shop
C. Manufacturer
D. Insurance company

_____ **3.** If a customer wants to know how much it will cost to replace a leaking intake manifold gasket on a particular vehicle, a technician can look up this procedure in the _____ to help estimate the cost.
A. Owner's manual
B. Labor guide
C. Service manual
D. Service information program

_____ **4.** When determining labor costs, every tenth of an hour equals _____ minutes.
  A. 4
  B. 5
  C. 6
  D. 7

_____ **5.** Initial information on a(n) _____ includes customer and vehicle details, along with a brief description of the customer's complaint(s).
  A. Repair order
  B. Service bulletin
  C. Insurance claim
  D. Recall notice

_____ **6.** Which of the following is NOT required to determine the total cost of service?
  A. Labor costs
  B. Tax amounts
  C. Cost of gas and consumables used to service the vehicle
  D. Odometer reading

_____ **7.** The _____ can provide potential new owners of used vehicles with an indication of how well the vehicle was maintained.
  A. Service history
  B. Owner's manual
  C. Repair order
  D. Service manual

_____ **8.** The vehicle identification number is designed for what type of motor vehicle?
  A. Trucks
  B. Motorcycles
  C. Cars
  D. All automotive vehicles

_____ **9.** The first digit of a North American vehicle identification number indicates which of the following?
  A. Manufacturer
  B. Country of origin
  C. Model
  D. Body type

_____ **10.** Which of the three Cs stands for understanding the reason that there is a fault?
  A. Connect
  B. Check
  C. Cause
  D. Control

_____ **11.** Usually, a factory service manual is specific to one year and make of vehicle.
  A. True
  B. False

_____ **12.** The information found in shop manuals provides a systematic procedure and identifies special tools, safety precautions, and specifications relevant to the task.
  A. True
  B. False

_____ **13.** A typical owner's manual contains step-by-step procedures and diagrams on how to identify if there is a fault and perform an effective repair.
  A. True
  B. False

_____ **14.** Online versions of labor guides can be updated as new models of vehicles are released or updates are made by the manufacturer.
  A. True
  B. False

_____ **15.** If you are replacing the brake pads on a vehicle with wheel locks, the customer may be charged for the extra time it takes to find the lock key and to remove and install the wheel locks.
  A. True
  B. False

_____ **16.** A labor guide is a computer application that may be used to identify part numbers for vehicle components.
  A. True
  B. False

_____ **17.** Most manufacturers store all service history performed in their dealerships.
  A. True
  B. False

_____ **18.** The VIN is a twelve-character identification code composed of letters and digits.
  A. True
  B. False

_____ **19.** The Vehicle Safety Certification label is used by technicians to identify engine and emission control information for the vehicle.
  A. True
  B. False

_____ **20.** A(n) _____ includes an overview of the controls and features of the vehicle; the proper operation, care, and maintenance of the vehicle; owner service procedures; and specifications or technical data.
  A. Repair order
  B. Owner's manual
  C. Emission control label
  D. Vehicle identification plate

_____ **21.** _____ can be mandatory and enforced by law, or voluntary in order to ensure the safe operation of the vehicle or prevent damage to their business and product image.
  A. Recalls
  B. Certification
  C. Warranties
  D. Labor guides

_____ **22.** When working on vehicles that have not been recalled, check to see if a _____ has been issued for that vehicle and type of repair.
  A. Warranty
  B. Technical service bulletin
  C. Parts program
  D. Lemon law

_____ **23.** _____ are essentially electronic versions of a parts manual, available via a CD/DVD, a computer network, or the Internet.
  A. Warranties
  B. Labor guides
  C. Parts programs
  D. Salvage titles

_____ **24.** A _____ is used by the technician to guide him or her to the concern and by the customer service staff to create the invoice when the work is completed.
  A. Salvage title
  B. Parts program
  C. Technical service bulletin
  D. Repair order

_____ **25.** The _____ section of a repair order contains information about the methods of payment, which can be cash, credit card, or account.
  A. Accounting
  B. Salvage
  C. Legal
  D. Service

_____ **26.** A _____ tells you that the vehicle has been wrecked and has suffered irreparable damage.
  A. Repair order
  B. Salvage title
  C. Vehicle safety certification
  D. Technical service bulletin

_____ **27.** The _____ label certifies that the vehicle meets the Federal Motor Vehicle Safety, Bumper, and Theft Prevention Standards in effect at the time of manufacture.

A. Refrigerant

B. Coolant

C. Belt routing

D. Vehicle safety certification

_____ **28.** The three Cs are an easy way to learn the fundamental steps in conducting repairs. They stand for _____, _____, and _____.

A. Care, concern, conduct

B. Concern, cause, correction

C. Cooling, cause, complete

D. Customer, cause, complete

_____ **29.** What VIN standards system is shown in the image?

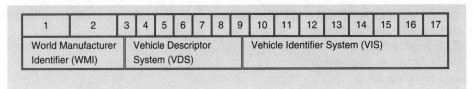

A. North American VIN system

B. ISO Standard 3779

C. Vehicle Emission Control system

D. Vehicle Safety Certification system

_____ **30.** What VIN standards system is shown in the image?

A. North American VIN system

B. Production Standard 3779

C. Vehicle Safety Certification system

D. ISO Standard 3779

_____ **31.** What vehicle information label is shown in the image?

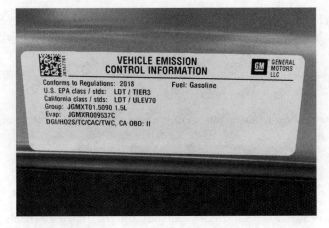

A. VECI label

B. VSC label

C. Refrigerant label

D. Coolant label

_____ **32.** What vehicle information label is shown in the image?

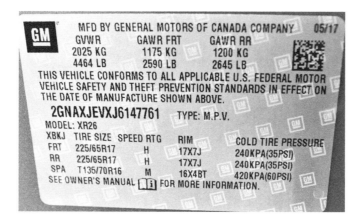

A. Coolant label
B. Refrigerant label
C. VSC label
D. Belt routing label

_____ **33.** What vehicle information label is shown in the image?

A. Refrigerant label
B. Coolant label
C. VSC label
D. VECI label

_____ **34.** What vehicle information label is shown in the image?

A. Refrigerant label
B. Coolant label
C. Belt routing label
D. VECI label

_____ **35.** What vehicle information label is shown in the image?

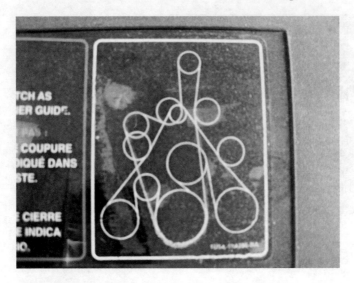

    A. Refrigerant label
    B. Coolant label
    C. Belt routing label
    D. VECI label

_____ **36.** Tech A says that the owner's manual will have the oil drain plug torque information. Tech B says that oil capacity information for that specific vehicle will be in the owner's manual. Who is correct?
    A. Tech A
    B. Tech B
    C. Both Tech A and Tech B
    D. Neither Tech A nor Tech B

_____ **37.** Tech A says that vehicle security personal identification number (PIN) codes can be found in the service manual. Tech B says that vehicle security PIN codes can be found in the owner's manual. Who is correct?
    A. Tech A
    B. Tech B
    C. Both Tech A and Tech B
    D. Neither Tech A nor Tech B

_____ **38.** Tech A says that paper service manuals are gone and electronic versions of the service manual are now available. Tech B says that online manuals are more current, as they can be updated periodically. Who is correct?
    A. Tech A
    B. Tech B
    C. Both Tech A and Tech B
    D. Neither Tech A nor Tech B

_____ **39.** Tech A says that service information programs are extremely helpful, as the technician can use a laptop at the repair for quick access to the procedure to perform a repair. Tech B says that service information programs allow a technician to know the labor guide and, with the support of the program, perform the task in the time allowed. Who is correct?
    A. Tech A
    B. Tech B
    C. Both Tech A and Tech B
    D. Neither Tech A nor Tech B

_____ **40.** Tech A says that TSBs are updates to the owner's manual. Tech B says that TSBs are generally updated information on model changes that do not affect the technician. Who is correct?
   A. Tech A
   B. Tech B
   C. Both Tech A and Tech B
   D. Neither Tech A nor Tech B

_____ **41.** Tech A says that labor guides are necessary for the service writer to quote prices for a customer on the repair bill. Tech B says that labor guides are what the customer pays and that the warranty pays more for labor using a different labor guide. Who is correct?
   A. Tech A
   B. Tech B
   C. Both Tech A and Tech B
   D. Neither Tech A nor Tech B

_____ **42.** Tech A says that locating the part number needed by the technician requires computer knowledge and mechanical knowledge. Tech B says that anybody can be a parts person. Who is correct?
   A. Tech A
   B. Tech B
   C. Both Tech A and Tech B
   D. Neither Tech A nor Tech B

_____ **43.** Tech A says that the repair order is just a piece of paper telling the technician what to do. Tech B says that the repair order is a legal and binding contract between the customer and the repair facility. Who is correct?
   A. Tech A
   B. Tech B
   C. Both Tech A and Tech B
   D. Neither Tech A nor Tech B

_____ **44.** Tech A says that customer authorization for a change in the repair order after initial authorization is the responsibility of the service writer or foreman. Tech B says that changes to the repair order can be dealt with after the repair is complete, as the customer will be satisfied if more repairs are completed. Who is correct?
   A. Tech A
   B. Tech B
   C. Both Tech A and Tech B
   D. Neither Tech A nor Tech B

_____ **45.** Tech A says that the VIN on a vehicle can help identify which engine is installed in the chassis. Tech B says that most digits in the VIN can be an identifier for information pertinent to that vehicle. Who is correct?
   A. Tech A
   B. Tech B
   C. Both Tech A and Tech B
   D. Neither Tech A nor Tech B

Hand and Measuring Tools

At the start of each chapter, you will find the Learning Objectives from the textbook. These are your objectives as you make your way through the exercises in this workbook and the chapter in your textbook. The following activities have been designed to help you refresh your knowledge of the material in this chapter.

Learning Objectives

After reading this chapter, you will be able to:

- LO 6-01 Identify basic hand tool safety.
- LO 6-02 Identify basic wrenches and sockets.
- LO 6-03 Identify other basic hand tools.
- LO 6-04 Identify basic hammers and struck tools.
- LO 6-05 Identify basic taps, dies, and specialty tools.
- LO 6-06 Measure precisely using measuring tools.
- LO 6-07 Measure precisely using other measuring tools.

Matching

Match the following terms with the correct description or example.

A. Bottoming tap

B. Cross-arm

C. Dead blow hammer

D. Dial bore gauge

E. Die stock

F. Intermediate tap

G. Micrometer

H. Parallax error

I. Peening

J. Pullers

K. Split ball gauge

L. Telescoping gauge

M. Vernier calipers

N. Wad punch

_____ 1. A gauge that is used to measure the inside diameter of bores with a high degree of accuracy and speed.

_____ 2. An accurate measuring device for internal, external, and depth measurements that incorporates fixed and adjustable jaws.

_____ 3. A thread-cutting tap designed to cut threads to the bottom of a blind hole.

_____ 4. A gauge that expands and locks to the internal diameter of bores; a caliper or outside micrometer is used to measure its size.

_____ 5. A type of hammer that has a cushioned head to reduce the amount of head bounce.

_____ 6. An accurate measuring device for internal and external dimensions.

_____ 7. A description for an arm that is set at right angles, or 90 degrees, to another component.

_____ 8. A type of punch that is hollow for cutting circular shapes in soft materials such as gaskets.

_____ 9. One of a series of taps designed to cut an internal thread. Also called a plug tap.

_____ 10. A term used to describe the action of flattening a rivet through a hammering action.

_____ 11. A tool used to grab and pull press-fit parts apart.

_____ 12. A tool used to hold a die during use.

_____ 13. A gauge that is good for accurately measuring small holes where telescoping gauges cannot fit.

_____ 14. A visual error caused by viewing measurement markers at an incorrect angle.

Multiple Choice

Read each item carefully, and then select the best response.

_____ **1.** Which of the following is a type of fastener?
   A. Bolt
   B. Nippers
   C. Speed brace
   D. Socket

_____ **2.** In the metric system, which of the following is measured by the distance between the peaks of threads in millimeters?
   A. Thread pitch
   B. Tensile strength
   C. Torque
   D. Thread count

_____ **3.** What type of wrench is also known as a tension wrench?
   A. Open-end wrench
   B. Ratcheting box-end wrench
   C. Torque wrench
   D. Pipe wrench

_____ **4.** As long as a bolt is not tightened too much, it will return to its original length when loosened; this is called _____.
   A. Torque
   B. Elasticity
   C. Yield
   D. Play

_____ **5.** A _____ is the fastest way to spin a fastener on or off a thread by hand, but it cannot apply much torque to the fastener.
   A. Breaker bar
   B. Sliding T-handle
   C. Speed brace
   D. Lug wrench

_____ **6.** What type of pliers are used for cutting wire and cotter pins?
   A. Diagonal
   B. Snap ring
   C. Flat-nosed
   D. Needle-nosed

_____ **7.** A screw or bolt with a cross-shaped recess requires a(n) _____.
   A. Flat blade screwdriver
   B. Offset screwdriver
   C. Phillips head screwdriver
   D. Allen wrench

_____ **8.** What type of screwdriver fits into spaces where a straight screwdriver cannot and is useful where there is not much room to turn it?
   A. Ratcheting screwdriver
   B. Impact driver
   C. Phillips head screwdriver
   D. Offset screwdriver

_____ **9.** What kind of tools are composed of a strong metal and used as a lever to move, adjust, or pry?
   A. Cold chisels
   B. Pry bars
   C. Drift punches
   D. Speed braces

_____ **10.** What type of file is thinner than other files, comes to a point, and is used for working in narrow slots?
   A. Warding file
   B. Triangular file
   C. Thread file
   D. Square file

_____ **11.** What type of file cleans clogged or distorted threads on bolts and studs?
   A. Triangular file
   B. Warding file
   C. Thread file
   D. Square file

_____ **12.** The name for this type of clamp comes from its shape; it can hold parts together while they are being assembled, drilled, or welded.
   A. J-clamp
   B. C-clamp
   C. D-clamp
   D. K-clamp

_____ **13.** What type of tap narrows at the tip to give it a good start in the hole where the thread is to be cut?
   A. Intermediate tap
   B. Taper tap
   C. Plug tap
   D. Bottoming tap

_____ **14.** What type of tool consists of three main parts: jaws, a cross-arm, and a forcing screw?
   A. Tap and die set
   B. Bench vice
   C. Flaring tool
   D. Gear pullers

_____ **15.** What tool is used to measure the gap between a straight edge and the surface being checked for flatness?
   A. Steel rule
   B. Caliper
   C. Feeler gauge
   D. Split ball gauge

_____ **16.** Thread pitch is a way of defining how much a fastener should be tightened.
   A. True
   B. False

_____ **17.** The higher the grade number of a fastener, the higher the tensile strength.
   A. True
   B. False

_____ **18.** Torque wrenches come in various types: beam style, clicker, dial, and electronic.
   A. True
   B. False

_____ **19.** The open-end wrench fits fully around the head of the bolt or nut and grips each of the six points at the corners just like a socket.
   A. True
   B. False

_____ **20.** Six- and 12-point sockets fit the heads of hexagonal-shaped fasteners.
   A. True
   B. False

_____ **21.** Locking jaw pliers, also called vice grips, are general-purpose pliers used to clamp and hold one or more objects.
   A. True
   B. False

_____ **22.** Allen wrenches are sometimes called hex keys.
   A. True
   B. False

_____ **23.** When a large chisel needs a really strong blow, it is time to use a dead blow hammer.
   A. True
   B. False

_____ **24.** When marks need to be drawn on an object like a steel plate to help locate a hole to be drilled, a drift punch can be used to mark the points so they will not rub off.
   A. True
   B. False

_____ **25.** A cold chisel gets its name from the fact that it is used to cut cold metals, rather than heated metals.
   A. True
   B. False

_____ **26.** A bottoming tap is used to tap a thread into a hole that does not come out the other side of the material.
   A. True
   B. False

_____ **27.** Always use a single flare if the tubing is to be used for higher pressures, such as in a brake system.
   A. True
   B. False

_____ **28.** A tubing cutter is more convenient and neater than a saw when cutting pipes and metal tubing.
   A. True
   B. False

_____ **29.** Depth micrometers are used to measure inside dimensions.
   A. True
   B. False

_____ **30.** Thread repair is used in situations where it is either not practical or possible to replace a damaged component.
   A. True
   B. False

_____ **31.** _____ is an umbrella term that describes a set of safety practices and procedures that are intended to reduce the risk of technicians inadvertently using tools, equipment, or materials that have been determined to be unsafe.
   A. Tagout/boxout
   B. Lockout/tagout
   C. Lockout/tap-out
   D. Box-end/open-end

_____ **32.** A _____ is a cylindrical piece of metal with a hexagonal head on one end and a thread cut into the shaft at the other end.
   A. Wrench
   B. Bolt
   C. Nut
   D. Socket

_____ **33.** Each bolt diameter in the standard system can have one of two thread pitches: _____ or _____.
   A. Rough, smooth
   B. Tight, loose
   C. Torque, yield
   D. Coarse, fine

_____ **34.** _____-to-yield means that a fastener is torqued to, or just beyond, its yield point.
   A. Torque
   B. Lockout
   C. Impact
   D. Depth

_____ **35.** A(n) _____ wrench has an open-end head on one end and a box-end head on the other end.
   A. Box-end
   B. Open-end
   C. Combination
   D. Flare nut

_____ **36.** _____ are a hand tool designed to hold, cut, or compress materials.
   A. Wrenches
   B. Pliers
   C. Sockets
   D. Screwdrivers

_____ **37.** End cutting pliers, also called _____, have a cutting edge at right angles to their length.
   A. Nippers
   B. Grippers
   C. Side cutters
   D. Vice grips

_____ **38.** A(n) _____ is used when a screw or a bolt is rusted/corroded in place or overtightened and needs a tool that can apply more force.
   A. Offset screwdriver
   B. Ratcheting screwdriver
   C. Rin snip
   D. Impact driver

_____ **39.** A _____ hammer is designed not to bounce back when it hits something.
   A. Hard rubber
   B. Dead blow
   C. Cold
   D. Brass

_____ **40.** _____ are used when the head of the hammer is too large to strike the object being hit without causing damage to adjacent parts.
   A. Wrenches
   B. Punches
   C. Socket drivers
   D. Screwdrivers

_____ **41.** A _____ has a hardened, sharpened blade and is designed to remove a gasket without damaging the sealing face of the component.
   A. Pry bar
   B. File
   C. Gasket scraper
   D. Screwdriver

_____ **42.** A screw _____ is a device designed to remove screws, studs, or bolts that have broken off in threaded holes.
   A. Driver
   B. File
   C. Scraper
   D. Extractor

_____ **43.** To cut a brand new thread on a blank rod or shaft, a die held in a _____ is used.
   A. Tap handle
   B. Die stock
   C. Screw extractor
   D. Bench vice

_____ **44.** A measuring _____ is a flexible type of ruler and a common measuring tool.
   A. Tape
   B. Micrometer
   C. Gauge
   D. Caliper

_____ **45.** A _____ can measure run out.
   A. Vernier caliper
   B. Micrometer
   C. Flaring tool
   D. Dial indicator

_____ **46.** A _____ is used to measure the width of gaps, such as the clearance between valves and rocker arms.
   A. Dial indicator
   B. Feeler gauge
   C. Dial bore gauge
   D. Micrometer

_____ **47.** The arrow in the image is pointing to what aspect of the socket?

   A. Depth
   B. Number of points
   C. Drive size
   D. Wall thickness

_____ **48.** The arrow in the image is pointing to what aspect of the socket?

   A. Standard or metric
   B. Number of points
   C. Depth
   D. Drive size

_____ **49.** The arrow in the image is pointing to what component of the flare tool?

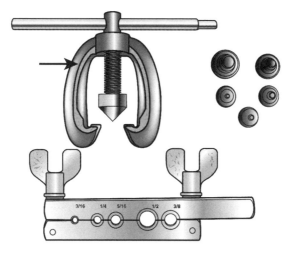

   A. T-handle
   B. Cone
   C. Yoke
   D. Clamp

_____ **50.** The arrow in the image is pointing to what component of the flare tool?

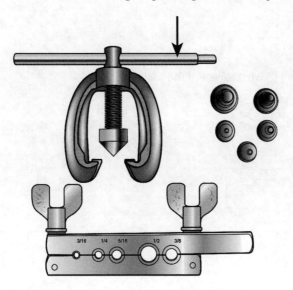

A. Yoke
B. T-handle
C. Die block
D. Clamp

_____ **51.** The arrow in the image is pointing to what component of the flare tool?

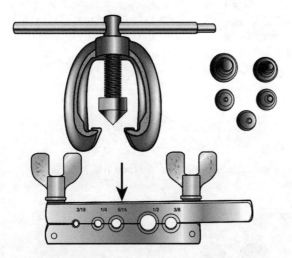

A. Die block
B. Yoke
C. Clamp
D. T-handle

_____ **52.** The arrow in the image is pointing to what part of the rivet?

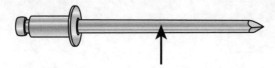

A. Mandrel head
B. Rivet body
C. Rivet head
D. Shank

_____ **53.** The arrow in the image is pointing to what part of the rivet?

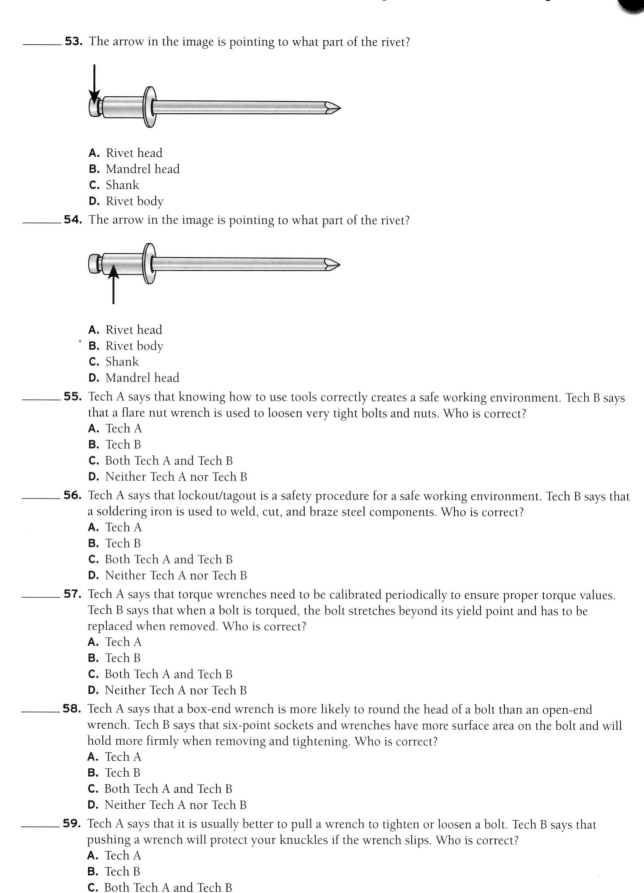

 A. Rivet head
 B. Mandrel head
 C. Shank
 D. Rivet body

_____ **54.** The arrow in the image is pointing to what part of the rivet?

 A. Rivet head
 B. Rivet body
 C. Shank
 D. Mandrel head

_____ **55.** Tech A says that knowing how to use tools correctly creates a safe working environment. Tech B says that a flare nut wrench is used to loosen very tight bolts and nuts. Who is correct?
 A. Tech A
 B. Tech B
 C. Both Tech A and Tech B
 D. Neither Tech A nor Tech B

_____ **56.** Tech A says that lockout/tagout is a safety procedure for a safe working environment. Tech B says that a soldering iron is used to weld, cut, and braze steel components. Who is correct?
 A. Tech A
 B. Tech B
 C. Both Tech A and Tech B
 D. Neither Tech A nor Tech B

_____ **57.** Tech A says that torque wrenches need to be calibrated periodically to ensure proper torque values. Tech B says that when a bolt is torqued, the bolt stretches beyond its yield point and has to be replaced when removed. Who is correct?
 A. Tech A
 B. Tech B
 C. Both Tech A and Tech B
 D. Neither Tech A nor Tech B

_____ **58.** Tech A says that a box-end wrench is more likely to round the head of a bolt than an open-end wrench. Tech B says that six-point sockets and wrenches have more surface area on the bolt and will hold more firmly when removing and tightening. Who is correct?
 A. Tech A
 B. Tech B
 C. Both Tech A and Tech B
 D. Neither Tech A nor Tech B

_____ **59.** Tech A says that it is usually better to pull a wrench to tighten or loosen a bolt. Tech B says that pushing a wrench will protect your knuckles if the wrench slips. Who is correct?
 A. Tech A
 B. Tech B
 C. Both Tech A and Tech B
 D. Neither Tech A nor Tech B

_____ **60.** Tech A says that a dead blow hammer reduces rebound of the hammer. Tech B says that a dead blow hammer should be used to cut the head of a bolt off with a chisel. Who is correct?
   A. Tech A
   B. Tech B
   C. Both Tech A and Tech B
   D. Neither Tech A nor Tech B

_____ **61.** Tech A says that gaskets can be removed quickly and safely with a portable grinder as long as the grinding wheel isn't too coarse. Tech B says that extreme care must be used when removing a gasket on an aluminum surface. Who is correct?
   A. Tech A
   B. Tech B
   C. Both Tech A and Tech B
   D. Neither Tech A nor Tech B

_____ **62.** Tech A says that when using a file, apply pressure to the file in the direction of the cut and no pressure when pulling the file back. Tech B says that file cards are used to file uneven surfaces. Who is correct?
   A. Tech A
   B. Tech B
   C. Both Tech A and Tech B
   D. Neither Tech A nor Tech B

Power Tools & Equipment

At the start of each chapter, you will find the Learning Objectives from the textbook. These are your objectives as you make your way through the exercises in this workbook and the chapter in your textbook. The following activities have been designed to help you refresh your knowledge of the material in this chapter.

Learning Objectives

After reading this chapter, you will be able to:

- LO 7-01 Identify and safely operate battery charging and jump-starting equipment.
- LO 7-02 Identify and safely operate air tools.
- LO 7-03 Use cutting and grinding tools.
- LO 7-04 Perform a solder repair.
- LO 7-05 Operate cleaning equipment.

Multiple Choice

Read each item carefully, and then select the best response.

_____ 1. What device is fitted to compressed air systems to remove the moisture or water from the compressed air that is condensed as a result of compressing air from the atmosphere?
   A. Spit filter
   B. Air drier
   C. Relief valve
   D. Strainer

_____ 2. What type of grinder uses disks rather than wheels?
   A. Bench grinder
   B. Angle grinder
   C. Straight grinder
   D. Pedestal grinder

_____ 3. What tool uses high pressure to blast small, abrasive particles to clean the surface of parts?
   A. Power washer
   B. Angle grinder
   C. Sand blaster
   D. Solvent tank

_____ 4. Serious, sometimes fatal, injuries can be caused by compressed air being injected into the body through the skin or into a body opening, such as your mouth or ear.
   A. True
   B. False

_____ 5. When soldering, always apply flux to the joint if cored solder is used.
   A. True
   B. False

_____ 6. Morse taper is a system for securing drill bits to drills.
   A. True
   B. False

_____ **7.** A solvent tank is a cleaning tank that is filled with a suitable solvent to clean parts by removing oil, grease, dirt, and grime.
A. True
B. False

_____ **8.** Batteries are filled with sulfuric acid, so if the hydrogen explodes, the battery case can then rupture and spray everything and everyone nearby with this dangerous and corrosive liquid.
A. True
B. False

_____ **9.** A(n) _____ is designed to regularly oil an air tool or air equipment so it does not have to be done manually, before or during its use.
A. Automatic oiler
B. Air drier
C. Air compressor
D. Rattle gun

_____ **10.** The most common air tool in an automotive shop is the air _____.
A. Impact wrench
B. Hammer
C. Drill
D. Nozzle

_____ **11.** A _____ incorporates microprocessors to monitor and control the charge rate so a battery receives the correct amount of charge, depending on its state of charge.
A. Surge protector
B. Memory saver
C. Memory minder
D. Smart charger

_____ **12.** The arrow in the image is pointing to what air tool?

A. Air ratchet
B. Air hammer
C. Air impact wrench
D. Air drill

_____ **13.** The arrow in the image is pointing to what air tool?

    A. Air ratchet
    B. Air nozzle
    C. Air impact wrench
    D. Air drill

_____ **14.** The arrow in the image is pointing to what air tool?

    A. Air impact wrench
    B. Air ratchet
    C. Air hammer
    D. Air nozzle

_____ **15.** The arrow in the image is pointing to what air tool?

    A. Air ratchet
    B. Air hammer
    C. Air impact wrench
    D. Air drill

_____ **16.** The arrow in the image is pointing to what air tool?

    A. Air nozzle
    B. Air hammer
    C. Air impact wrench
    D. Air drill

_____ **17.** Tech A says that when jump-starting a vehicle, a spark typically occurs when making the last jumper cable connection. Tech B says that all four of the jumper cable connections should be made at the battery terminals. Who is correct?
    A. Tech A
    B. Tech B
    C. Both Tech A and Tech B
    D. Neither Tech A nor Tech B

Fasteners and Thread Repair

At the start of each chapter, you will find the Learning Objectives from the textbook. These are your objectives as you make your way through the exercises in this workbook and the chapter in your textbook. The following activities have been designed to help you refresh your knowledge of the material in this chapter.

Learning Objectives

After reading this chapter, you will be able to:

- LO 8-01 Identify threaded fasteners.
- LO 8-02 Identify nonthreaded fasteners.
- LO 8-03 Replace threaded fasteners.
- LO 8-04 Repair damaged fastener threads.

Multiple Choice

Read each item carefully, and then select the best response.

_____ 1. To accomplish their job, fasteners come in a variety of hardnesses, which are defined in _____.
  A. Ranks
  B. Grades
  C. Sizes
  D. Shapes

_____ 2. Bolts are a cylindrical piece of metal with a(n) _____ head on one end and a thread cut into the shaft at the other end.
  A. Octagonal
  B. Square
  C. Hexagonal
  D. Circular

_____ 3. In the standard system, bolts, studs, and nuts are measured in _____.
  A. Threads per centimeter
  B. Threads per inch
  C. Threads per millimeter
  D. Threads per foot

_____ 4. Each bolt diameter in the standard system can have one of two thread pitches: _____ or UNF.
  A. ANC
  B. UNC
  C. ANF
  D. ENC

_____ 5. The higher the grade number of a bolt, the higher the _____.
  A. Flexibility of the bolt or nut
  B. Tensile strength of the bolt
  C. Weight of the bolt or nut
  D. Size of the bolt or nut

_____ **6.** Torque _____ is considered a more precise method to tighten torque-to-yield (TTY) bolts and is essentially a multistep process.
   A. Angle
   B. Limit
   C. Gauge
   D. Elasticity

_____ **7.** _____ are designed to provide a consistent clamping force when torqued to their yield point or just beyond.
   A. Torque wrenches
   B. Fasteners
   C. TTY bolts
   D. Threads

_____ **8.** The _____ method is a more precise way to tighten TTY bolts.
   A. Torque-to-yield
   B. Torque angle
   C. Torque assist
   D. Fastener

_____ **9.** A bolt has three hash marks on its head. What is its grade?
   A. 2
   B. 5
   C. 6
   D. 8

_____ **10.** Which nut is made with a deformed top thread?
   A. Lock nut
   B. J nut
   C. Castle nut
   D. Specialty nut

_____ **11.** When you want a nut or bolt to hold tight such that it needs heat to be removed with a socket and ratchet, you should use a(n) _____.
   A. Lubricant
   B. Blue thread locking compound
   C. Antiseize compound
   D. Red thread locking compound

_____ **12.** Which of the following are threaded fasteners?
   A. C-clips
   B. Trim screws
   C. Cotter pins
   D. Solid rivets

_____ **13.** Which of the following screws has a fluted tip to drill a hole into the base material without needing a pilot hole?
   A. Machine screw
   B. Self-tapping screw
   C. Trim screw
   D. Sheet metal screw

_____ **14.** Which of the following is NOT an example of a nonthreaded fastener?
   A. Snap ring
   B. Push clip
   C. Rivet
   D. Machine screw

_____ **15.** What type of pin is shown in the image?

    A. Spiral pin
    B. Roll pin
    C. Cotter pin
    D. Safety pin

_____ **16.** What type of nut is shown in the image?

    A. Deformed lock nut
    B. Nylon lock nut
    C. Sheet metal nut
    D. Castellated nut

_____ **17.** All of the following should be done in the process of removing a broken bolt, EXCEPT _____.
    A. Inspecting the site of the broken bolt
    B. Trying to remove the bolt with a pair of locking pliers
    C. Using a screw or bolt extractor
    D. Using a self-tapping screw to drill out the old fastener

_____ **18.** If a bolt is broken off flush with a surface, and the bolt is large enough in diameter, then you may be able to use a small _____ to turn the bolt by tapping on the outside diameter of the bolt, but in the reverse direction.
    A. Piston
    B. Center punch
    C. Fastener
    D. Stud

_____ **19.** Once a broken bolt is removed, run a lubricated tap or thread restoring tool of the correct size and thread pitch through the hole to clean up any rust or damage.
  A. True
  B. False

_____ **20.** The aim of thread repair is to restore the thread to a condition that restores the fastening integrity.
  A. True
  B. False

_____ **21.** To properly accomplish their given job, fasteners come in a variety of diameters for different sized loads and hardness, which are defined in grades.
  A. True
  B. False

_____ **22.** Tech A says that torque-to-yield head bolts are tightened to, or just past, their yield point. Tech B says that the yield point is the torque at which the bolt breaks. Who is correct?
  A. Tech A
  B. Tech B
  C. Both Tech A and Tech B
  D. Neither Tech A nor Tech B

_____ **23.** Tech A says that most metric fasteners can be identified by the grade number cast into the head of the bolt. Tech B says that standard bolts can generally be identified by hash marks cast into the head of the bolt indicating the bolt's grade. Who is correct?
  A. Tech A
  B. Tech B
  C. Both Tech A and Tech B
  D. Neither Tech A nor Tech B

_____ **24.** Tech A says that some locking nuts have a nylon insert that has a smaller diameter hole than the threads in the nut. Tech B says that castle nuts are used with cotter pins to prevent the nut from loosening. Who is correct?
  A. Tech A
  B. Tech B
  C. Both Tech A and Tech B
  D. Neither Tech A nor Tech B

_____ **25.** Thread repair is used in situations where it is not feasible to replace a damaged thread. Tech A states that this may be because the damaged thread is located in a large, expensive component. Tech B states that this may be because replacement parts are expensive or not available. Who is correct?
  A. Tech A
  B. Tech B
  C. Both Tech A and Tech B
  D. Neither Tech A nor Tech B

_____ **26.** Tech A says that when using a HeliCoil® to repair a damaged thread, the HeliCoil® is welded into the bolt hole. Tech B says that when using a solid sleeve type of insert, the damaged threads need to be drilled out with the correct size drill bit. Who is correct?
  A. Tech A
  B. Tech B
  C. Both Tech A and Tech B
  D. Neither Tech A nor Tech B

_____ **27.** Tech A says that if a bolt breaks off above the surface, you might be able to remove it using locking pliers or curved jaw Channellock® pliers. Tech B says that a hammer and punch can sometimes be used to remove a broken bolt. Who is correct?
  A. Tech A
  B. Tech B
  C. Both Tech A and Tech B
  D. Neither Tech A nor Tech B

_____ **28.** Tech A says that antiseize compound is a type of thread locking compound. Tech B says that thread locking compounds act like very strong glue. Who is correct?
   A. Tech A
   B. Tech B
   C. Both Tech A and Tech B
   D. Neither Tech A nor Tech B

_____ **29.** Tech A states that the aim of thread repair is to restore the thread to a condition that restores the fastening integrity. Tech B states that thread repairs can only be performed on internal threads. Who is correct?
   A. Tech A
   B. Tech B
   C. Both Tech A and Tech B
   D. Neither Tech A nor Tech B

Vehicle Protection and Jack and Lift Safety

At the start of each chapter, you will find the Learning Objectives from the textbook. These are your objectives as you make your way through the exercises in this workbook and the chapter in your textbook. The following activities have been designed to help you refresh your knowledge of the material in this chapter.

Learning Objectives

After reading this chapter, you will be able to:

- LO 9-01 Prepare vehicle for service, and return to customer.
- LO 9-02 Operate jacks, engine hoists, and stands.
- LO 9-03 Operate hoists and describe use of inspection pits.

Matching

Match the following terms with the correct description or example.

A. Four-post lift   **D.** Safe working load

B. Hydraulic jack   **E.** Single-post lift

C. Pneumatic jack   **F.** Two-post lift

_____ **1.** A type of vehicle jack that uses compressed gas or air to lift a vehicle.

_____ **2.** A type of vehicle jack that uses oil under pressure to lift vehicles.

_____ **3.** The maximum safe lifting load for lifting equipment.

_____ **4.** Vehicle lift with two centrally located lifting posts.

_____ **5.** Vehicle lift that uses a single central platform to lift a vehicle.

_____ **6.** Vehicle lift with four lifting posts, one near each corner of the vehicle.

Multiple Choice

Read each item carefully, and then select the best response.

_____ **1.** It is good practice for the service advisor to perform a(n) _____ with the customer to point out any existing damage or missing components on the vehicle.
   A. Detailed inspection
   B. Vehicle walk-around
   C. Audit
   D. Performance analysis

_____ **2.** The three main types of mechanisms that provide the lifting action for vehicle jacks include all of the following, EXCEPT _____.
   A. Hydraulic
   B. Pneumatic
   C. Electric
   D. Mechanical

_____ **3.** What type of jack is mounted on four wheels, two of which swivel to provide a steering mechanism?
   A. Bottle jack
   B. Sliding bridge jack
   C. Floor jack
   D. Scissor jack

_____ **4.** Often used on farms, what type of jack is designed to lift, winch, clamp, pull, and push?
  A. High-lift jack
  B. Scissor jack
  C. Sliding bridge jack
  D. Bottle jack

_____ **5.** What type of jack uses compressed air to either operate a large ram or inflate an expandable air bag to lift the vehicle?
  A. Bottle jack
  B. Pneumatic jack
  C. High-lift jack
  D. Floor jack

_____ **6.** Which lifting device is useful for raising a vehicle to a height that removes the need for the technician to bend down?
  A. Engine hoist
  B. Jack stand
  C. Farm jack
  D. Vehicle lift

_____ **7.** What type of lift comes in two configurations—symmetrical and asymmetrical?
  A. Four-post lifts
  B. Two-post lifts
  C. Farm lifts
  D. Single-post lifts

_____ **8.** The engine or component to be lifted is attached to the lifting arm of an engine hoist by which of the following?
  A. Vehicle lift
  B. Rope
  C. Lifting chain
  D. Safety lock

_____ **9.** The size of the vehicle jack you use is determined by _____.
  A. The length of the axles
  B. The length of the vehicle
  C. The weight of the vehicle
  D. The type of vehicle

_____ **10.** Customers expect their vehicles to be treated with care and respect while in your shop.
  A. True
  B. False

_____ **11.** Fender, carpet, seat, and steering wheel covers should be the first thing on and the last thing removed when working on vehicles.
  A. True
  B. False

_____ **12.** Never place tools in your back pocket.
  A. True
  B. False

_____ **13.** Only the most skilled and experienced drivers available should be allowed to test-drive higher performance vehicles.
  A. True
  B. False

_____ **14.** When multiple pieces of lifting equipment are used, the safe working load is determined by the highest rated piece of equipment.
  A. True
  B. False

_____ **15.** Vehicle jacks may be used to support the weight of the vehicle during any task that requires you to get underneath any part of the vehicle.
  A. True
  B. False

_____ **16.** High-lift jacks are usually fitted in pairs to four-post lifts as an accessory to allow the vehicle to be lifted off the drive-on lift runways.
    A. True
    B. False

_____ **17.** Tall jack stands are used to stabilize a vehicle up on a lift that is having a heavy component, such as a transaxle, removed or installed.
    A. True
    B. False

_____ **18.** Since the vehicle rests on its wheels on the four-post lift, the wheels cannot be removed unless the lift is fitted with sliding bridge jacks.
    A. True
    B. False

_____ **19.** Every vehicle lift in the shop must have a built-in mechanical locking device so that the vehicle lift can be secured at the chosen height after the vehicle is raised.
    A. True
    B. False

_____ **20.** _____ are a protective layer used to cover the fenders when work is conducted around the engine bay.
    A. Seat covers
    B. Fender bumpers
    C. Fender covers
    D. Fender mats

_____ **21.** If spills do occur, be sure to clean them up thoroughly using appropriate methods, which can usually be found in the _____ sheets for each material.
    A. Spill safety
    B. Safety data
    C. Spill disposal
    D. Chemical damage

_____ **22.** Special _____ materials in granular form can be used on some liquid spills, such as engine oil.
    A. Cleaning
    B. Waterproof
    C. Impermeable
    D. Absorbent

_____ **23.** When moving a vehicle in the shop, often _____ will be used by someone directing you to maneuver the vehicle.
    A. Hand signals
    B. Radio communications
    C. A flare gun
    D. Light signals

_____ **24.** Lifting equipment should be periodically _____ and _____ to make sure it is safe.
    A. Moved, operated
    B. Checked, lowered
    C. Checked, tested
    D. Cleaned, painted

_____ **25.** A _____ jack is a portable jack that usually has either a mechanical screw or a hydraulic ram mechanism that rises vertically from the center of the jack as the handle is operated.
    A. Floor
    B. Transmission
    C. Scissor
    D. Bottle

_____ **26.** _____ two-post lifts have arms that are of approximately equal length so that the vehicle is roughly centered lengthwise between the posts.
   A. Symmetrical
   B. Balanced
   C. Asymmetrical
   D. Compressed

_____ **27.** What type of jack is shown in the image?

   A. Transmission jack
   B. Floor jack
   C. Air jack
   D. Scissor jack

_____ **28.** What type of jack is shown in the image?

   A. Bottle jack
   B. Sliding bridge jack
   C. Transmission jack
   D. High-lift (or farm) jack

_____ **29.** What type of jack is shown in the image?

   A. Floor jack
   B. Sliding bridge jack
   C. Air jack
   D. Transmission jack

_____ **30.** What type of lift is shown in the image?

   A. Single-post lift
   B. Two-post lift
   C. Three-post lift
   D. Four-post lift

_____ **31.** What type of lift is shown in the image?

   A. Single-post lift
   B. Two-post lift
   C. Three-post lift
   D. Four-post lift

_____ **32.** Tech A says that you should always deal with the customer's valuables according to company policy. Tech B says that you should protect a customer's vehicle by washing it when you are finished with it. Who is correct?
  A. Tech A
  B. Tech B
  C. Both Tech A and Tech B
  D. Neither Tech A nor Tech B

_____ **33.** Tech A says that it is a good practice to perform a walk-around inspection of the vehicle with the customer. Tech B says that fender covers will protect the fenders from dents when working on the engine. Who is correct?
  A. Tech A
  B. Tech B
  C. Both Tech A and Tech B
  D. Neither Tech A nor Tech B

_____ **34.** Tech A says that a vehicle jack can be used to support the vehicle while working under it. Tech B says that a jack stand automatically adjusts to the vehicle's height. Who is correct?
  A. Tech A
  B. Tech B
  C. Both Tech A and Tech B
  D. Neither Tech A nor Tech B

_____ **35.** Tech A says that lifts should be inspected and certified periodically. Tech B says that safety locks do not need to be applied before working under the vehicle, unless you will be working for more than 10 minutes. Who is correct?
  A. Tech A
  B. Tech B
  C. Both Tech A and Tech B
  D. Neither Tech A nor Tech B

_____ **36.** Tech A says that an engine hoist can lift more weight when the legs and arms are extended. Tech B says that the bolts used to mount an engine to a stand should turn at least six turns into the engine. Who is correct?
  A. Tech A
  B. Tech B
  C. Both Tech A and Tech B
  D. Neither Tech A nor Tech B

_____ **37.** Tech A says that you should always ensure that the vehicle has enough ground clearance before driving on any type of lift. Tech B says that you should center the vehicle on the lift before raising it. Who is correct?
  A. Tech A
  B. Tech B
  C. Both Tech A and Tech B
  D. Neither Tech A nor Tech B

_____ **38.** Tech A says that you should always inspect a lifting device for leaks and operation before using it. Tech B says that all mechanical safety locks on a lift should be in place before getting under the vehicle. Who is correct?
  A. Tech A
  B. Tech B
  C. Both Tech A and Tech B
  D. Neither Tech A nor Tech B

_____ **39.** Tech A says that if you damage a customer's vehicle, you should insist the damage was there before you started working on it. Tech B says that you should always use seat and steering wheel covers, floor mats, and fender covers whenever a vehicle is being worked on. Who is correct?
  A. Tech A
  B. Tech B
  C. Both Tech A and Tech B
  D. Neither Tech A nor Tech B

_____ **40.** Tech A says that a person outside the vehicle should be used to supervise and guide a vehicle while it is being driven in the shop. Tech B says that as long as you are a student in the Auto Tech program, you can drive any of the vehicles on a test drive, even if you do not have a driver's license. Who is correct?

A. Tech A

B. Tech B

C. Both Tech A and Tech B

D. Neither Tech A nor Tech B

Vehicle Maintenance Inspection

At the start of each chapter, you will find the Learning Objectives from the textbook. These are your objectives as you make your way through the exercises in this workbook and the chapter in your textbook. The following activities have been designed to help you refresh your knowledge of the material in this chapter.

Learning Objectives

After reading this chapter, you will be able to:

- LO 10-01 Perform in-vehicle inspection.
- LO 10-02 Perform fluid inspection.
- LO 10-03 Perform belt, hose, and air filter/cabin air filter inspection.
- LO 10-04 Perform undervehicle inspection.
- LO 10-05 Perform exterior vehicle inspection.

Matching

Match the following terms with the correct description or example.

A. Body control module (BCM)
B. Coolant
C. CV joint
D. Diesel exhaust fluid (DEF)

E. Hydrometer
F. Hygroscopic
G. Pinion shaft

_____ 1. A property of a substance or liquid that causes it to attract and absorb moisture (water), as a sponge absorbs water.

_____ 2. A joint used to transmit torque through wider angles and without the change of velocity that occurs in U-joints.

_____ 3. On a drive axle using a ring-and-pinion gear assembly, the input component that drives the ring gear.

_____ 4. An onboard computer that controls many vehicle functions, including the vehicle interior and exterior lighting, horn, door locks, power seats, and windows.

_____ 5. The resulting mixture when antifreeze concentrate is mixed with water.

_____ 6. A tool that measures the specific gravity of a liquid.

_____ 7. A mixture of urea and water that is injected into the exhaust system of a late-model diesel-powered vehicle to reduce exhaust oxides of nitrogen emissions.

Multiple Choice

Read each item carefully, and then select the best response.

_____ 1. If the oil in an engine's lubrication system is too high, the oil will be struck by the crankshaft and will _____.
A. Burn
B. Smoke
C. Coagulate
D. Foam

_____ **2.** Do not remove the _____ when the engine is warm or hot.
    A. Oil dipstick
    B. Radiator cap
    C. Master cylinder cover
    D. Transmission dipstick

_____ **3.** The antifreeze protection level can be checked with which of the following?
    A. Hydrometer
    B. Hydroscope
    C. Hygroscope
    D. Refractoscope

_____ **4.** If an automatic transmission does not have a dipstick, the fluid level may be checked using a _____ on the side of the transmission.
    A. Master cylinder
    B. Reservoir
    C. Fill plug
    D. Sight glass

_____ **5.** Some late-model diesel-powered vehicles use a fluid called _____, which is injected into the exhaust stream to reduce oxides of nitrogen during certain driving conditions.
    A. Diesel exhaust fluid
    B. Oxide reducer
    C. Nitrous oxide
    D. Octane booster

_____ **6.** What type of drive belt has a flat profile with a number of grooves running lengthwise along the belt?
    A. V-type
    B. Serpentine type
    C. Variable diameter type
    D. Toothed

_____ **7.** The _____ is the most important component driven by the drive belt, and the engine will quickly overheat if the belt breaks or comes off.
    A. Alternator
    B. Air conditioning compressor
    C. Water pump
    D. Power steering pump

_____ **8.** What is the term used to describe the shininess on the surface of a belt where it comes in contact with the pulley?
    A. Shine
    B. Gloss
    C. Soaking
    D. Glazing

_____ **9.** What device on a multiport fuel-injected vehicle is typically located in a rectangular box within the air induction system?
    A. Fuel filter
    B. Air cleaner
    C. Serpentine belt
    D. Carburetor

_____ **10.** What type of fluid is normally green, orange, or yellow in color?
    A. DOT 4
    B. Power steering fluid
    C. Coolant
    D. Automatic transmission fluid

_____ **11.** Component damage or failure is often caused by a lack of service or low fluid level in the related system.
    A. True
    B. False

_____ **12.** Always check engine oil with the engine on.
    A. True
    B. False

_____ **13.** If the brake fluid level gets too low, air can be pulled into the hydraulic system.
    A. True
    B. False

_____ **14.** Most brake fluid is dark or black.
    A. True
    B. False

_____ **15.** Some automatic transmissions do not have a dipstick.
    A. True
    B. False

_____ **16.** Transmission fluid is added through the transmission reservoir.
    A. True
    B. False

_____ **17.** Many newer vehicles have transmissions that are considered sealed and lubricated for the life of the vehicle.
    A. True
    B. False

_____ **18.** There are two types of drive belts: the V-type and the serpentine type.
    A. True
    B. False

_____ **19.** The engine should be hot when inspecting radiator hoses.
    A. True
    B. False

_____ **20.** The warning lamps perform a self-check each time the ignition is switched on or the engine is cranked.
    A. True
    B. False

_____ **21.** Nearly all of the _____ in a vehicle, except some used in automatic transmissions/transaxles, get(s) old and wear(s) out, requiring replacement.
    A. Air filters
    B. Engine coolant
    C. Fluids
    D. Engine drive belts

_____ **22.** The level of _____ in an engine's lubrication system is critical to the engine's operation.
    A. Oil
    B. Coolant
    C. Brake fluid
    D. Power steering fluid

_____ **23.** Engine _____ must be controlled to prevent overheating and to maintain proper exhaust emission levels.
    A. Combustion
    B. Fluid
    C. Noise
    D. Temperature

_____ **24.** A hydraulic braking system depends on a special fluid called _____.
    A. Engine oil
    B. Engine coolant
    C. Brake fluid
    D. Power steering fluid

_____ **25.** When checking power steering fluid levels, the engine should be _____ and the fluid hot.
    A. Cold
    B. Off
    C. Driving
    D. Idling

_____ 26. Driving in dusty or wet conditions may require that the _____ be checked more often.
  A. Windshield washer fluid
  B. Power steering fluid
  C. Automatic transmission fluid
  D. Diesel exhaust fluid

_____ 27. Engine _____ are used to operate the various accessories on the engine, such as the water pump, power steering pump, air conditioner compressor, and alternator.
  A. Hoses
  B. Filters
  C. Drive belts
  D. Fluids

_____ 28. _____ that exceed a certain number per inch in a belt indicate that the belt may soon fail and should be replaced.
  A. Stains
  B. Cracks
  C. Grooves
  D. Oil spots

_____ 29. The _____ should flex at the hinge and be held firmly against the windshield by the wiper arm spring.
  A. Wiper arm
  B. Wiper blade
  C. Shock absorber
  D. Body control module

_____ 30. The _____ acts as a lever to increase the force applied to the brake assemblies by the driver.
  A. Parking brake
  B. Body control module
  C. Brake pedal
  D. Hydraulic clutch

_____ 31. What type of power steering fluid reservoir is shown in the image?

  A. Mounted on the engine-driven hydraulic pump
  B. Mounted on the automatic transmission
  C. Mounted on the engine drive belt
  D. Mounted separately

_____ **32.** What type of power steering fluid reservoir is shown in the image?

   A. Mounted on the engine-driven hydraulic pump
   B. Mounted on the automatic transmission
   C. Mounted on the engine drive belt
   D. Mounted separately

_____ **33.** Tech A says that during an exterior inspection of a vehicle, you should open and close doors to check that they are operating correctly. Tech B says you should never pull on bumpers or fenders. Who is correct?
   A. Tech A
   B. Tech B
   C. Both Tech A and Tech B
   D. Neither Tech A nor Tech B

_____ **34.** Tech A says wiper blade condition can cause a vehicle to fail a safety inspection in some states. Tech B says wiper blades should be checked for wear and tear. Who is correct?
   A. Tech A
   B. Tech B
   C. Both Tech A and Tech B
   D. Neither Tech A nor Tech B

_____ **35.** Tech A says that a low oil level is bad for the engine, but that it is OK for the level to be too high. Tech B says that overfilling the engine oil is bad for the engine. Who is correct?
   A. Tech A
   B. Tech B
   C. Both Tech A and Tech B
   D. Neither Tech A nor Tech B

_____ **36.** Tech A says that improper handling of a windshield wiper can lead to a broken windshield. Tech B says you should place a fender cover under the wiper arm to avoid any possible damage while working on windshield wipers. Who is correct?
   A. Tech A
   B. Tech B
   C. Both Tech A and Tech B
   D. Neither Tech A nor Tech B

_____ **37.** While servicing an automatic transaxle–equipped vehicle, the transaxle dipstick cannot be found. Tech A says that some of these vehicles do not have a transaxle dipstick. Tech B says that the dipstick may have fallen off, since all vehicles have a transaxle dipstick. Who is correct?
   A. Tech A
   B. Tech B
   C. Both Tech A and Tech B
   D. Neither Tech A nor Tech B

_____ **38.** Tech A says to test new wiper blades against a dry windshield to ensure they seat properly. Tech B says to wet the windshield and operate the wipers to check their performance. Who is correct?
  A. Tech A
  B. Tech B
  C. Both Tech A and Tech B
  D. Neither Tech A nor Tech B

_____ **39.** An engine serpentine belt has broken and come off the pulleys. Tech A says that it is OK to drive the vehicle without the belt. Tech B says that the belt only runs the charging system, so it is OK to drive a short distance without the belt. Who is correct?
  A. Tech A
  B. Tech B
  C. Both Tech A and Tech B
  D. Neither Tech A nor Tech B

_____ **40.** Tech A says to clean a windshield first, and then inspect the windshield for chips, scratches, or etching. Tech B says to always clean a windshield first and then inspect the windshield for cracks or signs of delamination. Who is correct?
  A. Tech A
  B. Tech B
  C. Both Tech A and Tech B
  D. Neither Tech A nor Tech B

_____ **41.** Tech A says that unlocking doors on most vehicles turns on the interior lights. Tech B says on some vehicles, the interior light will not turn on until a door is actually opened. Who is correct?
  A. Tech A
  B. Tech B
  C. Both Tech A and Tech B
  D. Neither Tech A nor Tech B

_____ **42.** Tech A says you should test the rear lights with the help of an assistant. Tech B says some shops have a mirror mounted in the service bay so the technician can check the rear lights. Who is correct?
  A. Tech A
  B. Tech B
  C. Both Tech A and Tech B
  D. Neither Tech A nor Tech B

Communication and Employability Skills

At the start of each chapter, you will find the Learning Objectives from the textbook. These are your objectives as you make your way through the exercises in this workbook and the chapter in your textbook. The following activities have been designed to help you refresh your knowledge of the material in this chapter.

Learning Objectives

After reading this chapter, you will be able to:

- LO 11-01 Demonstrate active listening skills.
- LO 11-02 Demonstrate effective speaking skills.
- LO 11-03 Demonstrate employability skills.
- LO 11-04 Demonstrate effective reading and researching skills.
- LO 11-05 Demonstrate effective writing skills.

Multiple Choice

Read each item carefully, and then select the best response.

_____ 1. Attempting to see a situation from someone else's point of view is known as which of the following?
   A. Sympathy
   B. Empathy
   C. Nonverbal communication
   D. Feedback

_____ 2. Phrases such as "I see" or "Tell me more" that indicate you are paying attention are examples of a(n) _____?
   A. Validating statement
   B. Supporting statement
   C. Closed question
   D. Open question

_____ 3. Statements like "Go on" and "Give me an example" that let the speaker know you would like more detail because you are genuinely interested in finding a solution are examples of a(n) _____?
   A. Validating statement
   B. Supporting statement
   C. Closed question
   D. Open question

_____ 4. What type of question usually begins with the words when or where?
   A. Open questions
   B. Yes/no questions
   C. Closed questions
   D. Good questions

_____ 5. Looking through the table of contents, introduction, conclusion, headings, and index until we find what we are looking for is the quickest way to use what type of reading?
   A. Absorbing
   B. Comprehending
   C. Selective
   D. Quick

_____ **6.** Which of the following is like conducting an investigation?
   A. Reading
   B. Researching
   C. Comprehending
   D. Skimming

_____ **7.** What online reference organization consists of over 75,000 active members, with more than 1.7 million years of combined experience, who share their knowledge with each other on over a dozen forums?
   A. International Automotive Technicians Network
   B. National Association for Stock Car Auto Racing
   C. Occupational Safety and Health Administration
   D. National Automotive Parts Association

_____ **8.** Which of the following should you do when you come across any defective equipment?
   A. Tag the defective item
   B. Wait to complete a defective equipment report
   C. Notify a coworker
   D. Unplug the equipment

_____ **9.** Learning and applying good communication skills will save you time and help you avoid or get through tricky situations.
   A. True
   B. False

_____ **10.** Most people tend to believe the actual words expressed over the nonverbal message you are sending.
   A. True
   B. False

_____ **11.** Your body position, eye contact, and facial expression can all set the direction of a conversation.
   A. True
   B. False

_____ **12.** Speaking is often referred to as a science.
   A. True
   B. False

_____ **13.** Closed questions allow individuals to answer with a simple yes or no.
   A. True
   B. False

_____ **14.** Nonverbal communication is not possible during a phone conversation.
   A. True
   B. False

_____ **15.** Being part of a team allows us to complement each other's strengths and weaknesses.
   A. True
   B. False

_____ **16.** If you have to take customers into the shop, always escort them, and be sure to keep them safe.
   A. True
   B. False

_____ **17.** Routinely texting your friends or taking personal calls during work hours is stealing from your employer.
   A. True
   B. False

_____ **18.** The customer satisfaction index (CSI) rating is reported each month and used to evaluate individual technicians and the entire dealership service facility.
   A. True
   B. False

_____ **19.** The person who wants the job done right will probably be very keen to have the repairs finished on schedule.
   A. True
   B. False

_____ **20.** The faster you try to read, the more you are likely to be able to concentrate on the meaning.
   A. True
   B. False

_____ **21.** A repair order can become a legal document that will be used by the court to determine if the shop has any liability in a lawsuit situation.
   A. True
   B. False

_____ **22.** Safety inspection forms are a way of keeping up on routine maintenance tasks such as changing the oil or replacing the belt of an air compressor.
   A. True
   B. False

_____ **23.** If a problem is noticed with a piece of equipment, the lockout/tagout procedure should be followed.
   A. True
   B. False

_____ **24.** Mental _____ are thoughts and feelings that interfere with our listening, such as our own assumptions, emotions, and prejudices.
   A. Practices
   B. Strategies
   C. Empathies
   D. Barriers

_____ **25.** As a person is speaking to you, you need to provide _____ feedback, which indicates to the speaker that you are engaged in what he or she is saying.
   A. Speaking
   B. Listening
   C. Writing
   D. Gathering

_____ **26.** _____ feedback includes very simple spoken signals that can enhance the conversation and let the person know you comprehend.
   A. Physical
   B. Mental
   C. Organizational
   D. Verbal

_____ **27.** Before speaking, take a moment to consider that your _____ reveals a lot about your feelings and adds significant meaning to your message.
   A. Tone of voice
   B. Choice of questions
   C. Choice of attire
   D. Selection of technicians

_____ **28.** A(n) _____ question encourages people to speak freely so we can gather facts, insights, and opinions from them.
   A. Closed
   B. Short
   C. Open
   D. Quiet

_____ **29.** When taking a _____ for someone else, make sure you have the caller's name and organization, contact details, the date and time of the call, and a summary of the caller's message.
   A. Commitment
   B. Violation
   C. Phone message
   D. Drug test

_____ **30.** _____ should contain information about who, what, when, where, and why, and direction on how a task should be completed.
   A. Conversations
   B. Customer service
   C. Readings
   D. Instructions

_____ **31.** Good _____ skills involve setting a clear vision of the goals, empowering each team member to contribute his or her best efforts, and recognizing each team member's strengths and weaknesses.
   A. Leadership
   B. Communication
   C. Training
   D. Reading

_____ **32.** Our _____ is the image we present of ourselves to the public.
   A. Appearance
   B. Communication
   C. Comprehension
   D. Education

_____ **33.** _____ means showing up on time (typically 5–10 minutes early to get ready to start work at the appointed time).
   A. Organizational
   B. Educational
   C. Punctual
   D. Technical

_____ **34.** _____ reading is reading only the parts we need to know. This method is useful when looking for a particular piece of information.
   A. Assistive
   B. Selective
   C. Technical
   D. Open

_____ **35.** Before you spend too much time researching information, it is important that you _____ the problem.
   A. Solve
   B. Define
   C. Diagram
   D. Test

_____ **36.** When completing a repair order, these elements constitute what are called the three Cs: _____, _____, and _____.
   A. Choice, creation, completion
   B. Consultation, creation, correction
   C. Concern, cause, correction
   D. Customer, complaint, correction

_____ **37.** If someone is injured in the workplace, a(n) _____ should be completed by those involved—both the victim and witnesses, if possible.
   A. Accident report
   B. Lockout
   C. Vehicle inspection form
   D. Supporting statement

_____ **38.** When completing a(n) _____, you need to check that all major components and systems are operational, secured, and safe in accordance with the vehicle manufacturer's recommendations.
   A. Accident report
   B. Customer satisfaction survey
   C. Customer concern sheet
   D. Inspection

_____ **39.** Tech A says that repair orders are legal documents, so they need to be filled out accurately and carefully. Tech B says that ensuring the repair order is well written, clear, and concise promotes a professional reputation. Who is correct?
   A. Tech A
   B. Tech B
   C. Both Tech A and Tech B
   D. Neither Tech A nor Tech B

_____ **40.** Tech A says that you should not waste time inspecting a vehicle in for a repair unless the customer requests it. Tech B says that performing an inspection in addition to a repair can lead to the discovery of additional concerns. Who is correct?
   A. Tech A
   B. Tech B
   C. Both Tech A and Tech B
   D. Neither Tech A nor Tech B

_____ **41.** Tech A says that the proper way to listen to a customer is to maintain eye contact with the customer in between taking notes. Tech B says that eye contact distracts you from taking good notes. Who is correct?
   A. Tech A
   B. Tech B
   C. Both Tech A and Tech B
   D. Neither Tech A nor Tech B

_____ **42.** Tech A says it is best to quietly listen to the customer and refrain from asking questions as much as possible. Tech B says it is constructive to ask questions throughout the conversation to obtain more details. Who is correct?
   A. Tech A
   B. Tech B
   C. Both Tech A and Tech B
   D. Neither Tech A nor Tech B

_____ **43.** Tech A says that maintaining an appearance of neatness is important, as it conveys to the customer the idea of careful, professional technicians. Tech B says that a dirty and cluttered shop indicates that the shop gets a lot of quality work done. Who is correct?
   A. Tech A
   B. Tech B
   C. Both Tech A and Tech B
   D. Neither Tech A nor Tech B

_____ **44.** Tech A says researching the service information is a waste of time. Tech B says that researching the service information saves time. Who is correct?
   A. Tech A
   B. Tech B
   C. Both Tech A and Tech B
   D. Neither Tech A nor Tech B

_____ **45.** Tech A says that an example of an open question is "What are the conditions like when your A/C is not working?" Tech B says that an example of an open question is "Does your A/C work at all?" Who is correct?
   A. Tech A
   B. Tech B
   C. Both Tech A and Tech B
   D. Neither Tech A nor Tech B

_____ **46.** Tech A says that one customer who has a bad experience stemming from miscommunication and a misdiagnosed repair due to a technician's failure to listen has more of an effect on the repair shop than several good and happy customers. Tech B says that it is only one customer and that since the happy ones paid their bills all is well. Who is correct?
   A. Tech A
   B. Tech B
   C. Both Tech A and Tech B
   D. Neither Tech A nor Tech B

_____ **47.** The first step in good communication is:

    A. Nonverbal communication

    B. Asking good questions

    C. Providing feedback

    D. Active listening

_____ **48.** Tech A says that the three Cs are "customer, complaint, and concern." Tech B says that the three Cs are "concern, cause, and correction." Who is correct?

    A. Tech A

    B. Tech B

    C. Both Tech A and Tech B

    D. Neither Tech A nor Tech B

Motive Power Theory—SI Engines

At the start of each chapter, you will find the Learning Objectives from the textbook. These are your objectives as you make your way through the exercises in this workbook and the chapter in your textbook. The following activities have been designed to help you refresh your knowledge of the material in this chapter.

Learning Objectives

After reading this chapter, you will be able to:

- LO 12-01 Demonstrate knowledge of heat engines.
- LO 12-02 Demonstrate knowledge of the physics of engine operation.
- LO 12-03 Demonstrate knowledge of force, work, and power.
- LO 12-04 Demonstrate knowledge of four-stroke engine arrangement, operation, and measurement.
- LO 12-05 Demonstrate knowledge of spark-ignition engine components.
- LO 12-06 Demonstrate knowledge of two-stroke and rotary engine operation.

Matching

Match the following terms with the correct description or example.

A. Base circle	**J.** Ignition
B. Bottom dead center (BDC)	**K.** Lobe
C. Cam	**L.** Piston clearance
D. Camshaft	**M.** Piston skirt
E. Column inertia	**N.** Power
F. Cylinder bore	**O.** Spark-ignition (SI) engine
G. Duration	**P.** Torque
H. Exhaust stroke	**Q.** Valve margin
I. Horsepower	**R.** Work

_____ **1.** The hole in the engine block that the piston fits into.

_____ **2.** An engine that relies on an electrical spark to ignite the air and fuel mixture.

_____ **3.** The stroke of the piston during which the exhaust valve is open and the piston is moving from bottom dead center (BDC) to top dead center (TDC) to push exhaust gas out of the cylinder.

_____ **4.** The clearance between the piston and the cylinder wall that allows for lubricating oil to reduce friction.

_____ **5.** The flat surface on the outer edge of the valve head between the valve head and the valve face.

_____ **6.** The principle that as a column of air flows, it creates inertia, which keeps air flowing until its inertia energy is spent; sometimes referred to as a "ram effect" when using tuned intake or exhaust systems.

_____ **7.** The amount of twisting force applied in a turning application, usually measured in foot-pounds.

_____ **8.** The egg-shaped lobe machined to a shaft used to cause opening and closing of the valves of a four-stroke cycle engine.

_____ **9.** The amount of time the valve stays open, given in degrees of rotation of the crankshaft.

_____ **10.** The position of the piston at the end of its stroke when it is closest to the crankshaft.

_____ **11.** The part of the engine that activates the valve train by using lobes riding against lifters.

_____ **12.** The lighting of the fuel and air mixture in the combustion chamber.

_____ **13.** The result of force creating movement.

_____ **14.** The area below the ring groove area of the piston that prevents the piston from cocking and becoming jammed in the cylinder bore.

_____ **15.** The rounded bottom part of the camshaft (off the lobe) where the valves remain closed or at rest.

_____ **16.** The rate or speed at which work is done.

_____ **17.** An amount of work performed in a given time.

_____ **18.** The raised portion on a camshaft; used to lift the lifter and open the valve.

Multiple Choice

Read each item carefully, and then select the best response.

_____ **1.** The steam engine and the Stirling engine are examples of what kind of engine?
   A. Internal combustion engine
   B. External combustion engine
   C. Rotary engine
   D. Reciprocating piston engine

_____ **2.** The effort to produce a push or pull action is referred to as _____.
   A. Power
   B. Work
   C. Force
   D. Torque

_____ **3.** The unit of measurement for torque in the metric system is _____.
   A. Ft-lb
   B. Newton meters
   C. Horsepower
   D. rpms

_____ **4.** How many radians are there in the circumference of a circle?
   A. 1
   B. 3.14
   C. 6.28
   D. Depends on the radius of the circle

_____ **5.** The _____ occurs as extreme force moves the piston from TDC to BDC with both valves remaining closed.
   A. Power stroke
   B. Compression stroke
   C. Intake stroke
   D. Exhaust stroke

_____ **6.** The distance the piston travels from TDC to BDC, or from BDC to TDC, is called the _____.
   A. Throw
   B. Piston stroke
   C. Displacement
   D. Compression ratio

_____ **7.** The Miller cycle engine and the Atkinson cycle engine are both variations on the traditional _____ engine.
   A. Two-stroke SI
   B. Four-stroke SI
   C. External combustion
   D. Rotary

_____ **8.** The _____ are the areas between the ring grooves that support the rings as the piston moves.
   A. Offsets
   B. Journals
   C. Ring lands
   D. Skirts

_____ **9.** The piston _____ prevents the piston from rocking and jamming in the cylinder bore.
  A. Skirt
  B. Pin boss
  C. Connecting rod
  D. Head

_____ **10.** What type of engines use pushrods to transfer the camshaft's lifting motion to the valves by way of rocker arms on top of the cylinder head?
  A. Flat-head engines
  B. Cam-in-block engines
  C. Overhead cam
  D. Dual overhead cam (DOHC) engines

_____ **11.** Which part rides on the camshaft lobes to actuate the pushrods, rocker arms, and valves?
  A. Clearance ramps
  B. B ramps
  C. Lifters
  D. Valve rotator

_____ **12.** The number of degrees between the centerline of the intake lobe and the centerline of the exhaust lobe is called the _____.
  A. Cam lobe centerline
  B. Cam lobe separation
  C. Cam lobe duration
  D. Cam lobe margin

_____ **13.** A machined surface on the back of the valve head is known as the valve _____.
  A. Face
  B. Margin
  C. Keeper
  D. Stem

_____ **14.** The valve train operates off the camshaft, and the part that rides against the cam lobe is the _____.
  A. Valve stem
  B. Valve lifter
  C. Rocker arm
  D. Fulcrum

_____ **15.** The _____ engine has enough clearance between the pistons and the valves, so in the event the timing belt breaks, any valve that is hanging all the way open will not contact the piston, thus preventing engine damage.
  A. Interference
  B. Rotary
  C. Freewheeling
  D. Two-stroke

_____ **16.** The _____ engine is also called the Wankel engine because it was improved upon by Felix Wankel for automotive use in the 1940s.
  A. Two-stroke
  B. Four-stroke
  C. Interference
  D. Rotary

_____ **17.** The internal combustion engine has almost completely replaced the external combustion engine and has been around for well over a century.
  A. True
  B. False

_____ **18.** Compression-ignition engines do not use spark plugs.
  A. True
  B. False

_____ **19.** Pressure and volume are inversely related; as one rises, the other falls.
  A. True
  B. False

_____ **20.** Force is measured in foot-pound (ft-lb) per second or ft-lb per minute.
   A. True
   B. False

_____ **21.** Movement must occur to produce torque.
   A. True
   B. False

_____ **22.** A turbocharger or supercharger will increase an engine's volumetric efficiency well above 100%.
   A. True
   B. False

_____ **23.** When the piston in the cylinder is at a position closest to the crankshaft, it is said to be at TDC.
   A. True
   B. False

_____ **24.** The power stroke starts with the exhaust valve closed, the intake valve(s) opening, and the piston moving from TDC to BDC.
   A. True
   B. False

_____ **25.** The block deck is the machined surface of the block farthest from the crankshaft.
   A. True
   B. False

_____ **26.** Increasing the diameter of the bore or increasing the length of the stroke will produce a larger piston displacement.
   A. True
   B. False

_____ **27.** The time during which both the intake and the exhaust valves are open is called valve overlap.
   A. True
   B. False

_____ **28.** A long block replacement includes the engine block from below the head gasket to above the oil pan.
   A. True
   B. False

_____ **29.** The piston body has a piston pinhole machined through a reinforced area called the piston pin boss.
   A. True
   B. False

_____ **30.** In many engines, the oil pan houses the oil pump, which is the heart of the engine in that it supplies critical lubrication and cooling for the internal moving parts of the engine.
   A. True
   B. False

_____ **31.** Intake and exhaust valves may all be actuated by a single camshaft, or there may be two camshafts per head, called dual overhead cam engines.
   A. True
   B. False

_____ **32.** The cam lobe ramp is where the rise of the cam lobe starts from the base circle to the top of the lobe.
   A. True
   B. False

_____ **33.** If valve clearance is too large, the valve can be held open longer than it should be.
   A. True
   B. False

_____ **34.** The rotary engine is not an internal combustion engine.
   A. True
   B. False

_____ **35.** The branch of physical science that deals with heat and its relation to other forms of energy, such as mechanical energy, is generally defined as _____.
   A. Thermodynamics
   B. Physics
   C. Thermology
   D. Relativity

_____ **36.** The gasoline _____ engine uses a crankshaft to convert the reciprocating movement of the pistons in their cylinder bores into rotary motion at the crankshaft.
A. Rotary
B. Stirling
C. Piston
D. Low-cylinder

_____ **37.** Heating a gas in a sealed container will _____ the pressure in the container.
A. Increase
B. Decrease
C. Equalize
D. Remove

_____ **38.** One _____ equals 33,000 ft-lb per minute.
A. Work
B. Kilowatt
C. Joule
D. Horsepower

_____ **39.** A _____ describes how many radius distances there are in the circumference of a circle.
A. Joule
B. Radian
C. Torque
D. Horsepower

_____ **40.** When the piston in a cylinder is at the position farthest away from the crankshaft, it is at _____.
A. Bottom dead center
B. Flat center
C. Top dead center
D. V configuration

_____ **41.** Internal combustion engines are designated by the amount of space their pistons displace as they move from TDC to BDC, which is called _____.
A. Piston stroke
B. Engine displacement
C. Bore
D. Piston displacement

_____ **42.** The process of using a column of moving air to create a low-pressure area behind it to assist in removing any remaining burned gases from the combustion chamber and replacing these gases with a new charge is called _____.
A. Intaking
B. Scavenging
C. Internal combustion
D. Valve overlapping

_____ **43.** The _____ is the single largest part of the engine.
A. Crankshaft
B. Flywheel
C. Connecting rod
D. Cylinder block

_____ **44.** A _____ is a weighted assembly that stores kinetic energy from each power stroke and helps keep the crankshaft turning through nonpower strokes.
A. Flywheel
B. Cylinder block
C. Cylinder head
D. Valve

_____ **45.** The _____ connects the piston to the crankshaft and transfers piston movement and combustion pressure to the crankshaft rod journals.
  A. Cylinder block
  B. Flywheel
  C. Connecting rod
  D. Overhead cam

_____ **46.** The upper two piston rings are compression rings that prevent combustion pressure, called _____, from leaking past the pistons into the crankcase.
  A. Blowby gas
  B. Intake stroke
  C. Engine vacuum
  D. Compressed air/fuel mixture

_____ **47.** Up until the 1950s, many engines had their valves installed in the engine block. Such engines are called _____ engines.
  A. Cam-in-block
  B. Overhead cam
  C. Flat-head
  D. Dual overhead cam

_____ **48.** Coil _____ occurs when the coils of the valve spring touch each other.
  A. Bind
  B. Block
  C. Compression
  D. Lift

_____ **49.** The opposite end of the valve stem has grooves machined into its end to receive locking pieces, sometimes referred to as valve _____, that retain a valve spring retainer and spring on the valve.
  A. Keepers
  B. Lockers
  C. Rotators
  D. Lifters

_____ **50.** In a _____ engine, the inlet and exhaust ports are opened and closed by the movement of the piston.
  A. Single-stroke
  B. Two-stroke
  C. Three-stroke
  D. Four-stroke

_____ **51.** A(n) _____ is the circular movement around the perimeter of another circle.
  A. Epicycloid curve
  B. Reciprocating bend
  C. Hypotrochoid arch
  D. Epitrochoid curve

_____ **52.** The arrow in the image is pointing to what part of the steam engine?

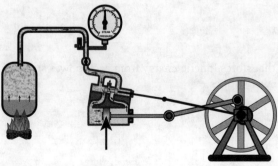

  A. Double acting piston
  B. Relief valve
  C. Throttle valve
  D. Inlet/exhaust valve

_____ **53.** The arrow in the image is pointing to what part of the steam engine?

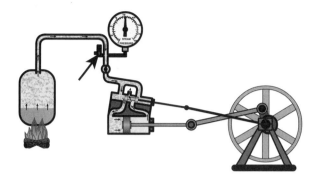

    A. Spent steam valve
    B. Throttle valve
    C. Double acting piston
    D. Relief valve

_____ **54.** The arrow in the image is pointing to what part of the steam engine?

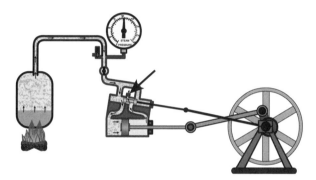

    A. Double acting piston
    B. Spent steam vent
    C. Throttle valve
    D. Inlet/exhaust valve

_____ **55.** The arrow in the image is pointing to what part of the Stirling engine?

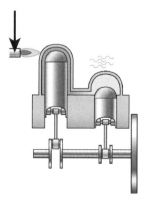

    A. Crankpin
    B. Heat source
    C. Cooling source
    D. Flywheel

_____ **56.** The arrows in the image are pointing to what part of the Stirling engine?

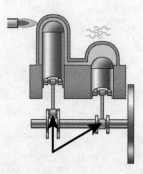

A. Crankpins
B. Heat sources
C. Cooling sources
D. Flywheels

_____ **57.** The arrow in the image is pointing to what part of the Stirling engine?

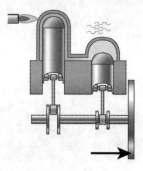

A. Piston
B. Heat source
C. Enclosed fluid
D. Flywheel

_____ **58.** The arrow in the image is pointing to what part of the engine cylinder?

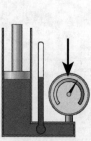

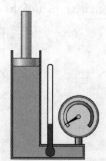

A. Thermometer
B. Air-tight container
C. Pressure gauge
D. Air-tight plunger

_____ **59.** The arrow in the image is pointing to what part of the engine cylinder?

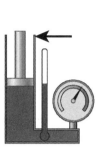 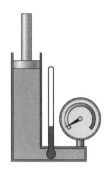

A. Pressure gauge
B. Thermometer
C. Air-tight container
D. Air-tight plunger

_____ **60.** The arrow in the image is pointing to what part of the engine cylinder?

 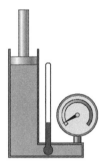

A. Pressure gauge
B. Air-tight plunger
C. Air-tight container
D. Thermometer

_____ **61.** The arrow in the image is pointing to what stage of the basic four-stroke combustion cycle in a compression-ignition engine?

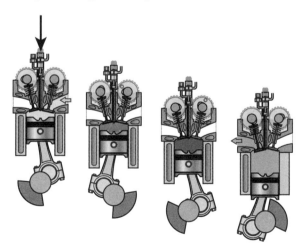

A. Compression
B. Power
C. Intake
D. Exhaust

_____ **62.** The arrow in the image is pointing to what stage of the basic four-stroke combustion cycle in a compression-ignition engine?

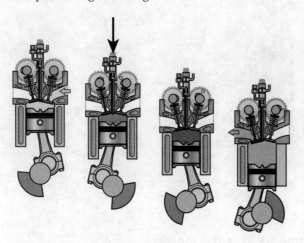

    A. Compression
    B. Intake
    C. Exhaust
    D. Power

_____ **63.** The arrow in the image is pointing to what stage of the basic four-stroke combustion cycle in a compression-ignition engine?

    A. Intake
    B. Compression
    C. Power
    D. Exhaust

_____ **64.** The arrows in the image are pointing to what part of the engine block?

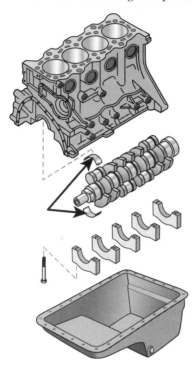

    A. Water jackets
    B. Crankcases
    C. Main bearings
    D. Cylinder bores

_____ **65.** The arrows in the image are pointing to what part of the engine block?

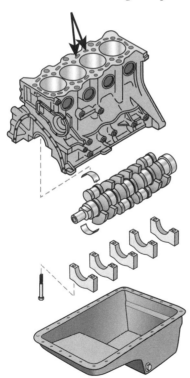

    A. Main bearing tunnels
    B. Water jackets
    C. Core plugs
    D. Oil pans

_____ **66.** The arrows in the image are pointing to what part of the engine block?

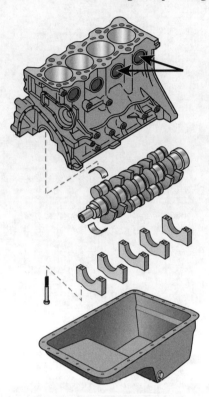

A. Main bearings
B. Core plugs
C. Crankcases
D. Crankshafts

_____ **67.** The arrows in the image are pointing to what part of the crankshaft?

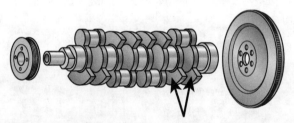

A. Counterweights
B. Main bearing journals
C. Harmonic balancers
D. Connecting rod journals

_____ **68.** The arrows in the image are pointing to what part of the crankshaft?

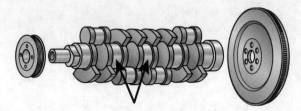

A. Flywheels
B. Webs
C. Main bearing journals
D. Flywheel mounting flanges

_____ **69.** The arrow in the image is pointing to what part of the crankshaft?

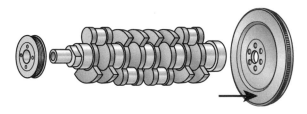

    A. Nose
    B. Oil pump drive
    C. Flywheel mounting flange
    D. Flywheel

_____ **70.** The arrow in the image is pointing to what part of the piston?

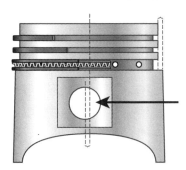

    A. Oil return hole
    B. Top compression ring
    C. Crown
    D. Pinhole

_____ **71.** The arrow in the image is pointing to what part of the piston?

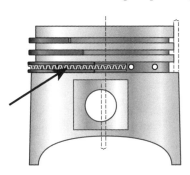

    A. Pin offset
    B. Expander
    C. Oil control ring side rail
    D. Ring groove

_____ **72.** The arrows in the image are pointing to what part of the piston?

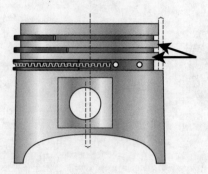

   A. Thrust skirts
   B. Top compression rings
   C. Ring lands
   D. Nonthrust skirts

_____ **73.** The arrow in the image is pointing to what part of the camshaft lobe?

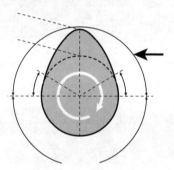

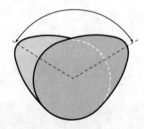

   A. Closing ramp
   B. Valve lift
   C. Lobe
   D. Opening ramp

_____ **74.** The arrow in the image is pointing to what part of the camshaft lobe?

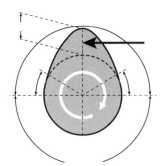

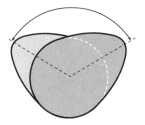

A. Base circle
B. Closing ramp
C. Valve lift
D. Lobe center line

_____ **75.** The arrow in the image is pointing to what part of the camshaft lobe?

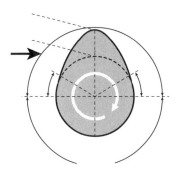

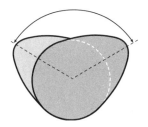

A. Opening ramp
B. Closing ramp
C. Valve opening duration
D. Base circle

_____ **76.** The arrow in the image is pointing to what part of the two-stroke engine?

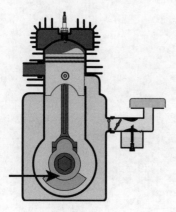

A. Crankcase
B. Crankshaft
C. Connecting rod
D. Reed valve

_____ **77.** The arrow in the image is pointing to what part of the two-stroke engine?

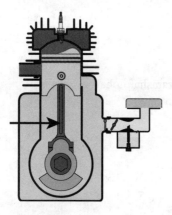

A. Piston
B. Crankshaft
C. Connecting rod
D. Reed valve

_____ **78.** The arrow in the image is pointing to what part of the two-stroke engine?

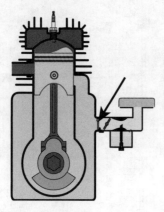

A. Reed valve
B. Piston
C. Crankshaft
D. Connecting rod

_____ 79. The arrow in the image is pointing to what part of the rotary engine?

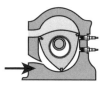

   A. Combustion chamber
   B. Exhaust manifold
   C. Stationary gear
   D. Rotor housing

_____ 80. The arrow in the image is pointing to what part of the rotary engine?

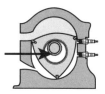

   A. Stationary gear
   B. Rotor
   C. Apex seal
   D. Intake manifold

_____ 81. The arrow in the image is pointing to what part of the rotary engine?

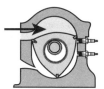

   A. Rotor housing
   B. Rotor
   C. Intake manifold
   D. Combustion chamber

_____ 82. Tech A says that engines using compression-ignition control timing by regulating when fuel is injected into the cylinder. Tech B says that engines using spark-ignition control timing by regulating when fuel is injected into the cylinder. Who is correct?
   A. Tech A
   B. Tech B
   C. Both Tech A and Tech B
   D. Neither Tech A nor Tech B

_____ 83. Tech A says that horsepower is a measurement of the amount of work being performed in a given amount of time. Tech B says that horsepower can be calculated by multiplying torque by rpm and dividing by 5252. Who is correct?
   A. Tech A
   B. Tech B
   C. Both Tech A and Tech B
   D. Neither Tech A nor Tech B

_____ 84. Tech A says that in a four-stroke engine, the piston is at TDC four times to complete the cycle. Tech B says that the air/fuel mixture is ignited once every two strokes. Who is correct?
   A. Tech A
   B. Tech B
   C. Both Tech A and Tech B
   D. Neither Tech A nor Tech B

_____ **85.** Tech A says that spark ignition typically occurs before TDC. Tech B says that spark ignition typically occurs after TDC. Who is correct?
   A. Tech A
   B. Tech B
   C. Both Tech A and Tech B
   D. Neither Tech A nor Tech B

_____ **86.** Tech A says that valve overlap occurs between the exhaust stroke and the intake stroke. Tech B says that valve overlap occurs to assist in scavenging the cylinder. Who is correct?
   A. Tech A
   B. Tech B
   C. Both Tech A and Tech B
   D. Neither Tech A nor Tech B

_____ **87.** Tech A says that compression ratio is the comparison of the volume above the piston at BDC to the volume above the piston at TDC. Tech B says that scavenging of the exhaust gases occurs once the exhaust valve closes. Who is correct?
   A. Tech A
   B. Tech B
   C. Both Tech A and Tech B
   D. Neither Tech A nor Tech B

_____ **88.** Tech A says that the design of an interference engine is such that the pistons can hit the valves if the timing belt breaks. Tech B says that the shape of the cam lobe determines how long and far the valves are held open. Who is correct?
   A. Tech A
   B. Tech B
   C. Both Tech A and Tech B
   D. Neither Tech A nor Tech B

_____ **89.** Tech A says that blowby gases occur when compression and combustion gases leak past the piston rings. Tech B says that a rotary engine uses reed valves to control airflow into the cylinder on intake. Who is correct?
   A. Tech A
   B. Tech B
   C. Both Tech A and Tech B
   D. Neither Tech A nor Tech B

Engine Mechanical Testing

At the start of each chapter, you will find the Learning Objectives from the textbook. These are your objectives as you make your way through the exercises in this workbook and the chapter in your textbook. The following activities have been designed to help you refresh your knowledge of the material in this chapter.

Learning Objectives

After reading this chapter, you will be able to:

- LO 13-01 Describe the overview of engine mechanical testing.
- LO 13-02 Isolate engine noises and vibrations.
- LO 13-03 Evaluate the results of engine vacuum tests.
- LO 13-04 Evaluate the results of cylinder power balance tests.
- LO 13-05 Evaluate the results of cylinder cranking compression tests.
- LO 13-06 Evaluate the results of cylinder leakage tests.

Matching

Match the following terms with the correct description or example.

A. Compression tester

B. Cylinder leakage tester

C. Data link connector (DLC)

D. Pressure transducer

E. Vacuum gauge

_____ **1.** The under-dash connector through which the scan tool communicates to the vehicle's computers; it displays the readings from the various sensors and can retrieve trouble codes, freeze-frame data, and system monitor data.

_____ **2.** A device used to measure the amount of pressure a cylinder can generate.

_____ **3.** A device used to measure the amount of vacuum an engine can generate during various operating conditions.

_____ **4.** A device that pumps air into the cylinder and measures the percentage of air that is leaking out of the cylinder.

_____ **5.** A device used to measure engine vacuum and display it graphically on a lab scope.

Multiple Choice

Read each item carefully, and then select the best response.

_____ **1.** What tool, available in both standard and electronic versions, is used by technicians to listen for unusual noises in a vehicle?
   A. Microphone
   B. Stethoscope
   C. Sound transducer
   D. Amplifier

_____ **2.** Many newer vehicles are programmed with a _____ capability that effectively shuts off the fuel injectors as long as the throttle is held to the floor before the ignition key is turned to the run or crank position.
   A. Time delay
   B. Cold start
   C. Crank mode
   D. Clear flood

_____ **3.** A low and steady gauge reading during a vacuum test indicates which of the following?
   A. Everything is good
   B. Possible late valve or ignition timing
   C. Possible worn rings
   D. Burned or leaking valve

_____ **4.** The typical vacuum reading for a properly running engine at idle is a steady _____ of mercury.
   A. 10 to 12 inches
   B. 17 to 21 inches
   C. 4 to 8 inches
   D. 20 to 23 inches

_____ **5.** Using a pressure transducer and _____ is a similar process to using a vacuum gauge, except it is much more accurate and allows you to look at the vacuum graphically.
   A. Lab scope
   B. Stethoscope
   C. Amp probe
   D. Monitor

_____ **6.** Which test identifies which cylinder(s) are not operating properly and gives a general indication of each cylinder's overall health?
   A. Vacuum gauge test
   B. Cylinder leakage test
   C. Pressure transducer test
   D. Power balance test

_____ **7.** In which test is a high-pressure hose hand threaded into the spark plug hole of the cylinder to be tested and then connected to the compression gauge?
   A. Power balance test
   B. Vacuum gauge test
   C. Cranking or running compression test
   D. Cylinder leakage test

_____ **8.** What test checks an engine's ability to move air into and out of the cylinder?
   A. Cylinder leakage test
   B. Vacuum gauge test
   C. Power balance test
   D. Running compression test

_____ **9.** Mechanical testing starts with a good visual inspection.
   A. True
   B. False

_____ **10.** A mechanic could diagnose a burned valve with just a piece of bubble gum and its wrapper.
   A. True
   B. False

_____ **11.** You can disable the ignition system so that it will crank but not start.
   A. True
   B. False

_____ **12.** A sharp oscillation back and forth in the needle or a dip in the vacuum gauge reading could relate to a problem such as a restricted exhaust system.
   A. True
   B. False

_____ **13.** Using a vacuum gauge is much more accurate than using a pressure transducer and lab scope.
   A. True
   B. False

_____ **14.** Disabling a cylinder that is not operating correctly will not produce much if any revolutions per minute (rpm) drop.
   A. True
   B. False

_____ **15.** A cranking compression test is performed when indications show a misfiring or dead cylinder that is not caused by an ignition or fuel problem.
   A. True
   B. False

_____ **16.** A rod bearing makes an evenly spaced single knock, while a main bearing generally makes a double knock.
   A. True
   B. False

_____ **17.** Engine mechanical testing uses a series of tests to assess the mechanical _____ of the engine.
   A. Motion
   B. Pump
   C. Condition
   D. Vacuum

_____ **18.** If _____ is not similar across all cylinders, a tune-up will not fix the problem.
   A. Compression
   B. Temperature
   C. Acceleration
   D. Intake

_____ **19.** A _____ can be used so that the vacuum gauge can be connected and, at the same time, vacuum can still be supplied to the existing component.
   A. Pressure transducer
   B. Lab scope
   C. Scan tool
   D. Vacuum tee

_____ **20.** A vacuum _____ reading shows the difference between outside atmospheric pressure and the amount of manifold pressure in the engine.
   A. Tee
   B. Gauge
   C. Scope
   D. Plug

_____ **21.** During an engine vacuum test, all vacuum gauges are calibrated to and all instructions and readings are referenced to _____.
   A. Sea level
   B. Flight level
   C. Operating temperature
   D. Terrain level

_____ **22.** The purpose of the _____ test is to see whether the cylinders are creating equal amounts of power and, if not, to isolate the problem to a particular cylinder or cylinders.
   A. Engine vacuum
   B. Pressure transducer
   C. Compression
   D. Power balance

_____ **23.** When a low-compression cylinder is found, you should put a couple of squirts of oil into the spark plug hole, crank the engine a few turns, and recheck the compression. If the compression rises significantly, the problem is typically worn _____.
   A. Cylinder walls
   B. Gaskets
   C. Piston rings
   D. Warning indicators

_____ **24.** The _____ test is performed by leaving all of the spark plugs in the engine except for the one in the cylinder that you are testing.
   A. Running compression
   B. Cylinder leakage
   C. Cylinder power balance
   D. Engine vacuum

_____ **25.** The _____ test is performed on a cylinder with low compression to determine the severity of the compression leak and where the leak is located.
   A. Intake manifold
   B. Cylinder leakage
   C. Engine vacuum
   D. Snap throttle

_____ **26.** What testing tool is shown in the image?

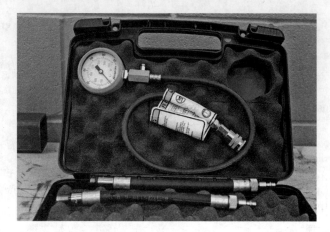

   A. Compression tester
   B. Vacuum gauge
   C. Pressure transducer and lab scope
   D. Cylinder leakage tester

_____ **27.** What testing tool is shown in the image?

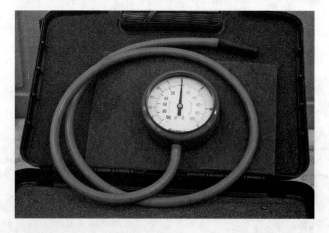

   A. Compression tester
   B. Cylinder leakage tester
   C. Stethoscope
   D. Vacuum gauge

_____ **28.** What testing tool is shown in the image?

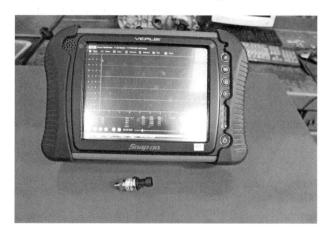

A. Pressure transducer and lab scope
B. Vacuum gauge
C. Digital multimeter (DMM)
D. Cylinder leakage tester

_____ **29.** What testing tool is shown in the image?

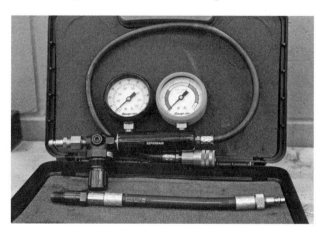

A. Vacuum gauge
B. Stethoscope
C. Cylinder leakage tester
D. Compression tester

_____ **30.** Tech A says that a cranking sound diagnosis can be used to diagnose problems in the ignition system. Tech B says that a cranking sound diagnosis can indicate differences in compression. Who is correct?
A. Tech A
B. Tech B
C. Both Tech A and Tech B
D. Neither Tech A nor Tech B

_____ **31.** Tech A says that a power balance test is a good way to narrow a misfire down to a particular cylinder or cylinders. Tech B says that a cylinder power balance test measures the volumetric efficiency of the cylinder being tested. Who is correct?
A. Tech A
B. Tech B
C. Both Tech A and Tech B
D. Neither Tech A nor Tech B

_____ **32.** Tech A says that a cranking compression wet test can indicate if the cylinder has worn piston rings. Tech B says that the throttle should be held wide open during a cranking compression check. Who is correct?
   A. Tech A
   B. Tech B
   C. Both Tech A and Tech B
   D. Neither Tech A nor Tech B

_____ **33.** A V-6 engine is being tested. Tech A says that low compression on a single cylinder will cause an engine not to start. Tech B says that low compression on a single cylinder will affect the engine's cranking sound. Who is correct?
   A. Tech A
   B. Tech B
   C. Both Tech A and Tech B
   D. Neither Tech A nor Tech B

_____ **34.** Tech A says that a cylinder leakage test is performed on a cylinder with low compression to determine the severity of the leak and where it is located. Tech B says that manufacturers will consider up to 50% cylinder leakage past the piston rings acceptable. Who is correct?
   A. Tech A
   B. Tech B
   C. Both Tech A and Tech B
   D. Neither Tech A nor Tech B

_____ **35.** Tech A says that a scan tool connected to the data link connector (DLC) will perform a cylinder power balance test and report cylinder pressures. Tech B says that a scan tool will perform a cylinder power balance test and report whether the rings or valves have failed. Who is correct?
   A. Tech A
   B. Tech B
   C. Both Tech A and Tech B
   D. Neither Tech A nor Tech B

_____ **36.** Tech A says that a vacuum gauge needle that dips 4 to 8 inches rhythmically can indicate a burned valve. Tech B says that a stethoscope can be used to determine the source of unusual engine noises. Who is correct?
   A. Tech A
   B. Tech B
   C. Both Tech A and Tech B
   D. Neither Tech A nor Tech B

_____ **37.** Tech A says that a vacuum test can determine exhaust restriction. Tech B says that when performing a cylinder power balance test, results should be 5% or less. Who is correct?
   A. Tech A
   B. Tech B
   C. Both Tech A and Tech B
   D. Neither Tech A nor Tech B

_____ **38.** Tech A says that a bad cam lobe or broken valve spring will show up during a running compression test. Tech B says that when performing a cylinder leakage test, the engine must be running to get proper results. Who is correct?
   A. Tech A
   B. Tech B
   C. Both Tech A and Tech B
   D. Neither Tech A nor Tech B

Lubrication System Theory

At the start of each chapter, you will find the Learning Objectives from the textbook. These are your objectives as you make your way through the exercises in this workbook and the chapter in your textbook. The following activities have been designed to help you refresh your knowledge of the material in this chapter.

Learning Objectives

After reading this chapter, you will be able to:

- LO 14-01 Describe the functions of lubricating oil.
- LO 14-02 Describe the common types of oil and their additives.
- LO 14-03 Identify and describe lubrication system components.
- LO 14-04 Describe oil-certifying bodies and their rating standards.
- LO 14-05 Describe oil indicators and warning systems.
- LO 14-06 Identify and describe the types of lubrication systems.

Matching

Match the following terms with the correct description or example.

A. Antifoaming agents
B. Bypass filter
C. Corrosion inhibitors
D. Crescent pump
E. Detergents
F. Dispersants
G. Extreme loading
H. Gelling
I. Hydrocracking
J. Hydrogenating

K. Lubrication system
L. Oil cooler
M. Oil slinger
N. Oxidation inhibitor
O. Polyalphaolefin (PAO) oil
P. Rotor-type oil pump
Q. Scavenge pump
R. Synthetic blend
S. Viscosity index improver
T. Viscosity

_____ 1. The thickening of oil to a point that it will not flow through the engine; it becomes close to a solid in extreme cold temperatures.

_____ 2. A device used to fling oil up onto moving engine parts.

_____ 3. Oil additives that keep contaminants held in suspension in the oil, to be removed by the filter or when the oil is changed.

_____ 4. The measurement of how easily a liquid flows.

_____ 5. A system of parts that work together to deliver lubricating oil to the various moving parts of the engine.

_____ 6. A pump used with a dry sump oiling system to pull oil from the dry sump pan and move it to an oil tank outside the engine.

_____ 7. Oil additives that help to reduce carbon deposits on parts such as piston rings and valves.

_____ 8. A blend of conventional engine oil and pure synthetic oil.

_____ 9. An oil filter system that only filters some of the oil.

_____ 10. Refining crude oil with hydrogen, resulting in a base oil that has the higher performance characteristics of synthetic oils.

_____ 11. An oil pump that uses a crescent-shaped part to separate the oil pump gears from each other, allowing oil to be moved from one side of the pump to the other.

_____ 12. An oil additive that resists a change in viscosity over a range of temperatures.

_____ 13. An oil additive that helps keep hot oil from combining with oxygen to produce sludge or tar.

_____ 14. Oil additives that keep acid from forming in the oil.

_____ 15. An oil pump that uses rounded gears to squeeze oil through.

_____ 16. Oil additives that keep oil from foaming.

_____ 17. An artificially made base stock (synthetic) that is not refined from crude oil. Oil molecules are more consistent in size, and no impurities are found in this oil, as it is made in a lab.

_____ 18. Large amount of pressure placed on two bearing surfaces to press oil from between them.

_____ 19. A process used during refining of crude oil. Hydrogen is added to crude oil to create a chemical reaction to take out impurities such as sulfur.

_____ 20. A device that takes heat away from engine oil by passing it near either engine coolant or outside air.

Multiple Choice

Read each item carefully, and then select the best response.

_____ 1. What kind of oil varies in color from a dirty yellow to dark brown to black and can be thin like gasoline or a thick oil- or tarlike substance?
A. Lubricating oil
B. Motor oil
C. Grease
D. Crude oil

_____ 2. _____ is a measure of how easily a liquid flows.
A. Pour point
B. Viscosity
C. Solidity
D. Reluctance

_____ 3. What kind of additive coats parts with a protective layer so that the oil resists being forced out under heavy load?
A. Extreme-pressure additives
B. Viscosity index improver
C. Pour point depressants
D. Dispersants

_____ 4. The _____ is a certifying body for engine oil.
A. American Petroleum Institute
B. American Society of Lubricant Standardization
C. International Automotive Engineers and Approval Committee
D. Japanese Society of Automotive Engineers

_____ 5. The letter W in a Society of Automotive Engineers (SAE) viscosity rating stands for _____.
A. Water content
B. Winter viscosity
C. Wax content
D. Weight

_____ 6. Which of the following organizations sets classification standards for motorcycle engines, both two-stroke and four-stroke?
A. American Society of Automotive Engineers
B. Association des Constructeurs Européens d'Automobiles
C. Japanese Automotive Standards Organization
D. International Lubricant Standardization and Approval Committee (ILSAC)

_____ 7. The American Petroleum Institute classifies oils into _____ groups.
A. Three
B. Four
C. Five
D. Six

_____ **8.** What type of oil can be man-made or highly processed petroleum?
  A. Group 1
  B. Synthetic
  C. Conventional
  D. Blended

_____ **9.** Lubrication oil is stored in the _____.
  A. Oil slinger
  B. Pickup tube
  C. Oil gallery
  D. Oil sump

_____ **10.** In a wet sump lubricating system, the _____ hold(s) the entire volume of the oil required to lubricate the engine.
  A. Drain plug
  B. Oil pan
  C. Crescents
  D. Barriers

_____ **11.** In a _____ oil pump, the driving gear meshes with a second gear; as both gears turn, their teeth separate, creating a low-pressure area.
  A. Geared
  B. Crescent
  C. Rotor-type
  D. Spline-type

_____ **12.** The spin-on type oil filter has a(n) _____ that fits into a groove in the base of the filter.
  A. Garter spring
  B. O-ring
  C. Plastic cap
  D. Compression ring

_____ **13.** Passageways called _____ allow oil to be fed to the crankshaft bearings first, then through holes drilled in the crankshaft to the connecting rods.
  A. Pickup tubes
  B. Sumps
  C. Oil jackets
  D. Galleries

_____ **14.** On horizontal-crankshaft engines, a _____ on the bottom of the connecting rod scoops up oil from the crankcase for the bearings.
  A. Crescent pump
  B. Dipper
  C. Rotor
  D. Strainer

_____ **15.** Clearances, such as those between the crankshaft journal and crankshaft bearing, fill with lubricating oil so engine parts move or float on layers of oil instead of directly on each other.
  A. True
  B. False

_____ **16.** Oxidation allows air bubbles to form in the engine oil, reducing the lubrication quality of oil and contributing to the breakdown of the oil.
  A. True
  B. False

_____ **17.** The International Lubricant Standardization and Approval Committee requires that the oil provide increased fuel economy over a base lubricant.
  A. True
  B. False

_____ **18.** If you are servicing a European vehicle, it is advised that you do not go by any American Petroleum Institute (API) recommendations; instead, make sure the oil meets the recommended Japanese Automotive Standards Organization (JASO) rating specified by the manufacturer.
   A. True
   B. False

_____ **19.** One of the impurities found in all crude oil is wax, which is removed during refining and is used for candle wax.
   A. True
   B. False

_____ **20.** Type 4 synthetic lubricating oil is not a true synthetic.
   A. True
   B. False

_____ **21.** On a wet sump lubricating system, the oil pan holds the entire volume of the oil required to lubricate the engine.
   A. True
   B. False

_____ **22.** The bypass filtering system is more common on diesel engines and is used in conjunction with a full-flow filtering system.
   A. True
   B. False

_____ **23.** The oil pressure sensor is also commonly called a sending unit because it sends a signal to the light, gauge, or message center in the dash.
   A. True
   B. False

_____ **24.** Oil would be good for thousands of miles longer in a vehicle driven in stop-and-go traffic with long periods of idling than if it were driven in moderate temperatures for long distances.
   A. True
   B. False

_____ **25.** Because there is no oil storage sump under the engine, the engine can be mounted much lower in a dry sump system.
   A. True
   B. False

_____ **26.** Most small four-stroke gasoline engines use only splash lubrication to lubricate all of the parts on the engine.
   A. True
   B. False

_____ **27.** _____ oil is distilled from crude oil and used as a base stock.
   A. Base
   B. Lubricating
   C. Oxidating
   D. Thickening

_____ **28.** _____ occurs between all surfaces that come into contact with each other.
   A. Corrosion
   B. Lubrication
   C. Contamination
   D. Friction

_____ **29.** Base stock derived from crude oil will not retain its viscosity if the temperature gets cold enough, so viscosity _____ improvers are added to the stock.
   A. Index
   B. Agent
   C. Corrosion
   D. Point

_____ **30.** The American _____ symbol is the donut symbol located on the back of the oil bottle.
   A. Society of Automotive Engineers
   B. Petroleum Institute
   C. Lubricant Standardization and Approval Committee
   D. Automotive Standards Organization

_____ **31.** _____ oils flow easily during cold engine start-up but do not thin out as much as the engine and oil come up to operating temperature.
   A. Ultra-low viscosity
   B. Two-stroke
   C. Multi-viscosity
   D. Corrosion

_____ **32.** Using the wrong _____ can void the customer's vehicle warranty, leaving the customer, or your shop, responsible for repairs.
   A. Oil
   B. Filter
   C. Pressure sensor
   D. Dipstick

_____ **33.** Group 2 and group 3 API classified oils are refined with _____ at much higher temperatures and pressures, in a process known as hydrocracking.
   A. Sulfur
   B. Nitrogen
   C. Oxygen
   D. Hydrogen

_____ **34.** Crude oil is broken down into _____ oil, which is then combined with additives to enhance the lubricating qualities.
   A. Ignition
   B. S-rated
   C. Mineral
   D. Synthetic

_____ **35.** True synthetic oils are based on man-made _____, commonly polyalphaolefin oil, which is a man-made oil base stock.
   A. Hydrocarbons
   B. Mineral oils
   C. Antifoaming agents
   D. Corrosion inhibitors

_____ **36.** Oil is drawn through the oil pump _____ from the oil sump by an oil pump.
   A. Slinger
   B. Inhibitor
   C. Cooler
   D. Strainer

_____ **37.** The oil pan is sealed to the engine with _____ or an oil pan gasket.
   A. Dispersants
   B. Extreme-pressure additives
   C. Silicone
   D. Detergents

_____ **38.** Between the oil sump and the oil pump is a(n) _____ with a flat cup and a wire mesh strainer immersed in the oil.
   A. Oil slinger
   B. Pickup tube
   C. Dipper
   D. Rotor lobe

_____ **39.** In a rotor-type oil pump, outside atmospheric pressure forces oil into the pump, and the oil fills the spaces between the _____.
   A. Oil spurt holes
   B. Rotor lobes
   C. Oil galleries
   D. Oil slingers

_____ **40.** An oil pressure _____ stops excess pressure from developing.
   A. Relief valve
   B. Inhibitor
   C. Dipper
   D. Pump

_____ **41.** The location of the _____ filter right after the oil pump ensures that all of the oil is filtered before it is sent to the lubricated components.
   A. Bypass
   B. Spin-on
   C. Cartridge
   D. Full-flow

_____ **42.** Some connecting rods have oil _____ holes that are positioned to receive oil from similar holes in the crankshaft.
   A. Slinger
   B. Pour
   C. Spurt
   D. Bypass

_____ **43.** Oil _____ systems are used to inform the driver when the oil needs to be changed.
   A. Filtering
   B. Monitoring
   C. Straining
   D. Lubricating

_____ **44.** Modern vehicle engines use a pressure, or _____, lubrication system where the oil is forced throughout the engine under pressure.
   A. Full-flow
   B. Force-feed
   C. Rotor-type
   D. Hydrocracking

_____ **45.** Diesel fuel has more _____ thermal units of heat energy than gasoline, so it produces more heat when it is ignited, placing more stress on the engine's moving parts.
   A. American
   B. Pressure-fed
   C. British
   D. Engine

_____ **46.** _____ can strip the threads on the oil pan, especially if it is made of aluminum, or can strip the threads on the drain plug itself.
   A. Overtightening
   B. Lubrication
   C. Hydrocracking
   D. Hydrogenating

_____ **47.** The arrow in the image is pointing to what part of the lubrication system?

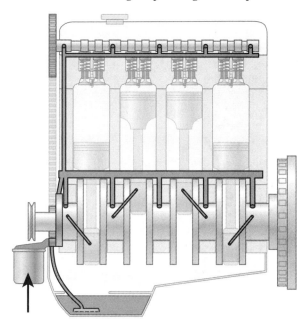

A. Camshaft
B. Oil pump
C. Oil filter
D. Main bearing

_____ **48.** The arrow in the image is pointing to what part of the lubrication system?

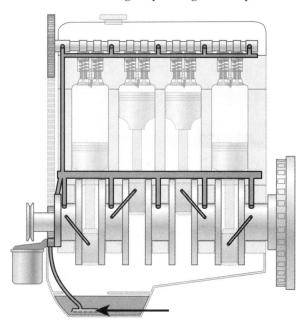

A. Cylinder head oil gallery
B. Oil pump pickup strainer
C. Oil pump
D. Oil filter

_____ **49.** The arrow in the image is pointing to what part of the lubrication system?

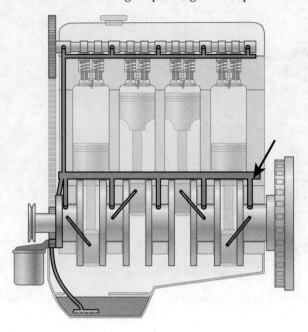

    A. Main oil gallery
    B. Cylinder head oil gallery
    C. Main bearing
    D. Big end bearing

_____ **50.** What type of oil pump is shown in the image?

    A. Geared oil pump
    B. Rotor-type oil pump
    C. Crescent pump
    D. Pressure relief pump

_____ **51.** What type of oil pump is shown in the image?

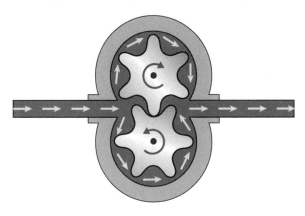

    A. Crescent pump
    B. Geared oil pump
    C. Rotor-type oil pump
    D. Pickup pump

_____ **52.** What type of oil pump is shown in the image?

    A. Crescent pump
    B. Rotor-type oil pump
    C. Scavenge pump
    D. Geared oil pump

_____ **53.** The arrow in the image is pointing to what part of the full-flow filtering system?

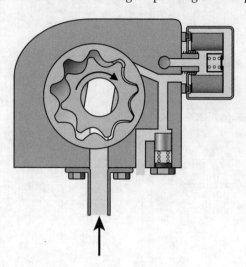

A. Oil pump pickup
B. Oil pump gear set
C. Oil pump discharge
D. Main oil gallery

_____ **54.** The arrow in the image is pointing to what part of the full-flow filtering system?

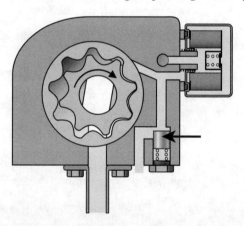

A. Relief valve spring
B. Oil pump gear set
C. Oil pump discharge
D. Relief valve

_____ **55.** The arrow in the image is pointing to what part of the full-flow filtering system?

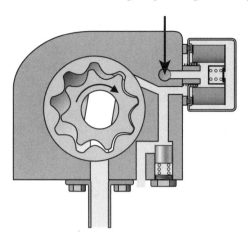

    A. Relief valve
    B. Oil pump discharge
    C. Main oil gallery
    D. Oil filter

_____ **56.** The arrow in the image is pointing to what part of the bypass filtering system?

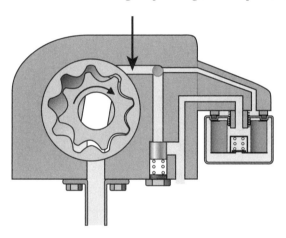

    A. Bypass gallery
    B. Oil filter
    C. Main oil gallery
    D. Oil pump discharge

_____ **57.** The arrow in the image is pointing to what part of the bypass filtering system?

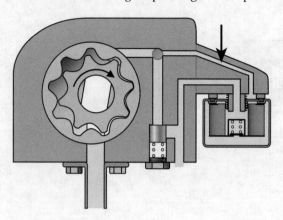

    A. Return to sump
    B. Bypass gallery
    C. Oil filter
    D. Oil pump discharge

_____ **58.** The arrow in the image is pointing to what part of the bypass filtering system?

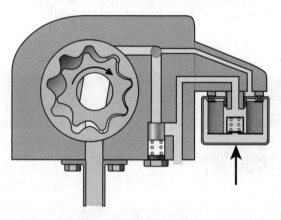

    A. Main oil gallery
    B. Bypass gallery
    C. Oil filter
    D. Oil pump discharge

_____ **59.** Tech A says that one function of oil is to clean. Tech B says that one function of oil is to cushion. Who is correct?
    A. Tech A
    B. Tech B
    C. Both Tech A and Tech B
    D. Neither Tech A nor Tech B

_____ **60.** Two techs are discussing 5W20 oil. Tech A says the W stands for "weight." Tech B says the W stands for "winter." Who is correct?
    A. Tech A
    B. Tech B
    C. Both Tech A and Tech B
    D. Neither Tech A nor Tech B

_____ **61.** Tech A says that the higher the viscosity number, the thicker the oil. Tech B says that most modern vehicles use single weight oil. Who is correct?
    A. Tech A
    B. Tech B
    C. Both Tech A and Tech B
    D. Neither Tech A nor Tech B

_____ **62.** Tech A says that spin-on oil filters need room temperature vulcanizing (RTV) gasket sealer to seal the gasket. Tech B says that some oil filters use a replaceable paper filter cartridge. Who is correct?
   A. Tech A
   B. Tech B
   C. Both Tech A and Tech B
   D. Neither Tech A nor Tech B

_____ **63.** Tech A says that most oil pumps are of the positive displacement style. Tech B says that oil pumps are designed to deliver more oil than is needed for an engine. Who is correct?
   A. Tech A
   B. Tech B
   C. Both Tech A and Tech B
   D. Neither Tech A nor Tech B

_____ **64.** Tech A says that oil pressure is reduced when bearing clearances increase. Tech B says that oil pressure is regulated by the pressure relief valve. Who is correct?
   A. Tech A
   B. Tech B
   C. Both Tech A and Tech B
   D. Neither Tech A nor Tech B

_____ **65.** Tech A says that a full-flow oil filter filters all of the oil going to the bearings. Tech B says that a bypass filter bypasses the pump, so that any particles in the oil will not damage the pump. Who is correct?
   A. Tech A
   B. Tech B
   C. Both Tech A and Tech B
   D. Neither Tech A nor Tech B

Servicing the Lubrication System

At the start of each chapter, you will find the Learning Objectives from the textbook. These are your objectives as you make your way through the exercises in this workbook and the chapter in your textbook. The following activities have been designed to help you refresh your knowledge of the material in this chapter.

Learning Objectives

After reading this chapter, you will be able to:

- LO 15-01 Check oil level and condition.
- LO 15-02 Change engine oil.
- LO 15-03 Diagnose engine lubrication system issues.

Multiple Choice

Match the following terms with the correct description or example.

_____ **1.** When the new O-ring in the new filter is installed over the old O-ring, this is called _____.
  A. Double gasketing
  B. Threaded filter
  C. Reluctance
  D. Hydrocracking

_____ **2.** Which of the following actions should be taken when installing a new oil filter?
  A. Wipe a small amount of contact cement on the gasket.
  B. Clean all oil from the mounting area and install the gasket dry.
  C. Wipe a small amount of oil on the new filter gasket.
  D. Wipe a small amount of gasket sealer on the new filter gasket.

_____ **3.** Which of the following statements is correct in regard to performing an oil change?
  A. Normal engine use and severe engine use require the oil to be changed at the same intervals.
  B. When draining oil, the oil should be cold.
  C. When changing oil, the drain plug should be tightened with the maximum force to prevent oil leakage.
  D. When draining oil, the engine should be fully warmed.

_____ **4.** When installing a spin-on oil filter, failure to lube the O-ring will lead to which of the following?
  A. Double gasketing
  B. Damage to the threads on the adapter
  C. The O-ring binding and rolling out of the oil filter groove
  D. Increased oil consumption

_____ **5.** Oil analysis is typically used in heavy vehicle applications that may use three or more gallons of engine oil.
  A. True
  B. False

_____ **6.** Checking the engine oil level regularly is necessary and should be performed during every fuel fill-up or every other fuel-up, depending on the age and condition of the vehicle.
  A. True
  B. False

_____ **7.** Some vehicles have a drain plug on the transmission/transaxle that can be mistaken for the oil drain plug.
  A. True
  B. False

_____ **8.** Manufacturers have returned to using cartridge oil filters more frequently because it is easier to properly dispose of the oil.
   A. True
   B. False

_____ **9.** When changing a cartridge oil filter, the only parts that get replaced are the paper filter cartridge and the _____ or seals.
   A. Filter housings
   B. Drain plugs
   C. O-rings
   D. End caps

_____ **10.** If the oil filter cartridge is on the top of the engine, the service information may direct you to pour a specified amount of oil into the filter cavity before the _____ is installed.
   A. O-ring
   B. End cap
   C. Housing
   D. Oil fill cap

_____ **11.** It may be necessary to top off the engine oil by adding a small quantity of oil to compensate for the amount _____ by the new filter.
   A. Indicated
   B. Released
   C. Absorbed
   D. Displaced

_____ **12.** If the oil pressure is good, turn the engine off and check underneath the vehicle to make sure no oil is leaking from the oil filter or the _____ plug.
   A. Oil pressure
   B. Drain
   C. Dipstick
   D. Spark

_____ **13.** Before draining the engine oil, a technician should make sure that _____.
   A. The engine is fully warmed up
   B. The fuel tank is empty
   C. The brakes are applied
   D. The steering system is locked

_____ **14.** During replacement, compared with a spin-on filter, a cartridge filter _____.
   A. Requires that all parts be replaced
   B. Makes it easier to properly dispose of the oil
   C. Generates a greater amount of waste
   D. Allows for minimal reuse

_____ **15.** What tool for lubrication repair is shown in the image?

   A. Digital multimeter
   B. Mirror and quality light
   C. Oil filter wrenches
   D. Pressure gauge

_____ **16.** What tool for lubrication repair is shown in the image?

    A. Sensor substitution box
    B. Digital multimeter
    C. Pressure gauge
    D. Mirror and quality light

_____ **17.** What tool for lubrication repair is shown in the image?

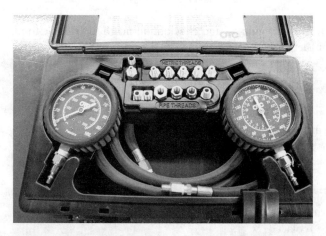

    A. Sensor substitution box
    B. Oil pressure sensor socket
    C. Digital multimeter
    D. Pressure gauge

_____ **18.** Tech A states that engine oil should be drained every time the filter is changed. Tech B states that over time, the oil becomes dirty and the additives wear out. Who is correct?
    A. Tech A
    B. Tech B
    C. Both Tech A and Tech B
    D. Neither Tech A nor Tech B

_____ **19.** Tech A says that an oil drain plug gasket made of aluminum, plastic, or fiber is reusable. Tech B says that a drain plug with an integrated silicone gasket is reusable as long as it is in good condition. Who is correct?
    A. Tech A
    B. Tech B
    C. Both Tech A and Tech B
    D. Neither Tech A nor Tech B

_____ 20. Tech A says that when installing a spin-on oil filter, it should be tightened with a filter wrench to prevent it from falling off. Tech B says that spin-on oil filters should be installed only by hand. Who is correct?
  A. Tech A
  B. Tech B
  C. Both Tech A and Tech B
  D. Neither Tech A nor Tech B

_____ 21. Tech A says that when filling an engine with oil, it should be filled until oil is coming out of the filler hole. Tech B says that it is a best practice to check the oil before and after running the vehicle. Who is correct?
  A. Tech A
  B. Tech B
  C. Both Tech A and Tech B
  D. Neither Tech A nor Tech B

_____ 22. Tech A says that when changing the spin-on oil filter, the old O-ring must be removed with the filter. Tech B says that the O-ring on the new filter should be dry when the filter is screwed on. Who is correct?
  A. Tech A
  B. Tech B
  C. Both Tech A and Tech B
  D. Neither Tech A nor Tech B

_____ 23. Tech A says that after changing the engine oil and filter, a static sticker should be applied to remind the owner when the next oil change is due. Tech B says that the maintenance reminder system should be reset to remind the owner when the next oil change is due. Who is correct?
  A. Tech A
  B. Tech B
  C. Both Tech A and Tech B
  D. Neither Tech A nor Tech B

_____ 24. Tech A says that when draining the engine oil, the engine should be cold for best results. Tech B says that the engine should be running when checking the engine oil level. Who is correct?
  A. Tech A
  B. Tech B
  C. Both Tech A and Tech B
  D. Neither Tech A nor Tech B

_____ 25. Tech A says that when removing oil, it can be easy to confuse the oil drain plug and the transmission drain plug. Tech B says that drain plugs should be tightened to the manufacturer-recommended torque specification. Who is correct?
  A. Tech A
  B. Tech B
  C. Both Tech A and Tech B
  D. Neither Tech A nor Tech B

_____ 26. Tech A says that oil changes are unnecessary as long as the oil is at the correct level. Tech B says that some engines have more than one oil drain plug. Who is correct?
  A. Tech A
  B. Tech B
  C. Both Tech A and Tech B
  D. Neither Tech A nor Tech B

_____ 27. Tech A says that the engine oil drain plug typically needs to be replaced at each oil change. Tech B says that engine oil drain plugs must be tightened with a torque wrench when they are installed. Who is correct?
  A. Tech A
  B. Tech B
  C. Both Tech A and Tech B
  D. Neither Tech A nor Tech B

_____ **28.** Tech A says that it takes about 1 pint of oil to raise the oil level from "add" to "full." Tech B says that it takes about 1 quart to raise it that much. Who is correct?

   A. Tech A

   B. Tech B

   C. Both Tech A and Tech B

   D. Neither Tech A nor Tech B

Cooling System Theory

At the start of each chapter, you will find the Learning Objectives from the textbook. These are your objectives as you make your way through the exercises in this workbook and the chapter in your textbook. The following activities have been designed to help you refresh your knowledge of the material in this chapter.

Learning Objectives

After reading this chapter, you will be able to:

- LO 16-01 Describe the methods of heat transfer, cooling system configurations, and their operation.
- LO 16-02 Describe engine coolant and its required properties.
- LO 16-03 Describe coolant flow in engines.
- LO 16-04 Describe the radiator and its associated components.
- LO 16-05 Describe the operation of the thermostat and water pump.
- LO 16-06 Describe the operation of cooling fans.
- LO 16-07 Describe hoses, belts, and tensioners.
- LO 16-08 Describe miscellaneous cooling system components.

Matching

Match the following terms with the correct description or example.

A. Coolant   **I.** Propylene glycol
B. Heater control valve   **J.** Radiator
C. Heater hoses   **K.** Radiator hoses
D. Cross flow radiator   **L.** Radiation
E. Downflow radiator   **M.** Surge tank
F. Ethylene glycol   **N.** Thermo-control switch
G. Heat dissipation   **O.** Thermostat
H. Overflow tank   **P.** Water jackets

_____ **1.** A device that transfers heat from a fluid within to a location outside.

_____ **2.** Flexible hoses that connect the stationary components of the cooling system, such as the heater core and radiator, to the engine, which is mounted on flexible mounts.

_____ **3.** Regulates coolant flow to the radiator. It opens at a predetermined temperature to allow coolant flow to the radiator for cooling. It also enables the engine to reach operating temperature more quickly for reduced emissions and wear.

_____ **4.** The movement of energy through space, such as the movement of energy from the sun to the earth.

_____ **5.** The passages surrounding the cylinders and head on the engine where coolant can flow to pick up excess heat. They are sealed by replaceable core plugs.

_____ **6.** Rubber hoses that connect the radiator to the engine. Because they are subject to pressure, they are reinforced with a layer of fabric, typically nylon.

_____ **7.** A fluid that contains antifreeze mixed with water.

_____ **8.** A temperature-sensitive switch that is mounted into the radiator or into a coolant passage on the engine to control electric fan operation.

_____ **9.** The spreading of heat over a large area to increase heat transfer.

_____ **10.** A valve that blocks off coolant flow to keep hot water from entering the heater core when less heat is requested by the operator.

_____ **11.** A chemical used as antifreeze. It is labeled as a nontoxic antifreeze.

_____ **12.** A type of radiator that has the coolant flow from left to right and is more conformable to the low hood designs of today's vehicles.

_____ **13.** A tank used to catch any coolant that is released from the radiator cap (works like a catch can).

_____ **14.** A chemical used as antifreeze that provides the lower freezing point of coolant and raises the boiling point. It is a toxic antifreeze.

_____ **15.** A pressurized tank that is piped into the cooling system. Coolant constantly moves through it. It is used when the radiator is not the highest part of the cooling system.

_____ **16.** A radiator in which the coolant flows from the top to the bottom.

Multiple Choice

Match the following terms with the correct description or example.

_____ **1.** Heat transfers through liquids and gases by a process called _____.
   A. Conduction
   B. Convection
   C. Radiation
   D. Transference

_____ **2.** Engines have an ideal operating temperature somewhere around _____, give or take 20° F, depending on the vintage of the vehicle.
   A. 200° F
   B. 212° F
   C. 250° F
   D. 325° F

_____ **3.** The circulation of coolant to the radiator is controlled by the _____.
   A. Water pump
   B. Radiator temperature sensor
   C. Air temperature
   D. Thermostat

_____ **4.** The best coolant is a _____ balance of water and antifreeze, making it an ideal coolant for both hot and cold climates and providing adequate corrosion protection.
   A. 60/40
   B. 40/60
   C. 50/50
   D. 70/30

_____ **5.** A force pulling outward on a rotating body is known as _____.
   A. Roll
   B. Convection
   C. Centrifugal force
   D. Spin

_____ **6.** Coolant absorbs heat via _____ from the engine and becomes hotter.
   A. Conduction
   B. Convection
   C. Radiation
   D. Induction

_____ **7.** As coolant moves through the hottest part of an engine with a reverse-flow design, steam tends to form and get stuck at the head cooling passages; this problem was fixed by adding a(n) _____.
   A. Overflow tank
   B. Cooling fan
   C. Surge tank
   D. Recirculation pump

_____ **8.** Timing belts are what type of belt?
   A. V-belt
   B. Serpentine belt
   C. Toothed belt
   D. Stretch-fit belt

_____ **9.** _____ are used around the radiator fan to help draw air through the entire radiator core and not just in front of the fan blades.
   A. Fins
   B. Baffles
   C. Shrouds
   D. Ducts

_____ **10.** The _____ is called a fan clutch because it can engage and disengage the cooling fan from the pulley.
   A. Torque connector
   B. Viscous coupler
   C. Thermo-control switch
   D. Electric solenoid

_____ **11.** The _____ is typically located on the water pump and connects to the intake manifold on many V-configured engines, such as a V6 or V8.
   A. Bypass hose
   B. Upper hose
   C. Lower hose
   D. Throttle body coolant line

_____ **12.** A _____ belt has a flat profile with a number of small V-shaped grooves running lengthwise along the inside of the belt.
   A. V-type
   B. Serpentine
   C. Stretch
   D. Toothed

_____ **13.** The _____ operates by using a signal sent from a coolant temperature sensor located on the engine in a coolant passage.
   A. Temperature gauge
   B. Thermostat
   C. Radiator
   D. Mechanical cooling fan

_____ **14.** The aluminum, brass, or steel plugs designed to seal the openings left from the casting process where the casting sand was removed are called _____.
   A. Core plugs
   B. Hard plugs
   C. Compression plugs
   D. Heater plugs

_____ **15.** What type of coolant contains a mixture of inorganic and organic additives?
   A. Organic acid technology
   B. Inorganic acid technology
   C. Hybrid organic acid technology
   D. Poly organic acid technology

_____ **16.** Heat is transferred from one solid to another by a process called radiation.
   A. True
   B. False

_____ **17.** No matter how efficiently fuel burning occurs, and no matter the size of the engine, the heat energy generated never completely transforms into kinetic energy.
   A. True
   B. False

_____ **18.** Some newer vehicles have a coolant heat storage system that uses a vacuum-insulated container similar to a Thermos bottle.
    A. True
    B. False

_____ **19.** Water alone is by far the best coolant there is, as it can absorb a larger amount of heat than most other liquids.
    A. True
    B. False

_____ **20.** Ethylene glycol is a nontoxic chemical that resists freezing and is used in nontoxic antifreezes.
    A. True
    B. False

_____ **21.** Because atmospheric pressure is lower at higher elevations, the boiling temperature of a liquid in an unsealed system is higher.
    A. True
    B. False

_____ **22.** In an automotive cooling system, electrolysis is possible when the coolant breaks down and becomes more acidic.
    A. True
    B. False

_____ **23.** Cooling systems, such as those used for turbocharger intercoolers/aftercoolers or hybrid high-voltage battery packs, may use air-to-air systems to cool the engine's intake air or high-voltage battery.
    A. True
    B. False

_____ **24.** All automobiles are now water cooled, although small engines can still be found in older automobiles that are air cooled.
    A. True
    B. False

_____ **25.** All engines operate best when they are at their full operating temperature.
    A. True
    B. False

_____ **26.** The hottest part of any engine is the cylinder head, since this is where the combustion chamber is located.
    A. True
    B. False

_____ **27.** Radiator hoses connect the water pump and engine to the heater core. They carry heated coolant to the heater core to be used to heat the passenger compartment.
    A. True
    B. False

_____ **28.** In both the cross flow and the downflow radiator designs, the core is built of the same components.
    A. True
    B. False

_____ **29.** The thermostat is a spring-loaded valve that is controlled by a wax pellet located inside the valve.
    A. True
    B. False

_____ **30.** Thermo-control switches often use a bimetallic strip that consists of two different metals or alloys laminated back to back that expand and contract at different rates.
    A. True
    B. False

_____ **31.** Some radiator hoses, especially lower hoses, have a spiral wire inside to keep the hose from collapsing during heavy acceleration when the water pump is drawing a lot of water from the radiator.
    A. True
    B. False

_____ **32.** The bottom radiator hose is typically attached to the heater core, which allows the heated coolant to enter the inlet side of the radiator.
   A. True
   B. False

_____ **33.** The heater control valve, if used, controls the flow of coolant to the heater core to control the temperature of the air desired by the operator.
   A. True
   B. False

_____ **34.** Coolant can be accurately identified according to its color, which may be anything from green or purple to yellow/gold, orange, blue, or pink.
   A. True
   B. False

_____ **35.** Poly organic acid technology coolant contains a proprietary blend of corrosion inhibitors and provides up to 7 years or 250,000 miles of protection.
   A. True
   B. False

_____ **36.** Heat is _____ energy, which cannot be destroyed, only transferred.
   A. Electrical
   B. Light
   C. Thermal
   D. Mechanical

_____ **37.** Heat is transferred through space by a process called _____.
   A. Conduction
   B. Radiation
   C. Convection
   D. Induction

_____ **38.** The principle that heat always moves from hot to cold is known as the second law of _____.
   A. Radiation
   B. Thermodynamics
   C. Convection
   D. Conduction

_____ **39.** Coolant is a mixture of water and _____, which is used to remove heat from the engine.
   A. Antifreeze
   B. Electrolytes
   C. Metal
   D. Oil

_____ **40.** _____ glycol is a chemical that resists freezing but is very toxic to humans and animals.
   A. Inorganic
   B. Propylene
   C. Ambient
   D. Ethylene

_____ **41.** A liquid under pressure higher than atmospheric pressure has _____ boiling point as that same liquid at atmospheric pressure.
   A. A higher
   B. A lower
   C. The same

_____ **42.** _____-cooled engines use heat-dissipating fins on the engine cylinders and heads to allow the movement of air to absorb heat and carry it away from the engine through special ducting.
   A. Water
   B. Air
   C. Absorption
   D. Rotary

_____ **43.** In a typical _____ design, coolant starts from the radiator and flows through the radiator outlet hose to the thermostat and then to the water pump.
   A. Normal-flow
   B. Slow-flow
   C. Reverse-flow
   D. Fast-flow

_____ **44.** The _____ is usually made of copper, brass, or aluminum tubes with copper, brass, aluminum, or plastic tanks on the sides or top for coolant to collect in.
   A. Actuator
   B. Water jacket
   C. Rotor housing
   D. Radiator

_____ **45.** _____ coolers are used to cool automatic transmission fluid, power steering fluid, exhaust gas recirculation gases, and compressed intake air.
   A. Auxiliary
   B. Viscous
   C. Ambient
   D. Surge

_____ **46.** One way to prevent coolant from boiling is to use a radiator _____ cap.
   A. Pressure
   B. Thermal
   C. Conduction
   D. Overflow

_____ **47.** The coolant _____ system consists of an overflow bottle, a sealed radiator pressure cap, and a small hose connecting the bottle to the radiator neck.
   A. Response
   B. Management
   C. Recovery
   D. Mitigation

_____ **48.** The _____ is usually belt driven from a pulley on the front of the crankshaft.
   A. Actuator
   B. Water pump
   C. Thermostat
   D. Surge tank

_____ **49.** Engine-driven _____ may be located on the water pump shaft or, in a few cases, may be attached directly to the engine crankshaft.
   A. Actuators
   B. Thermostats
   C. Surge tanks
   D. Cooling fans

_____ **50.** The bottom or lower _____ is connected between the outlet of the radiator and the inlet of the water pump.
   A. Actuator
   B. Thermostat
   C. Radiator hose
   D. Water jacket

_____ **51.** A _____ belt looks like an ordinary serpentine belt but is found on vehicles without a tensioner.
   A. Stretch-fit
   B. X-type
   C. Toothed
   D. V-type

_____ **52.** _____ are used to keep the drive belt tight around the pulleys to ensure the least amount of slippage without causing damage to component bearings.
    A. Tensioners
    B. Hoses
    C. Water jackets
    D. Core plugs

_____ **53.** Temperature _____ can come in two forms: a temperature gauge or a temperature light located in the instrument cluster.
    A. Regulators
    B. Actuators
    C. Indicators
    D. Control valves

_____ **54.** The _____ is simply a small radiator that is mounted inside the heater box in the passenger compartment.
    A. Temperature indicator
    B. Heater core
    C. Actuator
    D. Water jacket

_____ **55.** Air doors are moved by one of three methods: cable, vacuum actuator, or an electric _____, called a stepper motor.
    A. Hose
    B. Jacket
    C. Indicator
    D. Actuator

_____ **56.** The arrow in the image is pointing to what type of heat transfer?

    A. Conduction
    B. Convection
    C. Radiation
    D. Collection

_____ **57.** The arrow in the image is pointing to what type of heat transfer?

    A. Radiation
    B. Convection
    C. Deduction
    D. Conduction

_____ **58.** The arrows in the image are pointing to what type of heat transfer?

    A. Convection
    B. Induction
    C. Radiation
    D. Conduction

_____ **59.** What percentage of heat loss occurs in a gasoline engine owing to the cooling system?
    A. 5%
    B. 33%
    C. 25%
    D. 70%

_____ **60.** What percentage of heat loss occurs in a gasoline engine owing to internal friction and from radiating off hot engine components?
    A. 1%
    B. 5%
    C. 25%
    D. 50%

_____ **61.** What type of radiator is shown in the image?

    A. Downflow radiator
    B. Slow-flow radiator
    C. Fast-flow radiator
    D. Cross flow radiator

_____ **62.** What type of radiator is shown in the image?

    A. Downflow radiator
    B. Slow-flow radiator
    C. Fast-flow radiator
    D. Cross flow radiator

_____ **63.** The arrow in the image is pointing to what part of the coolant recovery system?

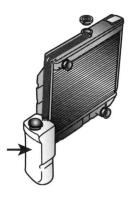

    A. Radiator neck
    B. Overflow bottle
    C. Connecting hose
    D. Radiator cap

_____ **64.** The arrow in the image is pointing to what part of the coolant recovery system?

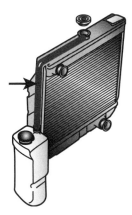

    A. Radiator cap
    B. Radiator neck
    C. Overflow bottle
    D. Connecting hose

_____ **65.** The arrow in the image is pointing to what part of the coolant recovery system?

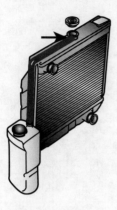

    A. Connecting hose
    B. Radiator cap
    C. Radiator neck
    D. Overflow bottle

_____ **66.** The arrow in the image is pointing to what part of the water pump?

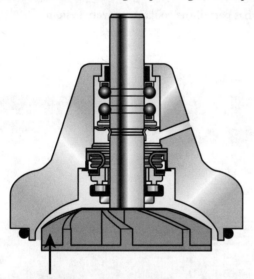

    A. Impeller
    B. Bearing
    C. Weep hole
    D. Housing

_____ **67.** The arrow in the image is pointing to what part of the water pump?

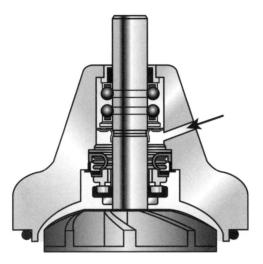

A. Housing
B. Impeller
C. Bearing
D. Weep hole

_____ **68.** The arrow in the image is pointing to what part of the water pump?

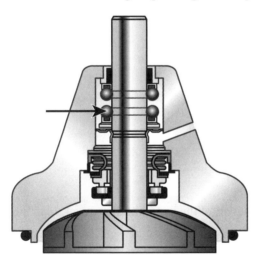

A. Bearing
B. Impeller
C. Housing
D. Weep hole

_____ **69.** The arrow in the image is pointing to what part of the hydraulically operated cooling fan?

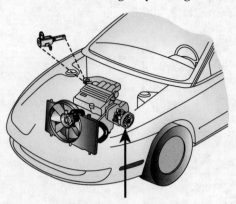

    A. Control valve assembly
    B. Cooling fan
    C. Hydraulic motor
    D. Hydraulic pump

_____ **70.** The arrow in the image is pointing to what part of the hydraulically operated cooling fan?

    A. Control valve assembly
    B. Hydraulic pump
    C. Cooling fan
    D. Hydraulic motor

_____ **71.** The arrow in the image is pointing to what part of the hydraulically operated cooling fan?

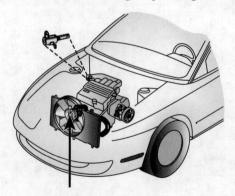

    A. Hydraulic motor
    B. Control valve assembly
    C. Hydraulic pump
    D. Cooling fan

_____ **72.** The arrow in the image is pointing to what part of the cooling system?

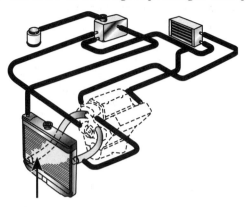

A. Heater
B. Upper radiator hose
C. Lower radiator hose
D. Header tank

_____ **73.** The arrow in the image is pointing to what part of the cooling system?

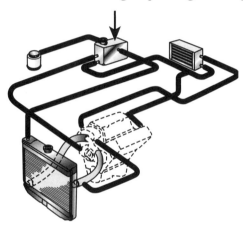

A. Overflow
B. Header tank
C. Heater
D. Upper radiator hose

_____ **74.** The arrow in the image is pointing to what part of the cooling system?

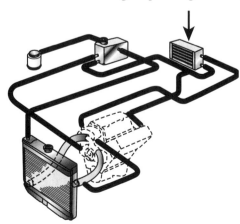

A. Heater core
B. Lower radiator hose
C. Header tank
D. Overflow

_____ **75.** Tech A says that the cooling system is designed to keep the engine as cool as possible. Tech B says that the heater core can remove heat from the cooling system. Who is correct?
  A. Tech A
  B. Tech B
  C. Both Tech A and Tech B
  D. Neither Tech A nor Tech B

_____ **76.** Tech A says that the thermostat is open until the engine warms up, and then it closes. Tech B says that a faulty radiator cap can be the cause of boiling coolant. Who is correct?
  A. Tech A
  B. Tech B
  C. Both Tech A and Tech B
  D. Neither Tech A nor Tech B

_____ **77.** Tech A says that when pure antifreeze is used in the cooling system, the protection level is –70° F. Tech B says that pure antifreeze will cool the engine better than water. Who is correct?
  A. Tech A
  B. Tech B
  C. Both Tech A and Tech B
  D. Neither Tech A nor Tech B

_____ **78.** Tech A says that some drive belts are a stretch-fit design. Tech B says that the thermostat may have a bleed valve that should be accurately positioned when the thermostat is replaced. Who is correct?
  A. Tech A
  B. Tech B
  C. Both Tech A and Tech B
  D. Neither Tech A nor Tech B

_____ **79.** Tech A says that the design of a surge tank system helps to purge air from the cooling system. Tech B says that overflow tanks are pressurized. Who is correct?
  A. Tech A
  B. Tech B
  C. Both Tech A and Tech B
  D. Neither Tech A nor Tech B

_____ **80.** Tech A says that when you find coolant hoses collapsed, the radiator pressure cap has failed. Tech B says that overflow tanks are reservoirs designed to hold coolant and allow flow back to the radiator. Who is correct?
  A. Tech A
  B. Tech B
  C. Both Tech A and Tech B
  D. Neither Tech A nor Tech B

_____ **81.** Tech A says that electric cooling fans operate whenever the engine is running. Tech B says that electric cooling fans are used to cause a large airflow over the radiator at low vehicle speeds. Who is correct?
  A. Tech A
  B. Tech B
  C. Both Tech A and Tech B
  D. Neither Tech A nor Tech B

_____ **82.** Tech A says that there are a number of coolant types, each with its own life span. Tech B says that mixing of coolants is generally OK, as all coolants use the same chemical base. Who is correct?
  A. Tech A
  B. Tech B
  C. Both Tech A and Tech B
  D. Neither Tech A nor Tech B

Servicing the Cooling System

At the start of each chapter, you will find the Learning Objectives from the textbook. These are your objectives as you make your way through the exercises in this workbook and the chapter in your textbook. The following activities have been designed to help you refresh your knowledge of the material in this chapter.

Learning Objectives

After reading this chapter, you will be able to:

- LO 17-01 Test engine coolant.
- LO 17-02 Service engine coolant.
- LO 17-03 Service drive belts.
- LO 17-04 Service coolant hoses.
- LO 17-05 Service thermostat and bypass.
- LO 17-06 Service fan, clutch, shroud, and water pump.
- LO 17-07 Diagnose cooling system performance.
- LO 17-08 Diagnose cooling system leaks and verify engine operating temperature.

Matching

Match the following steps with the correct sequence for pressure testing the cooling system.

A. Step 1		**D.** Step 4	
B. Step 2		**E.** Step 5	
C. Step 3		**F.** Step 6	

_____ **1.** Install the radiator cap on the pressure tester, and pressurize the cap to the correct pressure.

_____ **2.** Install the tester on the radiator. Pressurize the system to the specified cap pressure.

_____ **3.** Verify specified cooling system pressure.

_____ **4.** Check hoses, soft plugs, and any heater cores; determine necessary action.

_____ **5.** Watch the pressure reading for a drop while performing a visual check for any leaks.

_____ **6.** Top off the radiator with coolant or water.

Match the following steps with the correct sequence for using a hydrometer to test the freeze point of the coolant.

A. Step 1		**C.** Step 3
B. Step 2		

_____ **1.** Hold the tool vertical. Read the scale to verify the freeze protection of the coolant. Return the coolant sample to the radiator or surge tank.

_____ **2.** Release the bulb to pull a coolant sample into the hydrometer. Verify it is above the minimum level in the tester.

_____ **3.** Be sure the cooling system is cool. Remove the pressure cap. Determine the type of antifreeze, and verify that the hydrometer is designed to be used with it. Place the hydrometer tube in the coolant, and squeeze the bulb on top.

Match the following steps with the correct sequence for draining and refilling coolant.

A. Step 1   **C.** Step 3

B. Step 2

_____ **1.** Remove the radiator cap. Locate the radiator drain plug, if equipped, and place a catch pan marked for coolant underneath the drain plug. Drain the radiator into the catch pan.

_____ **2.** Refill the cooling system with the proper coolant mix after closing the drain plugs. Start the engine and verify the proper level. Dispose of the coolant in an approved way.

_____ **3.** Remove the block drain plugs, and allow the coolant to drain into the pan.

Match the following steps with the correct sequence for checking and replacing a coolant hose.

A. Step 1   **C.** Step 3

B. Step 2

_____ **1.** Refill coolant, verify correct level, and pressure check for leaks.

_____ **2.** Inspect the hoses by squeezing them, and visually inspect the clamps. Remove the hose clamp. Carefully pull or cut the hose off the fittings.

_____ **3.** Verify that the replacement hose is correct. Cut to length if necessary. Reinstall the new hose all the way into position. Install new clamps.

Multiple Choice

Read each item carefully, and then select the best response.

_____ **1.** What are the two types of belt tensioners?
  A. V type and serpentine
  B. Vertical- and horizontal-mounted
  C. Viscous and nonviscous
  D. Automatic and manual

_____ **2.** Which of the following statements is correct?
  A. Belts should be inspected when the engine is running.
  B. Torn or split belts are serviceable.
  C. A bottomed-out belt will slip and squeal.
  D. Belts with cracks should always be replaced.

_____ **3.** What spring clamp is not adjustable?
  A. Gear clamp
  B. Worm clamp
  C. Screw clamp
  D. Wire clamp

_____ **4.** What is the best way to remove coolant hoses?
  A. Twist the hose off.
  B. Pry the hose off with a screwdriver.
  C. Slit the hose and carefully peel it off if it is stuck.
  D. Remove the hose as quickly as possible using any tool.

_____ **5.** Which of the following is a nonadjustable clamp?
  A. Gear clamp
  B. Worm clamp
  C. Screw clamp
  D. Spring clamp

_____ **6.** Which of the following is a clamp that can be adjusted using a nut driver?
  A. Screw clamp
  B. Gear clamp
  C. Wire clamp
  D. Spring clamp

_____ **7.** What is the best way to remove hoses from heater cores?
  A. Slit the hose and peel it off.
  B. Pry the hose off with a screwdriver.
  C. Twist the hose off.
  D. Heat the hose mildly.

_____ **8.** When replacing the thermostat, what percentage of the coolant in the system should be drained to avoid spills?
  A. At least 40%
  B. At least 50%
  C. At least 60%
  D. At least 70%

_____ **9.** Which of the following statements is true with respect to removing and replacing a thermostat?
  A. Do not bleed air from the cooling system.
  B. Tighten the housing bolts to maximum torque.
  C. Drain at least 50% of the coolant to avoid spills.
  D. Faulty thermostats should be replaced rather than repaired.

_____ **10.** When inspecting the _____, clean any gasket material from the mating surfaces.
  A. Thermostat housing
  B. O-rings
  C. Clamps
  D. Hoses

_____ **11.** A hose that has become very weak and is in danger of ballooning or bursting is known as
_____.
  A. Soft hose
  B. Swollen hose
  C. Cracked hose
  D. Hardened hose

_____ **12.** A hose that has lost its elasticity and is swelling under pressure is known as _____.
  A. Soft hose
  B. Swollen hose
  C. Cracked hose
  D. Hardened hose

_____ **13.** When a hose deteriorates, all of the following can happen, EXCEPT _____.
  A. Bursting of the hose
  B. Coolant being pumped out of the engine
  C. Overheating of the engine
  D. Overcooling of the engine

_____ **14.** With _____, a belt that has been soaked in oil will not grip properly on the pulleys and will slip.
  A. Glazing
  B. Bottoming out
  C. Oil contamination
  D. Pulley wear

_____ **15.** _____ is shininess on the surface of the belt that comes in contact with the pulley.
  A. Oil contamination
  B. Cracking
  C. Bottoming out
  D. Glazing

_____ **16.** _____ belts are unserviceable and should be replaced immediately.
  A. Cracked
  B. Torn or split
  C. Glazed
  D. Contaminated

_____ 17. In a serpentine belt system, the pulley edges should be within _____ of alignment with each other.
   A. 2 mm
   B. 1.6 mm
   C. 3 mm
   D. 0.5 mm

_____ 18. A _____ can tell the proportions of antifreeze and water in coolant mix (or the level of freeze protection) by measuring a liquid's specific gravity.
   A. Thermometer
   B. Voltmeter
   C. Scan tool
   D. Refractometer

_____ 19. One drawback to _____ is that they are typically antifreeze specific.
   A. Refractometers
   B. Hydrometers
   C. Scan tools
   D. Thermometers

_____ 20. As corrosion inhibitors break down over time, the solution of water and antifreeze will become _____ acidic.
   A. Slightly less
   B. Much less
   C. More
   D. The acidity will not change.

_____ 21. If voltages are over _____ volts when the load or loads are activated, then electrical diagnosis will have to be performed on the circuit being tested.
   A. 0.2
   B. 0.3
   C. 0.5
   D. 1.0

_____ 22. A refractometer is used to test a coolant's _____.
   A. pH
   B. Freeze protection
   C. Electrolysis
   D. Level

_____ 23. Draining and refilling the coolant is necessary in all of the following cases, EXCEPT _____.
   A. When a customer requests it
   B. When the pH is out of specification
   C. When the coolant is contaminated
   D. When the coolant level is low

_____ 24. Antifreeze has a higher _____ than water, so the higher the float rises in the liquid, the greater the percentage of antifreeze in the mix.
   A. Boiling point
   B. Specific gravity
   C. Density
   D. Freezing point

_____ 25. Radiator hose problems include all of the following, EXCEPT _____.
   A. Swollen hose
   B. Hardened hose
   C. Elongated hose
   D. Cracked hose

_____ 26. All of the following are types of clamps that secure the hose to the component it is connected to, EXCEPT _____.
   A. Worm clamp
   B. Screw clamp
   C. Spring clamp
   D. C clamp

_____ **27.** It is necessary to remove and replace a thermostat in all of the following situations, EXCEPT _____.
  A. When the thermostat is found to be faulty
  B. When the thermostat is creating an overheating concern
  C. When the thermostat is creating an underheating concern
  D. When the thermostat has been in use for a long time

_____ **28.** When installing a thermostat, a new gasket and/or O-ring seal should be installed _____.
  A. Every time
  B. If it is faulty
  C. If it cannot be repaired
  D. With every alternate thermostat replacement

_____ **29.** Before starting a repair or service task on the cooling system, allow sufficient time for the _____.
  A. Coolant to drain
  B. System to cool
  C. Pressure to decrease
  D. Engine to cool

_____ **30.** Tech A says that coolant leaking out of the water pump weep hole is normal. Tech B says that a refractometer measures the specific gravity of a liquid. Who is correct?
  A. Tech A
  B. Tech B
  C. Both Tech A and Tech B
  D. Neither Tech A nor Tech B

_____ **31.** Tech A says that a cooling system pressure tester can be used to find leaks in the cooling system. Tech B says that the cooling system pressure tester can be used to test the pressure at which the radiator cap vents. Who is correct?
  A. Tech A
  B. Tech B
  C. Both Tech A and Tech B
  D. Neither Tech A nor Tech B

_____ **32.** Tech A says that most coolants are designed to last the life of the vehicle. Tech B says that a hydrometer can be used to test the freeze point of coolant. Who is correct?
  A. Tech A
  B. Tech B
  C. Both Tech A and Tech B
  D. Neither Tech A nor Tech B

_____ **33.** Tech A says that belts are an important part of the cooling system if they are used to drive the fan or water pump. Tech B says that since the belt is made of rubber, it cannot be overtightened, so making it extra tight is a good insurance policy to be sure it will not slip. Who is correct?
  A. Tech A
  B. Tech B
  C. Both Tech A and Tech B
  D. Neither Tech A nor Tech B

_____ **34.** Tech A says the bypass hose is used to regulate cooling system pressure. Tech B says that radiator hoses should be checked during routine maintenance. Who is correct?
  A. Tech A
  B. Tech B
  C. Both Tech A and Tech B
  D. Neither Tech A nor Tech B

_____ **35.** Tech A states that coolant hoses should be checked anytime the vehicle is in the shop for a maintenance inspection. Tech B states that most technicians generally replace both radiator hoses at once as a sensible precaution. Who is correct?
  A. Tech A
  B. Tech B
  C. Both Tech A and Tech B
  D. Neither Tech A nor Tech B

_____ **36.** Tech A states that hardening can be verified by squeezing the hose and comparing it to a known good hose. Tech B states that a cracked hose can be verified by a visual inspection. Who is correct?
  A. Tech A
  B. Tech B
  C. Both Tech A and Tech B
  D. Neither Tech A nor Tech B

_____ **37.** Tech A says that when replacing cooling system hoses, it may be best to slit the hose rather than trying to twist it off. Tech B says that cooling system hoses need to be changed only when they start leaking. Who is correct?
  A. Tech A
  B. Tech B
  C. Both Tech A and Tech B
  D. Neither Tech A nor Tech B

_____ **38.** Tech A states that before starting a repair or service task on the cooling system, sufficient time should be allowed for the system to cool adequately before opening the pressurized system. Tech B states that the presence of coolant does not impact the replacement of the thermostat. Who is correct?
  A. Tech A
  B. Tech B
  C. Both Tech A and Tech B
  D. Neither Tech A nor Tech B

_____ **39.** Tech A states that once the thermostat has been removed, any old gasket material and corrosion that has built up on both sealing surfaces where the thermostat seats should be cleaned. Tech B states that it is fine to let a faulty thermostat run until it completely breaks down. Who is correct?
  A. Tech A
  B. Tech B
  C. Both Tech A and Tech B
  D. Neither Tech A nor Tech B

_____ **40.** Tech A states that all thermostats are equipped with an air bleed valve. Tech B states that thermostats may or may not be equipped with an air bleed valve. Who is correct?
  A. Tech A
  B. Tech B
  C. Both Tech A and Tech B
  D. Neither Tech A nor Tech B

_____ **41.** Tech A states that the thermostat should be fully seated in the groove before the housing is installed. Tech B states that it is important that the thermostat be installed so the sensing bulb is toward the engine. Who is correct?
  A. Tech A
  B. Tech B
  C. Both Tech A and Tech B
  D. Neither Tech A nor Tech B

_____ **42.** The thermostat falls out of its recessed groove. Tech A states that one of the thermostat housing ears will break off when the bolts are tightened, requiring replacement. Tech B states that the thermostat itself is likely to be damaged. Who is correct?
  A. Tech A
  B. Tech B
  C. Both Tech A and Tech B
  D. Neither Tech A nor Tech B

_____ **43.** Tech A states that the manufacturer's procedure should be followed to properly bleed all air from the cooling system. Tech B states that it is necessary to remove and replace a thermostat in the event that the thermostat is found to be faulty. Who is correct?
  A. Tech A
  B. Tech B
  C. Both Tech A and Tech B
  D. Neither Tech A nor Tech B

_____ **44.** Tech A states that a thermostat replacement should be suggested to the customer during a cooling system flush. Tech B states that a thermostat replacement should be suggested to the customer during a water pump replacement. Who is right?

    A. Tech A

    B. Tech B

    C. Both Tech A and Tech B

    D. Neither Tech A nor Tech B

_____ **45.** Tech A says that the thermostat may have an air bleed that must be positioned properly during replacement. Tech B says that the thermostat housing must be flush against the mating surface or the ear may break off when the bolts are tightened. Who is correct?

    A. Tech A

    B. Tech B

    C. Both Tech A and Tech B

    D. Neither Tech A nor Tech B

Automatic Transmission Fundamentals

At the start of each chapter, you will find the Learning Objectives from the textbook. These are your objectives as you make your way through the exercises in this workbook and the chapter in your textbook. The following activities have been designed to help you refresh your knowledge of the material in this chapter.

Learning Objectives

After reading this chapter, you will be able to:

- LO 18-01 Describe the functions and types of automatic transmissions.
- LO 18-02 Describe torque converter construction.
- LO 18-03 Describe torque converter operation.
- LO 18-04 Describe lockup converters and heat exchangers.
- LO 18-05 Describe gear train operation.
- LO 18-06 Describe methods of holding–driving gears.
- LO 18-07 Describe automatic transmission components.

Matching

Match the following terms with the correct description or example.

A. Extension housing   **E.** One-way clutch

B. Fluid coupler   **F.** Ring gear

C. Gear ratio   **G.** Spur gear

D. Helical-cut gear   **H.** Transaxle

_____ **1.** A gear with teeth cut parallel to its axis of rotation.

_____ **2.** A type of hydraulic coupling used on vintage vehicles to connect and transfer power from the engine to the transmission.

_____ **3.** A type of gear that is cut on a helix or spiral.

_____ **4.** A component of the automatic transmission housing that covers the output shaft of the transmission. It also supports the end of the driveshaft and may hold components such as the vehicle speed sensor, speedometer drive assembly, and governor assembly.

_____ **5.** The outer gear in a planetary gear. Also the gear that meshes with the pinion gear in a final drive.

_____ **6.** The relationship between two gears in mesh as a comparison to input versus output.

_____ **7.** A type of transmission, typically used in front-wheel drive vehicles, in which the transmission also includes the differential and final drive gear assembly.

_____ **8.** A type of holding device that allows free rotation in one direction but will lock up in the opposite direction; also called an over-running clutch.

Multiple Choice

Read each item carefully, and then select the best response.

_____ **1.** Transmissions that use two pulleys that change diameter in response to vehicle load and speed are known as _____.

     A. Dual-clutch transmissions

     B. Continuously variable transmissions

     C. Power-splitting transmissions

     D. Automatic transmissions

_____ **2.** With which type of hybrid drive train does the engine have two or more methods for the power to flow through the transmission?
A. Series hybrid
B. Parallel hybrid
C. Dual-clutch
D. Continuously variable

_____ **3.** Early automatic transmissions used a _____ rather than a torque converter.
A. Fluid coupler
B. Flywheel
C. Transaxle
D. Countershaft

_____ **4.** In its simplest form, a single-stage torque converter has three elements. Which of the following is NOT one of them?
A. Turbine
B. Impeller
C. Fluid coupler
D. Stator

_____ **5.** The _____ has a small set of curved blades attached to a central hub and is positioned between the impeller and the turbine.
A. Fluid coupler
B. Pinion
C. Servo
D. Stator

_____ **6.** Automatic transmission vehicles use a(n) _____, sometimes called a transmission cooler, in one tank of the radiator.
A. Evaporator
B. Heat exchanger
C. Blend box
D. Condenser

_____ **7.** The difference in diameter between the driving gear and the driven gear is known as _____.
A. Leverage
B. Torque conversion
C. The gear ratio
D. Reduction

_____ **8.** What type of gears are often used to change the direction of power flow by 90 degrees?
A. Helical-cut gears
B. Sun gears
C. Planetary gears
D. Hypoid gears

_____ **9.** A friction-lined steel belt that wraps around the outside of a drum inside the automatic transmission is called a _____.
A. Band
B. Clutch
C. Pinion
D. Strap

_____ **10.** Which of the following should never be used to lubricate an internal transmission seal?
A. ATF fluid
B. Grease
C. Petroleum jelly
D. Automatic transmission assembly lubricant

_____ **11.** An automatic transmission can select and shift gears without input from the driver.
A. True
B. False

_____ **12.** In a manual transmission, engine torque is controlled by two, typically wet, clutches.
A. True
B. False

_____ **13.** Modern transmissions no longer use fluid couplers.
   A. True
   B. False

_____ **14.** The stator has a small set of curved blades attached to a central hub and is positioned between the impeller and the turbine.
   A. True
   B. False

_____ **15.** Most automatic transmission vehicles use a heat exchanger, sometimes called a transmission cooler, in one tank of the radiator.
   A. True
   B. False

_____ **16.** Planetary gears are held in place in a planet carrier.
   A. True
   B. False

_____ **17.** Helical-cut gears tend to be much louder and do not offer as much strength as the other types of gears.
   A. True
   B. False

_____ **18.** Automatic transmission fluids are typically dyed purple for easy identification.
   A. True
   B. False

_____ **19.** In a manual transmission, there is no flywheel; there is a thin, lightweight steel flexplate.
   A. True
   B. False

_____ **20.** Teflon seals come in several styles—continuous, butt cut, scarf-cut, and step joint.
   A. True
   B. False

_____ **21.** _____ transmissions are a newer type of automatic transmission that uses two, typically wet, clutches.
   A. Conventional
   B. Continuously variable
   C. Dual-clutch
   D. Power-splitting

_____ **22.** In a _____ hybrid drive train, the electric motor is not typically able to propel the vehicle on its own. This electric motor is often placed between the engine and the transmission.
   A. Parallel
   B. Power-splitting
   C. Series-parallel
   D. Series

_____ **23.** A series-parallel hybrid drive train uses what is called a _____ transmission.
   A. Power-splitting
   B. Dual-shaft
   C. Dual-clutch
   D. Continuously variable

_____ **24.** In the automatic transmission, the _____ is mounted between the engine and the transmission, in the same place as a manual transmission clutch.
   A. Lockup converter
   B. Gear train
   C. Flexplate
   D. Torque converter

_____ **25.** A _____ is basically two fans facing each other.
   A. Torque converter
   B. Fluid coupler
   C. Gasket
   D. Parking pawl

_____ **26.** The _____ has a large number of vanes attached to the torque converter housing to form the driving member.
   A. Impeller
   B. Turbine
   C. Stator
   D. Dual clutch

_____ **27.** Planetary gears revolve inside a larger _____ gear that wraps around the outside of the whole planetary gear set.
   A. Hypoid
   B. Spur
   C. Sun
   D. Ring

_____ **28.** _____ gears were used in early transmissions due to ease of manufacturing and lower cost.
   A. Sun
   B. Ring
   C. Spur
   D. Planet

_____ **29.** When the driver shifts the transmission into park, a lever, called a _____, is forced into notches cut into a hardened steel drum on the output shaft of the transmission.
   A. Flexplate
   B. Parking pawl
   C. Parking pan
   D. Ring gear

_____ **30.** One-way clutches can be either one-way _____ or _____.
   A. Splines, drums
   B. Rollers, splines
   C. Gaskets, splines
   D. Rollers, sprags

_____ **31.** The arrow in the image is pointing to what part of the torque converter mounted between the engine and transmission?

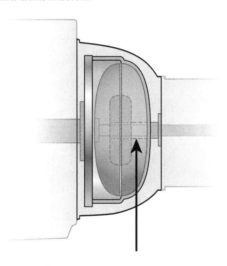

   A. Flexplate
   B. Crankshaft
   C. Stator
   D. Turbine

_____ **32.** The arrow in the image is pointing to what part of the torque converter mounted between the engine and transmission?

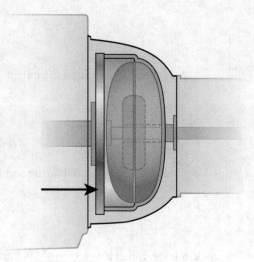

A. Impeller
B. Flexplate
C. Crankshaft
D. Turbine

_____ **33.** The arrow in the image is pointing to what part of the torque converter mounted between the engine and transmission?

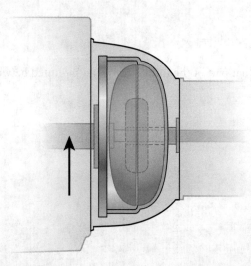

A. Engine
B. Torque converter
C. Crankshaft
D. Transmission

_____ **34.** The arrow in the image is pointing to what part of the simple planetary gear set?

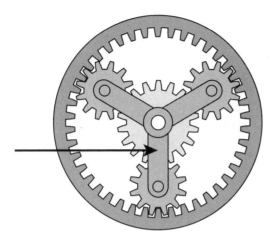

    A. Planet carrier
    B. Sun gear
    C. Ring gear
    D. Planet gear

_____ **35.** The arrow in the image is pointing to what part of the simple planetary gear set?

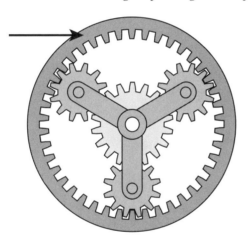

    A. Planet gear
    B. Planet carrier
    C. Ring gear
    D. Sun gear

_____ **36.** The arrow in the image is pointing to what part of the simple planetary gear set?

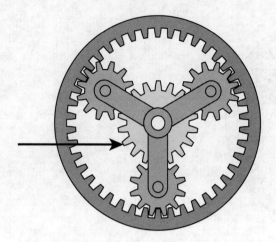

    A. Ring gear
    B. Planet gear
    C. Planet carrier
    D. Sun gear

_____ **37.** What type of gear set style is shown in the image?

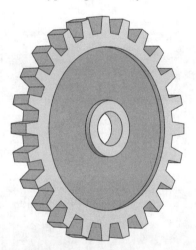

    A. Spur gear
    B. Helical gear
    C. Hypoid gear
    D. Sun gear

_____ **38.** What type of gear set style is shown in the image?

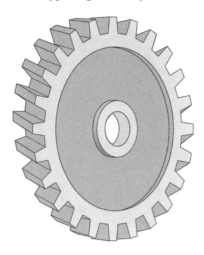

    A. Spur gear
    B. Ring gear
    C. Helical gear
    D. Hypoid gear

_____ **39.** What type of gear set style is shown in the image?

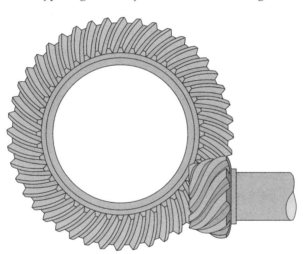

    A. Planet gear
    B. Helical gear
    C. Spur gear
    D. Hypoid gear

_____ **40.** The arrow in the image is pointing to what part of the transmission band?

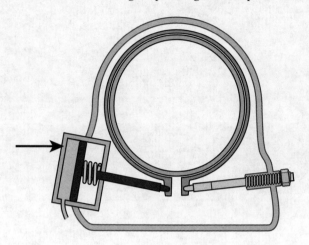

    A. ATF
    B. Servo
    C. Drum
    D. Locknut

_____ **41.** The arrow in the image is pointing to what part of the transmission band?

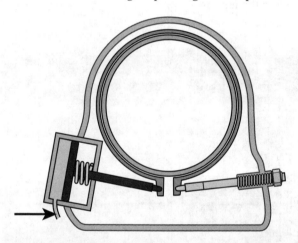

    A. Adjustment screw
    B. ATF
    C. Band
    D. Drum

_____ **42.** The arrow in the image is pointing to what part of the transmission band?

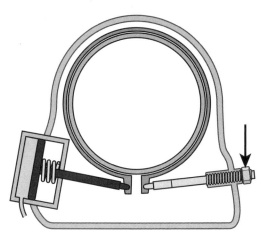

A. ATF
B. Adjustment screw
C. Housing
D. Locknut

_____ **43.** Tech A says that the higher the numerical gear ratio (4:1), the more torque that will be applied to the wheels. Tech B says that the lower the numerical gear ratio (2:1), the more torque that will be applied to the wheels. Who is correct?
A. Tech A
B. Tech B
C. Both Tech A and Tech B
D. Neither Tech A nor Tech B

_____ **44.** Tech A says that helical-cut gears are stronger than straight-cut gears. Tech B says that helical-cut gears are noisier than straight-cut gears. Who is correct?
A. Tech A
B. Tech B
C. Both Tech A and Tech B
D. Neither Tech A nor Tech B

_____ **45.** Tech A says that a transaxle must have a type of differential assembly to allow the front wheels to turn at different speeds while driving around a curve. Tech B says that a conventional rear-wheel drive vehicle will typically use a differential assembly that is located in the rear axle assembly. Who is correct?
A. Tech A
B. Tech B
C. Both Tech A and Tech B
D. Neither Tech A nor Tech B

_____ **46.** Tech A says that a parallel hybrid transmission has two or more different devices to propel the vehicle. Tech B says that hybrid power-splitting transmissions use two adjustable pulleys and a heavy metal belt. Who is correct?
A. Tech A
B. Tech B
C. Both Tech A and Tech B
D. Neither Tech A nor Tech B

_____ **47.** Tech A says that planetary gear sets are used in all automatic transmissions. Tech B says that planetary gear sets are often combined together to create the needed gear ratios for a transmission. Who is correct?
  A. Tech A
  B. Tech B
  C. Both Tech A and Tech B
  D. Neither Tech A nor Tech B

_____ **48.** Tech A says that the impeller is connected directly to the flex plate. Tech B says that the turbine shaft drives the front pump. Who is correct?
  A. Tech A
  B. Tech B
  C. Both Tech A and Tech B
  D. Neither Tech A nor Tech B

_____ **49.** Tech A says that most automatic transmissions have a ring gear for the starter as part of the flexplate. Tech B says that the ring gear is typically replaceable (interference fit) on a flexplate. Who is correct?
  A. Tech A
  B. Tech B
  C. Both Tech A and Tech B
  D. Neither Tech A nor Tech B

_____ **50.** Tech A says that a knocking noise in the engine/transmission area is typically caused by a bad front pump in the transmission. Tech B says that a knocking noise in the engine/transmission area is more likely to be caused by loose flexplate bolts. Who is correct?
  A. Tech A
  B. Tech B
  C. Both Tech A and Tech B
  D. Neither Tech A nor Tech B

_____ **51.** Tech A says that painting a rebuilt transmission helps to identify it as being a rebuilt transmission. Tech B says that painting a rebuilt transmission prevents the housing from leaking due to case porosity. Who is correct?
  A. Tech A
  B. Tech B
  C. Both Tech A and Tech B
  D. Neither Tech A nor Tech B

_____ **52.** Tech A says that the torque converter multiplies the amount of torque transmitted from the engine to the transmission. Tech B says that the torque converter couples and uncouples the engine and transmission as the vehicle stops and starts in traffic. Who is correct?
  A. Tech A
  B. Tech B
  C. Both Tech A and Tech B
  D. Neither Tech A nor Tech B

Maintaining the Automatic Transmission/Transaxle

At the start of each chapter, you will find the Learning Objectives from the textbook. These are your objectives as you make your way through the exercises in this workbook and the chapter in your textbook. The following activities have been designed to help you refresh your knowledge of the material in this chapter.

Learning Objectives

After reading this chapter, you will be able to:

- LO 19-01 Check automatic transmission fluid level and inspect for leaks.
- LO 19-02 Perform automatic transmission fluid service.
- LO 19-03 Perform the in-vehicle transmission service tasks.

Multiple Choice

Read each item carefully, and then select the best response.

_____ 1. How much fluid does it take to raise the level from the bottom of the cross-hatched area or add mark to the full mark on most transmission dipsticks?
   A. Half pint
   B. Quart
   C. Pint
   D. Gallon

_____ 2. A test tool that can help pinpoint noises in a vehicle while the vehicle is driven is called a(n) _____.
   A. NVH analyzer
   B. Stethoscope
   C. Scan tool
   D. PID tester

_____ 3. What feature makes a high-quality aftermarket scan tool or factory scan tool better than a code reader?
   A. Views live data
   B. Reads diagnostic trouble codes (DTCs)
   C. Gives omnidirectional control
   D. Writes DTCs

_____ 4. Replacement of which of the following components can be performed without the removal of the transmission?
   A. Torque converter
   B. Transmission band
   C. Powertrain mounts
   D. Pump assembly

_____ 5. Replacement of which of the following components requires the removal of the driveshaft and the extension housing?
   A. Speed sensor seal
   B. Extension housing bushing
   C. Pan gasket
   D. Shift solenoid

_____ 6. A loud thump when the accelerator is applied in drive or reverse and again when the brake pedal is applied may indicate a _____.
   A. Broken powertrain mount
   B. Dry extension housing bushing
   C. Transmission fluid leak
   D. Bad shift sensor

_____ 7. If the _____ is not adjusted properly, the transmission position indicator will not be set correctly, and the customer will not know which gear the vehicle is in.
   A. Speed sensor
   B. Shift sensor
   C. Shift linkage
   D. Lockup torque converter

_____ 8. Some vehicles do not have a transmission dipstick.
   A. True
   B. False

_____ 9. All transmissions are checked with the engine at operating temperature, idling, and the transmission in park.
   A. True
   B. False

_____ 10. Some modern transmission fluids are a darker red when they are brand new and even have a slightly burnt smell to them.
   A. True
   B. False

_____ 11. The presence of a small amount of clutch material or metal in the bottom of the pan indicates a serious transmission problem.
   A. True
   B. False

_____ 12. Scanning the powertrain control module for trouble codes will indicate the specific transmission problem.
   A. True
   B. False

_____ 13. Bidirectional control allows the technician to command the chassis dynamometer to operate in "forward" or "reverse" without moving the shift selector in the vehicle.
   A. True
   B. False

_____ 14. If a transmission suffers catastrophic failure, metal particles and old clutch material can become stuck inside the fluid cooler or lines.
   A. True
   B. False

_____ 15. When fluids leak, they travel _____ due to gravity and typically toward the _____ of the vehicle.
   A. Upward, front
   B. Upward, back
   C. Downward, front
   D. Downward, back

_____ 16. If a transmission leak is hard to locate, place a leak detection _____ in the transmission fluid.
   A. Sensor
   B. Degreaser
   C. Pan
   D. Dye

_____ 17. The new way to check the quality of transmission fluid is to take a few drops and place them on a clean _____.
   A. Paper towel
   B. Plate
   C. Pan
   D. Actuator

_____ **18.** Many high-quality or factory scan tools give technicians _____ control of the transmission, allowing them to command different solenoids and actuators "on" and "off" to check their operation.
   A. Unidirectional
   B. Bidirectional
   C. Tridirectional
   D. Complete

_____ **19.** The powertrain _____ hold the engine and transmission in the proper position in the vehicle.
   A. Seals
   B. Sensors
   C. Solenoids
   D. Mounts

_____ **20.** With worn powertrain mounts, the _____ can become bound, and the driver might be able to start a vehicle while it is in gear or have the vehicle in the wrong gear.
   A. Manual valve
   B. Shift linkage
   C. Transmission position indicator
   D. Transmission dipstick

_____ **21.** Most powertrain mounts are of the _____ style and subject to cracks and tears that occur over time due to their constant flexing.
   A. Metal
   B. Plastic
   C. Rubber
   D. Hydraulic

_____ **22.** Some _____ mounts use a computer-controlled electronic control valve that can open an alternate passageway or vary the size of the orifice.
   A. Hydraulic
   B. Rubber
   C. Metal
   D. Plastic

_____ **23.** A transmission fluid _____ that is leaking could cause the new transmission to fail from low fluid level or could cause antifreeze to be drawn into the cooler from the radiator, contaminating the fluid.
   A. Heater
   B. Cooler
   C. Radiator
   D. Solenoid

_____ **24.** Tech A says that you should check the fluid level in many automatic transmissions with the engine idling and the transmission in park. Tech B says that some transmissions are made without dipsticks. Who is correct?
   A. Tech A
   B. Tech B
   C. Both Tech A and Tech B
   D. Neither Tech A nor Tech B

_____ **25.** Tech A says that increasing the transmission fluid level from the add mark to the full mark will require a quart of fluid, just as with engine oil. Tech B says that transmission solenoids should be tested electrically and mechanically. Who is correct?
   A. Tech A
   B. Tech B
   C. Both Tech A and Tech B
   D. Neither Tech A nor Tech B

_____ **26.** Tech A says that transmission pan gaskets are always replaced. Tech B says that some newer transmissions have reusable pan gaskets. Who is correct?
   A. Tech A
   B. Tech B
   C. Both Tech A and Tech B
   D. Neither Tech A nor Tech B

_____ **27.** Tech A says that the cooler and cooler lines should be flushed as part of a transmission rebuild. Tech B says that when checking the fluid level on the dipstick, always read the highest level on either side. Who is correct?
   A. Tech A
   B. Tech B
   C. Both Tech A and Tech B
   D. Neither Tech A nor Tech B

_____ **28.** Tech A says that using a scan tool to activate the torque converter clutch (TCC) with the engine idling in drive and the brakes applied is a typical troubleshooting task. Tech B says that diagnosis is a waste of time on a faulty transmission since it will be rebuilt anyway. Who is correct?
   A. Tech A
   B. Tech B
   C. Both Tech A and Tech B
   D. Neither Tech A nor Tech B

_____ **29.** Tech A says you should check fluid level in all types of automatic transmissions with the engine idling and the transmission in park. Tech B says that a transmission fluid leak can result in failure of the transmission due to a low fluid level. Who is correct?
   A. Tech A
   B. Tech B
   C. Both Tech A and Tech B
   D. Neither Tech A nor Tech B

_____ **30.** Tech A says that if the linkage is not adjusted properly, the transmission position indicator (PRNDL indicator) will not be set correctly, and the driver may not know which gear the vehicle is in. Tech B says that even if the linkage is not adjusted properly, the backup or reverse lights will still work normally. Who is correct?
   A. Tech A
   B. Tech B
   C. Both Tech A and Tech B
   D. Neither Tech A nor Tech B

_____ **31.** Tech A says that often in the event of a broken powertrain mount, customers complain of a loud thump when they apply the accelerator in drive or reverse. Tech B says that the loud thump may be the engine and/or transmission being pulled up from the broken mount due to the engine's torque, and then being set back down when the brake is applied. Who is correct?
   A. Tech A
   B. Tech B
   C. Both Tech A and Tech B
   D. Neither Tech A nor Tech B

_____ **32.** Tech A says that a transmission fluid cooler that is leaking could cause the new transmission to fail from low fluid level. Tech B says that such a leak could cause antifreeze to be drawn into the cooler from the radiator and contaminate the transmission fluid. Who is correct?
   A. Tech A
   B. Tech B
   C. Both Tech A and Tech B
   D. Neither Tech A nor Tech B

_____ **33.** Tech A says that some manufacturers install a replaceable in-line transmission filter in the cooler line. Tech B says that some transmissions have external filters that look similar to engine oil filters. Who is correct?
   A. Tech A
   B. Tech B
   C. Both Tech A and Tech B
   D. Neither Tech A nor Tech B

_____ **34.** On some vehicles, _____ and springs are held in place by the transmission filter.
   A. Seals
   B. Check balls
   C. Check valves
   D. Gaskets

_____ **35.** Tech A says that transmission cooler lines should be flushed only in the direction of flow. Tech B says that a transmission cooler should be flushed in both directions. Who is correct?
   A. Tech A
   B. Tech B
   C. Both Tech A and Tech B
   D. Neither Tech A nor Tech B

_____ **36.** Tech A says that some clutch material in the bottom of the pan is abnormal when servicing a transmission filter. Tech B says that a small amount of metal in the pan is abnormal. Who is correct?
   A. Tech A
   B. Tech B
   C. Both Tech A and Tech B
   D. Neither Tech A nor Tech B

_____ **37.** Which of the following is NOT an example of an in-vehicle transmission repair?
   A. Replacing external seals
   B. Replacing external gaskets
   C. Replacing a servo or band
   D. Replacing the valve body

_____ **38.** Tech A says that some powertrain mounts are computer controlled. Tech B says that some powertrain mounts are hydraulically controlled, similarly to a shock absorber. Who is correct?
   A. Tech A
   B. Tech B
   C. Both Tech A and Tech B
   D. Neither Tech A nor Tech B

_____ **39.** Tech A says that a misaligned range sensor may cause reverse lights to not operate. Tech B says that a misadjusted neutral safety switch may allow a vehicle to start in drive. Who is correct?
   A. Tech A
   B. Tech B
   C. Both Tech A and Tech B
   D. Neither Tech A nor Tech B

_____ **40.** Tech A says that some manufacturers advise against changing the transmission fluid for the life of the vehicle. Tech B says that some manufacturers recommend changing the transmission fluid every 5,000 miles if used for towing or operated in dusty conditions. Who is correct?
   A. Tech A
   B. Tech B
   C. Both Tech A and Tech B
   D. Neither Tech A nor Tech B

Hybrid and Continuously Variable Transmissions

At the start of each chapter, you will find the Learning Objectives from the textbook. These are your objectives as you make your way through the exercises in this workbook and the chapter in your textbook. The following activities have been designed to help you refresh your knowledge of the material in this chapter.

Learning Objectives

After reading this chapter, you will be able to:

- LO 20-01 Describe hybrid vehicle functions.
- LO 20-02 Describe various hybrid vehicle models.
- LO 20-03 Describe continuously variable transmission (CVT) operation.

Matching

Match the following terms with the correct description or example.

A. Belt alternator starter (BAS)

B. CVT

C. Electronic continuously variable transmission (ECVT)

D. Hybrid drive system

E. Integrated motor assist (IMA)

F. Regenerative braking

G. Torque assist

H. Torque smoothing

I. Variable-diameter pulley (VDP)

_____ **1.** A pulley that can change its diameter by moving closer or further apart.

_____ **2.** A drive system that uses two or more propulsion systems, such as electric motors and an internal combustion engine (ICE).

_____ **3.** A type of hybrid drive system that uses a belt-driven alternator/starter that operates on 42 volts.

_____ **4.** A process that uses an electric motor to smooth out engine power pulses when an ICE is operating at low rpm or when the vehicle is using fuel management techniques such as cylinder deactivation.

_____ **5.** A type of hybrid transmission that often uses two electric motors in combination with an ICE. The two electric motors and the ICE transfer power through a planetary gear set, allowing an infinite amount of gear ratios.

_____ **6.** A type of braking in which the kinetic energy of the vehicle's motion is captured, rather than being lost to heat, as it is in a conventional braking system.

_____ **7.** A type of transmission that has no fixed gears, as in a conventional transmission, but rather can adjust gear ratios infinitely within the design of the transmission.

_____ **8.** Use of an electric motor to supplement the engine's torque whenever additional torque is needed, allowing for a smaller ICE to be used.

_____ **9.** A Honda hybrid drive system that uses a moderate-sized electric motor installed between the engine and the transmission.

Multiple Choice

Read each item carefully, and then select the best response.

_____ 1. When the powertrain control module shuts off the hybrid engine at a stoplight, this is an example of _____.
   A. Torque smoothing
   B. Regenerative braking
   C. Torque assist
   D. Idle stop

_____ 2. What ability can a hybrid employ to smooth out the power pulses of the ICE and create a flatter torque output curve?
   A. Torque assist
   B. Torque smoothing
   C. Idle stop
   D. Regenerative braking

_____ 3. When a conventional vehicle is being stopped, most of the kinetic energy of the vehicle's movement is converted into _____.
   A. Heat
   B. Electricity
   C. Electromagnetic energy
   D. Chemical energy

_____ 4. What ability of a hybrid engine reduces the need to use the ICE at lower rpms by using an electric motor to help propel the vehicle from a stop?
   A. Torque smoothing
   B. Kinetic braking
   C. Torque assist
   D. Integrated propulsion

_____ 5. What hybrid system is classified as a parallel hybrid because the electric motor is operating at the same time as the gasoline engine?
   A. Two-mode
   B. IMA
   C. BAS
   D. Continuously variable

_____ 6. What type of transaxle cannot be used in combination with the IMA system?
   A. Automatic transmission
   B. Manual transmission
   C. CVT
   D. Dual-clutch transmission

_____ 7. The Toyota and Lexus hybrids are classified as _____ because either the ICE or the electric motor/generator can propel the vehicle, or both can be used together.
   A. Series-parallel hybrids
   B. Parallel hybrids
   C. Series hybrids
   D. Integrated hybrids

_____ 8. A CVT has _____ fixed gear ratios.
   A. Three
   B. Four
   C. Five
   D. Zero

_____ **9.** All of the following are basic types of CVTs commonly used in production vehicles, EXCEPT _____.
  A. VDP
  B. IMA
  C. Reeves drive
  D. Electronic continuously variable

_____ **10.** The _____ type of CVT can be found in both Toyota and Ford hybrid vehicles.
  A. ECVT
  B. Toroidal
  C. VDP
  D. Reeves drive

_____ **11.** Automobile manufacturers have had hybrid vehicles and vehicles with CVTs in mass production for more than 20 years.
  A. True
  B. False

_____ **12.** Mechanical brakes can be used to recapture kinetic energy while braking.
  A. True
  B. False

_____ **13.** Hybrid vehicles are able to maintain a higher gas mileage rating during highway driving than in stop-and-go traffic.
  A. True
  B. False

_____ **14.** An electric motor is capable of creating its maximum torque as soon as it begins spinning.
  A. True
  B. False

_____ **15.** In many full hybrids, the vehicle can operate at low speeds using the electric motor only.
  A. True
  B. False

_____ **16.** Currently, Honda IMA hybrids cannot drive using the electric motor only.
  A. True
  B. False

_____ **17.** During electric-only mode, the ICE on Toyota and Lexus hybrids is still used to continually charge the batteries.
  A. True
  B. False

_____ **18.** Some of the early CVTs in Europe used a stiff rubber belt, as do many snowmobiles and all-terrain vehicles.
  A. True
  B. False

_____ **19.** As the vehicle gains speed, the input-pulley diameter of the VDP increases, while the output-pulley diameter decreases.
  A. True
  B. False

_____ **20.** CVTs require special transmission fluid.
  A. True
  B. False

_____ **21.** Many hybrid vehicles that use idle stop have a small, electric _____ that is used to prevent a delay in the engagement of the transmission when the vehicle is restarted.
  A. Generator
  B. Hydraulic pump
  C. Brake
  D. Transaxle

_____ **22.** On a hybrid vehicle, when the driver initiates a stop, the electric motor becomes a(n) _____.
   A. Generator
   B. Alternator
   C. Starter
   D. IMA

_____ **23.** The _____ system uses a 42-volt battery.
   A. IMA
   B. Intelligent multi-mode drive
   C. CVT
   D. BAS

_____ **24.** The IMA system uses a thin electric _____ in place of a conventional flywheel or flexplate.
   A. Generator
   B. Starter
   C. Motor
   D. Alternator

_____ **25.** Toyota and Lexus hybrids use voltages from _____ to _____ volts, depending on the application.
   A. 200, 650
   B. 10, 100
   C. 300, 900
   D. 200, 450

_____ **26.** The only major difference between the Ford hybrid and the Toyota hybrid is that on the Ford system, rather than having the two electric motor/generators directly connected to the ring gear and the sun gear, they are attached through a set of _____ gears.
   A. Planet
   B. Transfer
   C. Spur
   D. Helical

_____ **27.** A CVT has a transmission that can infinitely change the _____ within its operational design.
   A. Gear ratio
   B. Fuel ratio
   C. Speeds
   D. Voltage

_____ **28.** Each of the VDP pulleys has two movable drive faces called _____.
   A. Discs
   B. Belts
   C. Pivots
   D. Sheaves

_____ **29.** The most common type of CVT is called a VDP or _____ drive CVT.
   A. Toroidal
   B. Reeves
   C. Standard
   D. Roller

_____ **30.** The _____ is made up of hundreds of transversely mounted steel plates that are held in place with several steel bands running longitudinally around the edge of the plates.
   A. Rubber belt
   B. Steel belt
   C. Steel pulley
   D. Sheave

_____ **31.** The arrow in the image is pointing to what part of the Ford hybrid transmission?

   A. Traction motor
   B. Generator motor
   C. Output to wheels
   D. Engine input

_____ **32.** The arrow in the image is pointing to what part of the Ford hybrid transmission?

   A. Traction motor
   B. Transfer gear from engine
   C. Transfer gear from traction motor
   D. Generator motor

_____ **33.** The arrow in the image is pointing to what part of the Ford hybrid transmission?

   A. Generator motor
   B. Traction motor
   C. Transfer gear from traction motor
   D. Engine input

_____ **34.** The arrow in the image is pointing to what part of the input and output pulleys?

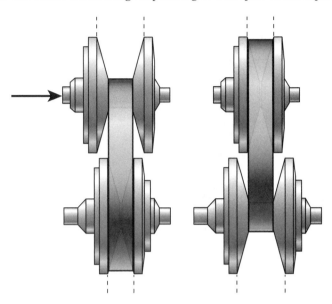

    A. Input shaft
    B. Sheaves
    C. Belt
    D. Output shaft

_____ **35.** The arrows in the image are pointing to what part of the input and output pulleys?

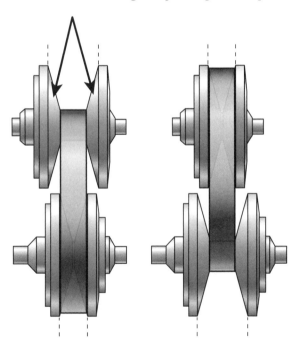

    A. Belts
    B. Input shafts
    C. Output shafts
    D. Sheaves

_____ **36.** The arrow in the image is pointing to what part of the input and output pulleys?

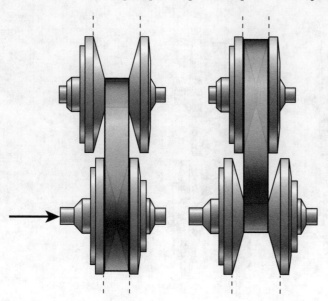

   A. Belt
   B. Output shaft
   C. Input shaft
   D. Sheaves

_____ **37.** Tech A says that hybrid vehicles have an ICE and an electric motor. Tech B says that most hybrid vehicles utilize regenerative braking to help improve fuel economy. Who is correct?
   A. Tech A
   B. Tech B
   C. Both Tech A and Tech B
   D. Neither Tech A nor Tech B

_____ **38.** Tech A says that hybrid vehicles will typically use the same transmission as a non-hybrid vehicle. Tech B says that some hybrid transmissions have a small electric fluid pump for when the transmission is in idle/stop mode. Who is correct?
   A. Tech A
   B. Tech B
   C. Both Tech A and Tech B
   D. Neither Tech A nor Tech B

_____ **39.** Tech A says that BAS vehicles use an alternator to help slow the vehicle and act as a starter. Tech B says that the BAS alternator can be used to propel the vehicle without the ICE. Who is correct?
   A. Tech A
   B. Tech B
   C. Both Tech A and Tech B
   D. Neither Tech A nor Tech B

_____ **40.** Tech A says that the transmission in a Toyota hybrid has a separate gear to provide reverse. Tech B says that Toyota hybrids use a VDP CVT transmission. Who is correct?
   A. Tech A
   B. Tech B
   C. Both Tech A and Tech B
   D. Neither Tech A nor Tech B

_____ **41.** Tech A says that some VDP CVTs use a steel belt to transfer power from pulley to pulley. Tech B says that regeneration in a hybrid vehicle is highest during acceleration. Who is correct?
   A. Tech A
   B. Tech B
   C. Both Tech A and Tech B
   D. Neither Tech A nor Tech B

_____ **42.** Tech A says that some CVTs require a heating system to warm the transmission fluid during cold weather operation. Tech B says that CVTs use conventional automatic transmission fluid. Who is correct?
   A. Tech A
   B. Tech B
   C. Both Tech A and Tech B
   D. Neither Tech A nor Tech B

_____ **43.** Tech A says that on a VDP CVT, each pulley has a movable sheave. Tech B says that most CVTs in vehicles use a rubber drive belt between two movable pulleys. Who is correct?
   A. Tech A
   B. Tech B
   C. Both Tech A and Tech B
   D. Neither Tech A nor Tech B

_____ **44.** Tech A says that regeneration occurs during braking. Tech B says that regeneration charges the main battery pack. Who is correct?
   A. Tech A
   B. Tech B
   C. Both Tech A and Tech B
   D. Neither Tech A nor Tech B

_____ **45.** Tech A says that the Honda IMA system uses a thin electric motor between the engine and the CVT transmission. Tech B says that the Honda IMA system uses voltages between 144 and 158 volts. Who is correct?
   A. Tech A
   B. Tech B
   C. Both Tech A and Tech B
   D. Neither Tech A nor Tech B

Manual Transmission/ Transaxle Principles

At the start of each chapter, you will find the Learning Objectives from the textbook. These are your objectives as you make your way through the exercises in this workbook and the chapter in your textbook. The following activities have been designed to help you refresh your knowledge of the material in this chapter.

Learning Objectives

After reading this chapter, you will be able to:

- LO 21-01 Describe the history of manual transmissions.
- LO 21-02 Describe manual transmission fundamentals.
- LO 21-03 Describe manual transmission drivetrain layout and operation.
- LO 21-04 Describe manual transmission gears, shafts, and bearings.
- LO 21-05 Describe the purpose of clutches, transmission/transaxles, and transfer cases.
- LO 21-06 Describe the purpose of differentials, final drives, and axles.
- LO 21-07 Perform drivetrain fluid maintenance.

Matching

Match the following terms with the correct description or example.

A. Accelerator pedal
B. Axial load
C. Dead axle
D. CV drive axle
E. Drivetrain
F. Gear set
G. Helical gears
H. Independent rear axle
I. Live axle
J. Power flow

K. Radial load
L. Rotational speed
M. Shaft
N. Solid axle
O. Splined
P. Spur gears
Q. Thrust washers
R. Transfer case
S. Transmission

_____ 1. Gears that have teeth set on an angle to the gear face; they operate more quietly than spur gears.

_____ 2. An assembly that houses a variety of gear sets that allow the vehicle to be driven at a wider range of speeds and terrain conditions than would be possible without a transmission.

_____ 3. The load that is perpendicular to a shaft, usually controlled by bearings or bushings.

_____ 4. Gears with straight-cut gear teeth.

_____ 5. Flat, washer-shaped bearings that provide a wear surface between two rotating components that are loaded axially.

_____ 6. An axle that provides power to one wheel.

_____ 7. Typically, a shaft and gear that have parallel grooves machined in them so they mate with each other and lock together rotationally.

_____ 8. An axle that provides power to the wheels.

_____ 9. The load applied in line with a shaft. It can be controlled with thrust bearings.

_____ 10. An assembly used in four-wheel-drive vehicles to transmit power to either two wheels only or all four wheels.

_____ 11. The component assemblies that transmit power from the engine all the way to the drive wheels.

_____ **12.** The path that power takes from the beginning of an assembly to the end. In a transmission, power flow changes as different gears are selected by the driver.

_____ **13.** The foot-operated pedal used by the driver to increase and decrease the amount of power the engine develops.

_____ **14.** The speed at which an object rotates, measured in revolutions per minute (rpm).

_____ **15.** An axle that supplies no power to the wheels.

_____ **16.** A type of axle that is not flexible, with splines on one end to fit the final drive unit and a flange on the other end to power the wheel.

_____ **17.** Two or more gears that are in mesh with each other.

_____ **18.** A type of rear-suspension system that allows each wheel on the axle to move independently of the other.

_____ **19.** The long, narrow component that carries one or more gears or has gears machined into it.

Match the following steps with the correct sequence for checking the fluid level of a manual transmission.

   A. Step 1   **B.** Step 2   **C.** Step 3

_____ **1.** If the transmission fluid begins to run out as the filler plug is removed, let the transmission fluid seek its own level before reinstalling the filler plug. The transmission fluid level should be at the bottom of the filler plug hole.

_____ **2.** If the fluid level is low, refill with the specified fluid, reinstall the filler plug, and wipe the area around the filler plug hole with a clean shop towel. Tighten the filler plug to the specified torque.

_____ **3.** Safely raise and support the vehicle on the lift so that it is level. Inspect the transmission for leaks. Remove the filler plug, using the proper tool. Inspect the filler plug and fill the hole for thread damage, and replace or repair if necessary.

Match the following steps with the correct sequence for checking and adjusting the differential/transfer case fluid level.

   A. Step 1   **B.** Step 2   **C.** Step 3

_____ **1.** Safely raise and support the vehicle. Inspect the differential and transfer case for leaks. Position a clean drain pan under the filler plug. Remove the filler plug, using the proper tool. Inspect the filler plug threads for thread, and replace if necessary.

_____ **2.** If the fluid level is low, refill with the specified fluid, reinstall the filler plug, and wipe the area around the filler plug hole with a clean shop towel. Tighten the filler plug to the specified torque.

_____ **3.** If the fluid begins to run out as the filler plug is removed, let the fluid seek its own level before reinstalling the filler plug. The fluid level should be at the bottom of the filler plug hole.

Match the following steps with the correct sequence for identifying the cause of fluid loss in a transmission.

   A. Step 1   **C.** Step 3   **E.** Step 5

   B. Step 2   **D.** Step 4   **F.** Step 6

_____ **1.** Look for leaks at the drain and fill plugs.

_____ **2.** Look for leaks in the rear tail shaft seal.

_____ **3.** Check the gear fluid level. If excess level is found, let the excess drain into a container. If the fluid is more than ¼″ (6 mm) below the bottom of the threaded hole, add new fluid of the correct viscosity and type.

_____ **4.** Safely raise and support the vehicle. Look for leaks in the transmission bell housing to the engine block (front seal).

_____ **5.** Look for leaks in the transmission breather outlet. Look for leaks in all case gasket areas.

_____ **6.** Inspect for transmission case defects, such as cracks and porosity.

Multiple Choice

Read each item carefully, and then select the best response.

_____ **1.** In 1894, Louis René Panhard and Émile Levassor designed a(n) _____.
   A. Rear-wheel drive carriage
   B. Synchromesh transmission
   C. Automatic transmission
   D. Multi-gear manual transmission

_____ **2.** Who connected an engine to a transmission and created a live rear axle by using a metal axle shaft supported by bushings in 1898?
   A. C. E. Duryea
   B. Louis Renault
   C. Henry Ford
   D. Ferdinand Porsche

_____ **3.** In 1928, Cadillac introduced the first _____.
   A. Synchromesh transmission
   B. Front-wheel drive automobile
   C. Planetary gear set
   D. Multi-gear transmission

_____ **4.** The amplification of the input force by trading distance moved for greater output force is the definition of _____.
   A. Ratio
   B. Mechanical advantage
   C. Power flow
   D. Rotational speed

_____ **5.** Given a 4:1 mechanical advantage, if a person pushes the lever down with a force of 100 lb, the force the lever generates against the rock is _____.
   A. 140 lb
   B. 40 lb
   C. 400 lb
   D. 4 lb

_____ **6.** If the drive gear has 15 teeth and the driven gear has 30 teeth, then the gear ratio is _____.
   A. 2:1
   B. 800 lb
   C. 1:4
   D. 4 lb

_____ **7.** The path in which power is transmitted through a series of components is called _____.
   A. Linkage
   B. Power flow
   C. Drivetrain
   D. Transmission

_____ **8.** What term relates to layouts where the transmission and final drive are integrated into a common assembly and usually used on front-wheel drive vehicles?
   A. Differential
   B. Drivetrain
   C. Torque converter
   D. Transaxle

_____ **9.** What transmits power from the transmission to the final drive assembly on rear-wheel drive vehicles?
   A. Clutch plate
   B. Transaxle
   C. Drive shaft
   D. Live axle

_____ **10.** Flexible drive axles are called _____.
- **A.** Radial axles
- **B.** Dead axles
- **C.** Half shafts
- **D.** Shafts

_____ **11.** What provides front-wheel drive axles the ability to change length as the vehicle goes over bumps and dips and allows for the wheels to be steered?
- **A.** Constant velocity joints
- **B.** Universal joints
- **C.** Drive shafts
- **D.** Synchronizers

_____ **12.** Which of the following components allows the axles to turn at different speeds when the vehicle is cornering or turning?
- **A.** Constant velocity joints
- **B.** Universal joints
- **C.** Differential assembly
- **D.** Transfer case

_____ **13.** If one wheel is stuck in the snow or ice, a(n) _____ allows both of the rear wheels to supply power to the ground in order to continue forward motion.
- **A.** Transfer case
- **B.** Limited slip differential
- **C.** Transaxle
- **D.** Open differential

_____ **14.** What term refers to grease's ability to normally be a solid, but while under stress to flow or become thin in order to lubricate properly?
- **A.** Thixotropy
- **B.** Viscosity
- **C.** Viscidity
- **D.** Glutinousness

_____ **15.** Fluid leaks may be a result of _____.
- **A.** Too little fluid in the transmission
- **B.** Proper fluid type
- **C.** Transmission case porosity or cracks
- **D.** Tight seals

_____ **16.** Prior to 1898, vehicles were either belt or chain driven.
- **A.** True
- **B.** False

_____ **17.** If the output force is four times greater than the input force, the input distance moved is two times greater than the output distance.
- **A.** True
- **B.** False

_____ **18.** As the gear ratio decreases, the output speed increases.
- **A.** True
- **B.** False

_____ **19.** The clutch pedal is used for acceleration or deceleration of the engine.
- **A.** True
- **B.** False

_____ **20.** The final drive assembly incorporates a set of differential gears arranged so they sit between the two axles.
- **A.** True
- **B.** False

_____ **21.** Shafts are used to support gears and are machined precisely to accommodate bearings and individual gears.
- **A.** True
- **B.** False

_____ 22. Splines allow shafts and gears to have greater rotational speeds, while at the same time minimizing metal-to-metal friction, thus contributing to increased fuel economy and longer transmission life.
   A. True
   B. False

_____ 23. The main difference between the transmission and the transaxle is that the transaxle incorporates the final drive assembly in its construction.
   A. True
   B. False

_____ 24. Transmissions/transaxles are rated by the manufacturers for how much twisting force, measured in foot-pounds, they can handle.
   A. True
   B. False

_____ 25. All manual transmissions use the same type of fluid.
   A. True
   B. False

_____ 26. A(n) _____ applies a friction device between the gear and the shaft to match the gear speed to the shaft speed.
   A. Accelerator pedal
   B. Clutch system
   C. Final drive assembly
   D. Gear synchronizer

_____ 27. A gear set in which the _____ (drive) gear has half as many teeth as the _____ (driven) gear has a gear ratio of 2:1.
   A. Input, output
   B. Output, input
   C. Forward, reverse
   D. Reverse, forward

_____ 28. Driving a large gear with a smaller gear results in a gear _____.
   A. Alignment
   B. Reduction
   C. Release
   D. Rotation

_____ 29. The _____ system is the medium by which the driver can connect and disconnect the engine from the transmission, resulting in the vehicle's forward or rearward movement.
   A. 4WD
   B. Clutch
   C. Traction control
   D. Brake

_____ 30. The final _____ gives the final gear reduction to the drivetrain and powers the drive wheels through axles.
   A. Gear set
   B. Driveshaft
   C. Transfer case
   D. Drive assembly

_____ 31. The drive shaft uses _____, which allow the drive shaft to change angles due to the movement of the suspension relative to the body.
   A. Universal joints
   B. Constant velocity joints
   C. Brake pedals
   D. Differential gears

_____ **32.** Both front-wheel drive and rear-wheel drive vehicles use _____ to power the wheels.
  A. Solid axles
  B. Dead axles
  C. Drive axles
  D. Gear sets

_____ **33.** Depressing the clutch pedal closes the contacts of the _____ switch.
  A. Input
  B. Clutch safety
  C. Dimmer
  D. Backup light

_____ **34.** _____ are round parts with teeth cut on the outside perimeter.
  A. Gears
  B. Pedals
  C. Axles
  D. Drivetrains

_____ **35.** In a manual transmission vehicle, the component that locks the engine and transmission together is the _____ system.
  A. Input
  B. Axle
  C. Shaft
  D. Clutch

_____ **36.** In a four-wheel drive vehicle, the _____ takes the power from the transmission and directs it to one or both axles, depending on the mode selected.
  A. Gear set
  B. Gear synchronizer
  C. Half shaft
  D. Transfer case

_____ **37.** With the _____ assembly, if one wheel is stuck in the snow or ice, the other wheel cannot supply power.
  A. Clutch
  B. Drive axle
  C. Open differential
  D. Final drive

_____ **38.** The _____ drive axle uses one half-shaft axle for each of the two wheels.
  A. Independent suspension
  B. Solid
  C. Open
  D. Nonflexible

_____ **39.** A _____ allows the wheels to rotate freely on the axle assembly and to not drive the wheels.
  A. Solid axle
  B. Dead axle
  C. Live axle
  D. Semi-floating axle

_____ **40.** _____ maintenance is designed to extend the service life of the transmission, improve the reliability of the vehicle, and help prevent vehicle breakdowns that could leave the driver and passengers stranded.
  A. Extension
  B. Assistive
  C. Preventive
  D. Annual

_____ **41.** The arrow in the image is pointing to what part of the transmission?

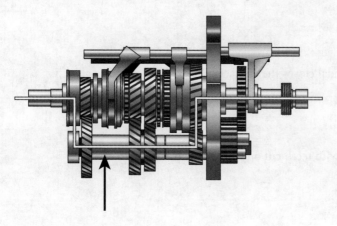

   A. Counter shaft
   B. Input shaft
   C. First gear
   D. Output shaft

_____ **42.** The arrow in the image is pointing to what part of the transmission?

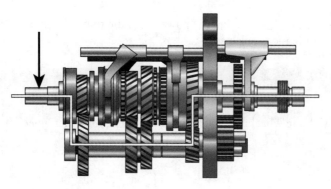

   A. Counter shaft
   B. First gear
   C. Input shaft
   D. Output shaft

_____ **43.** The arrow in the image is pointing to what part of the transmission?

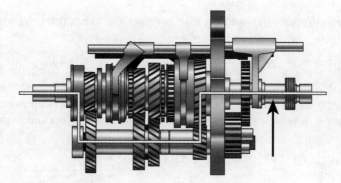

   A. First gear
   B. Counter shaft
   C. Output shaft
   D. Input shaft

_____ **44.** The arrow in the image is pointing to what part of the final drive assembly?

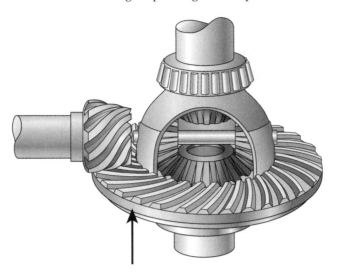

    A. First gear
    B. Ring gear
    C. Pinion gear
    D. Spider gear

_____ **45.** The arrows in the image are pointing to what part of the final drive assembly?

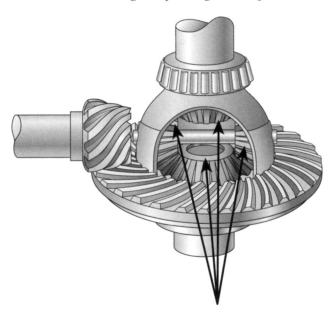

    A. First gears
    B. Pinion gears
    C. Ring gears
    D. Spider gears

_____ **46.** The arrow in the image is pointing to what part of the final drive assembly?

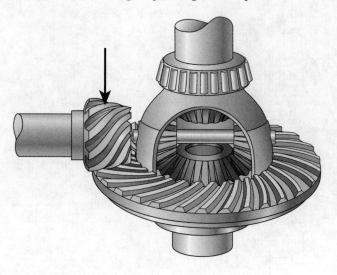

A. Spider gear
B. Pinion gear
C. First gear
D. Ring gear

_____ **47.** The arrow in the image is pointing to what part of the drivetrain?

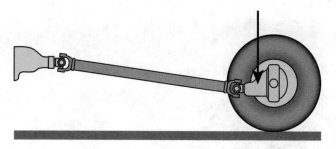

A. Universal joint
B. Transmission
C. Drive shaft
D. Final drive assembly

_____ **48.** The arrow in the image is pointing to what part of the drive train?

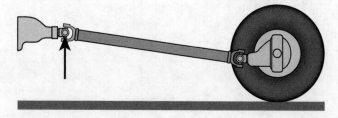

A. Drive shaft
B. Universal joint
C. Final drive assembly
D. Transmission

_____ **49.** The arrow in the image is pointing to what part of the drivetrain?

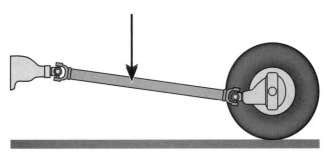

- **A.** Drive shaft
- **B.** Final drive assembly
- **C.** Transmission
- **D.** Universal joint

_____ **50.** The arrow in the image is pointing to what part of the axial load?

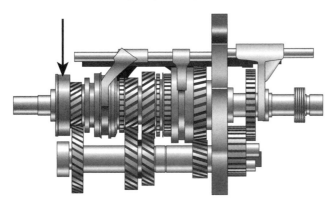

- **A.** Thrust bearing
- **B.** Spur gear
- **C.** Thrust washer
- **D.** Helical gear

_____ **51.** The arrows in the image are pointing to what part of the axial load?

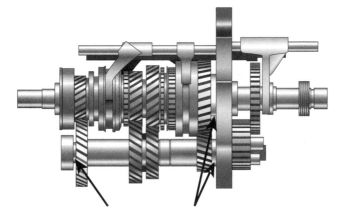

- **A.** Spur gears
- **B.** Helical gears
- **C.** Thrust bearings
- **D.** Thrust washers

_____ **52.** The arrow in the image is pointing to what part of the clutch components?

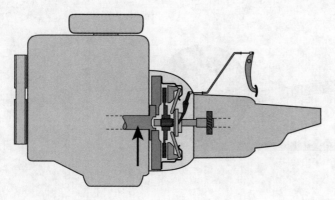

    A. Clutch
    B. Crankshaft
    C. Engine
    D. Clutch pedal

_____ **53.** The arrow in the image is pointing to what part of the clutch components?

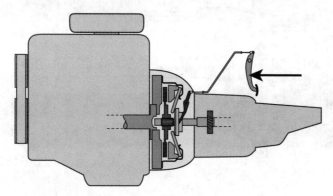

    A. Crankshaft
    B. Engine
    C. Clutch pedal
    D. Transmission input shaft

_____ **54.** The arrow in the image is pointing to what part of the clutch components?

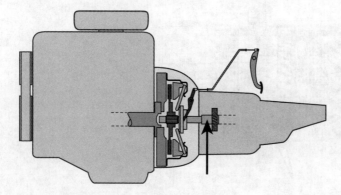

    A. Clutch
    B. Transmission
    C. Transmission input shaft
    D. Crankshaft

_____ **55.** Tech A says that friction bearings are made up of balls and rollers. Tech B says that nonfriction bearings are in sliding contact between moving surfaces. Who is correct?
   A. Tech A
   B. Tech B
   C. Both Tech A and Tech B
   D. Neither Tech A nor Tech B

_____ **56.** Tech A says that gear ratios are all the same in any five-speed transmission. Tech B says that gear ratios vary from transmission to transmission. Who is correct?
   A. Tech A
   B. Tech B
   C. Both Tech A and Tech B
   D. Neither Tech A nor Tech B

_____ **57.** Tech A says that gear lube can be used in all manual transmissions. Tech B says that some manual transmissions use engine oil as a lubricant. Who is correct?
   A. Tech A
   B. Tech B
   C. Both Tech A and Tech B
   D. Neither Tech A nor Tech B

_____ **58.** Tech A says that the final drive is used to provide an increase in the rotational speed of the axles. Tech B says that the final drive is used to provide an increase in twisting force to the axles. Who is correct?
   A. Tech A
   B. Tech B
   C. Both Tech A and Tech B
   D. Neither Tech A nor Tech B

_____ **59.** Tech A says that a gear set that has a drive gear with nine teeth and a driven gear with 27 teeth has a gear ratio of 3:1. Tech B says that the drive gear is also called the output gear. Who is correct?
   A. Tech A
   B. Tech B
   C. Both Tech A and Tech B
   D. Neither Tech A nor Tech B

_____ **60.** Tech A says that transmission fluid levels are critical to the life of the transmission. Tech B says that transmission oil should be changed when engine oil is changed. Who is correct?
   A. Tech A
   B. Tech B
   C. Both Tech A and Tech B
   D. Neither Tech A nor Tech B

_____ **61.** Tech A says that the clutch uses a cone-style synchronizer to match the speed of the engine to the manual transmission. Tech B says that the fill plug hole is where the fluid level should be checked in most manual transmissions. Who is correct?
   A. Tech A
   B. Tech B
   C. Both Tech A and Tech B
   D. Neither Tech A nor Tech B

_____ **62.** Tech A says that the differential assembly provides a means for the inside and outside wheels to turn at different speeds when going around a corner. Tech B says that the differential assembly provides smooth shifts and reduces gear "grinding" by matching gear speeds. Who is correct?
   A. Tech A
   B. Tech B
   C. Both Tech A and Tech B
   D. Neither Tech A nor Tech B

The Clutch System

At the start of each chapter, you will find the Learning Objectives from the textbook. These are your objectives as you make your way through the exercises in this workbook and the chapter in your textbook. The following activities have been designed to help you refresh your knowledge of the material in this chapter.

Learning Objectives

After reading this chapter, you will be able to:

- LO 22-01 Describe clutch principles.
- LO 22-02 Describe the purpose and design of flywheels.
- LO 22-03 Describe pressure plates and clutch discs.
- LO 22-04 Describe throw-out bearing, clutch fork, and pilot bearing.
- LO 22-05 Describe clutch operating mechanisms.
- LO 22-06 Perform clutch maintenance.

Matching

Match the following terms with the correct description or example.

A. Carrier
B. Clutch disc
C. Clutch fork
D. Coefficient of friction
E. Coil-spring pressure plate
F. Diaphragm pressure plate
G. Flywheel
H. Free play
I. Friction facing
J. Fulcrum ring

K. Pilot bearing
L. Push-type clutch
M. Pressure plate
N. Quadrant ratchet
O. Release mechanisms
P. Single-plate clutch
Q. Slave cylinder
R. Throw-out bearing
S. Torsional vibrations
T. Transmission input shaft

_____ 1. The bearing or bushing that supports the front of the transmission input shaft.

_____ 2. A heavy metal disc bolted to the crankshaft that is used to smooth out the engine's power pulses and keep the engine moving through the non-power strokes.

_____ 3. A type of pressure plate that uses coil springs to provide the clamping force.

_____ 4. The part of the clutch release mechanism that imparts clutch pedal force to the rotating pressure plate levers.

_____ 5. The material riveted to each side of the clutch disc that mates to the flywheel and pressure plate. Used to provide friction and a wear surface for the clutch assembly.

_____ 6. The speeding up and slowing down of a shaft, which happen at a relatively high frequency.

_____ 7. The device used in some cable-operated clutches to provide self-adjustment as the clutch disc wears.

_____ 8. The component in a hydraulically operated clutch that converts hydraulic pressure to mechanical movement at the clutch fork.

_____ 9. The center component of the clutch assembly, with friction material riveted on each side. Also called a clutch plate or friction disc.

_____ 10. Components that operate the clutch. Usually included are the throw-out bearing and the clutch fork. Some manufacturers include the operating system.

_____ 11. A slightly conical, spring steel plate used to provide the clamping force for the clutch assembly.

_____ 12. The part of the throw-out bearing assembly that holds the bearing.

_____ 13. A clutch assembly that uses only one plate to transfer torque from the engine to the transmission. This is the most common type of light vehicle clutch.

_____ 14. The part of the clutch linkage that operates the throw-out bearing.

_____ 15. The amount of clearance in the clutch release mechanism as measured at the clutch pedal.

_____ 16. The shaft that brings engine torque into the transmission.

_____ 17. A steel ring that is used as a pivot point for the diaphragm spring in the pressure plate.

_____ 18. The amount of resistance to movement between any two surfaces that are in contact with each other.

_____ 19. The assembly that applies and removes the clamping force on the clutch disc.

_____ 20. A typical clutch system used in modern vehicles where the clutch fork pushes the release bearing forward to release the friction facing from the pressure plate.

Multiple Choice

Read each item carefully, and then select the best response.

_____ 1. The amount of torque a clutch can transmit is dependent on all of the following, EXCEPT _____.
   A. Diameter of the clutch
   B. Coefficient of friction
   C. Total spring force
   D. Thickness of the clutch material

_____ 2. The _____ is able to connect to and disconnect from the engine's flywheel through the operation of the clutch.
   A. Crankshaft
   B. Input shaft
   C. Pilot bearing
   D. Output shaft

_____ 3. The main purpose of the _____ is to smooth out the power pulses from the pistons during the power strokes.
   A. Pressure plate
   B. Throw-out bearing
   C. Flywheel
   D. Clutch disc

_____ 4. Refinishing the flywheel moves the pressure plate toward the engine and away from the throw-out bearing, increasing _____.
   A. Free play
   B. Clutch chatter
   C. Runout
   D. Fuel economy

_____ 5. The snout of the input shaft rides on the _____.
   A. Throw-out bearing
   B. Hub
   C. Pilot bearing
   D. Flywheel

_____ 6. Which of the following is a type of clutch operating mechanism?
   A. Linkage style
   B. Multi-plate
   C. Throw-out style
   D. Diaphragm pressure

_____ 7. The pilot bearing is the front bearing, and at the other end of the input shaft is the _____ bearing.
   A. Passenger
   B. Transmission input
   C. Clutch release
   D. Transmission output

_____ **8.** If the clutch pedal has too little _____, the throw-out bearing could remain in contact with the pressure plate levers.
   A. End play
   B. Spring pressure
   C. Free play
   D. Diaphragm pressure

_____ **9.** The _____ method is used to properly bleed the hydraulic clutch system.
   A. Upward bleeding
   B. Forward bleeding
   C. Pressure bleeding
   D. Assistive bleeding

_____ **10.** Automotive manual transmission clutches are wet clutches.
   A. True
   B. False

_____ **11.** Both output shaft torque and speed can be increased at the same time.
   A. True
   B. False

_____ **12.** The output shaft is connected directly to the wheels through the drivetrain components and cannot be disconnected.
   A. True
   B. False

_____ **13.** The function of the dual mass flywheel is to absorb torsional crankshaft vibrations.
   A. True
   B. False

_____ **14.** Coil-spring pressure plate clutches require less pedal effort to operate.
   A. True
   B. False

_____ **15.** The clutch throw-out bearing and clutch fork work together to compress the pressure plate springs when the clutch pedal is pressed.
   A. True
   B. False

_____ **16.** Some vehicles do not require a pilot bearing.
   A. True
   B. False

_____ **17.** In a cable-style clutch control system, the slave cylinder is located directly behind the throw-out bearing and pushes directly on it.
   A. True
   B. False

_____ **18.** When initially placing the transmission into gear, the transmission input shaft must be disconnected from the engine to prevent gear grinding.
   A. True
   B. False

_____ **19.** Old fluid in the hydraulic system can cause master cylinder and slave cylinder damage.
   A. True
   B. False

_____ **20.** The _____ allows the driver to engage and disengage the engine from the transmission while operating the vehicle.
   A. Cable
   B. Clutch
   C. Pressure plate
   D. Starter ring gear

_____ **21.** Two or more clutch plates can be used to form a _____ clutch, increasing the number of facings and the torque capacity.
   A. Multi-plate
   B. Push-type
   C. Release
   D. Starter

_____ **22.** The _____ bolts onto the flywheel.
   A. Clutch disc
   B. Driven center plate
   C. Pressure plate
   D. Fulcrum ring

_____ **23.** The flywheel _____ enables the starter motor drive gear to crank the engine over.
   A. Carrier
   B. Ring gear
   C. Pressure plate
   D. Bearing

_____ **24.** The _____ flywheel improves the engine's fuel economy by smoothing out the power pulses and focusing them in the direction of engine rotation.
   A. One-piece
   B. Throw-out
   C. Dual mass
   D. Steel ring

_____ **25.** The diaphragm pressure plate is located inside the clutch cover on two _____, held in place by a number of rivets passing through the diaphragm.
   A. Flywheels
   B. Throw-out bearings
   C. Discs
   D. Fulcrum rings

_____ **26.** The clutch disc is also called a _____ or a friction disc.
   A. Driven center plate
   B. Coil-spring pressure plate
   C. Diaphragm pressure plate
   D. Transmission input plate

_____ **27.** The throw-out bearing carrier slides on the sleeve of the front _____ that extends from the front of the transmission.
   A. Bearing retainer
   B. Clutch disc
   C. Clutch fork
   D. Pressure plate

_____ **28.** The clutch fork _____ is generally screwed into the bell housing and is usually replaceable.
   A. Disc
   B. Pivot
   C. Plate
   D. Sleeve

_____ **29.** While the vehicle is in motion, the transmission of engine _____ must be interrupted for shifting of the gears.
   A. Pressure
   B. Friction
   C. Power
   D. Torque

_____ **30.** Some vehicles use a quadrant _____, which automatically adjusts the clutch pedal free play as needed when the pedal is lifted by the driver's toe.
  A. Disc
  B. Plate
  C. Ratchet
  D. Sleeve

_____ **31.** In hydraulic clutch release mechanisms, the clutch pedal acts on a master cylinder connected by a hydraulic tube and flexible hose to a _____ cylinder mounted on (external type) or in (internal type) the transmission bell housing.
  A. Slave
  B. Spring
  C. Front bearing
  D. Center

_____ **32.** Older technology clutch operating systems used a series of levers with an equalizing mechanism called a _____.
  A. Coil spring
  B. Pressure plate
  C. Fulcrum ring
  D. Bell crank

_____ **33.** As the clutch wears, the _____ becomes thinner.
  A. Clutch fork
  B. Starter disc
  C. Friction disc
  D. Fulcrum disc

_____ **34.** The arrow in the image is pointing to what part of the light vehicle clutch?

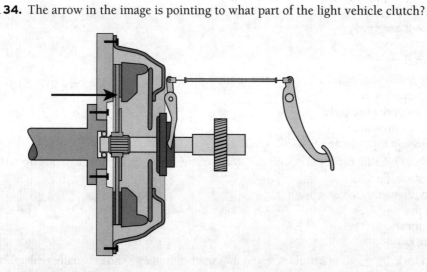

  A. Central hub
  B. Clutch disc
  C. Clutch cover
  D. Bolt

_____ **35.** The arrow in the image is pointing to what part of the light vehicle clutch?

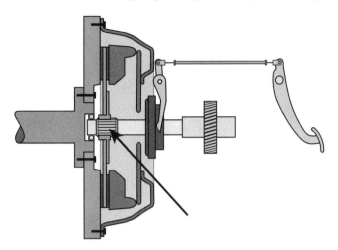

    A. Friction facings
    B. Diaphragm springs
    C. Throw-out bearings
    D. Splines

_____ **36.** The arrow in the image is pointing to what part of the light vehicle clutch?

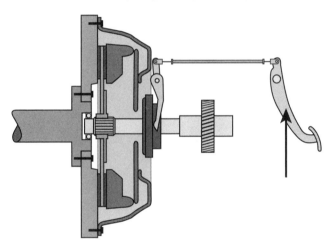

    A. Crankshaft
    B. Pressure plate
    C. Clutch fork
    D. Clutch pedal

_____ **37.** The arrows in the image are pointing to what part of the dual mass flywheel?

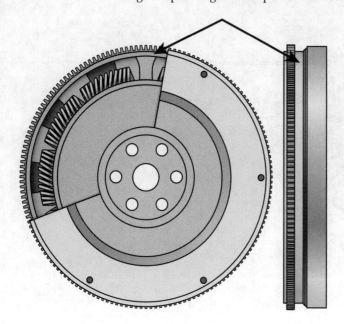

   A. Secondary flywheels
   B. Primary flywheels
   C. Planetary gears
   D. Ring gears

_____ **38.** The arrows in the image are pointing to what part of the dual mass flywheel?

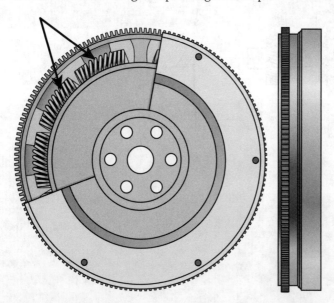

   A. Planetary gears
   B. Torsion springs
   C. Ring gears
   D. Cushions

_____ **39.** The arrow in the image is pointing to what part of the dual mass flywheel?

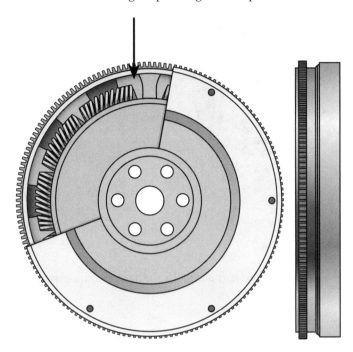

A. Secondary flywheel
B. Planetary gear
C. Ring gear
D. Cushion

_____ **40.** The arrow in the image is pointing to what part of the diaphragm pressure plate?

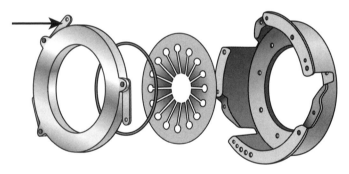

A. Drive strap
B. Machined surface
C. Pressure plate
D. Cover

_____ **41.** The arrow in the image is pointing to what part of the diaphragm pressure plate?

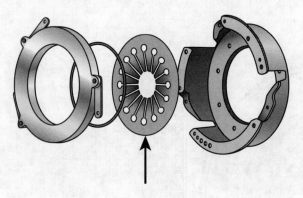

A. Pressure plate
B. Fulcrum ring
C. Diaphragm spring
D. Cover

_____ **42.** The arrow in the image is pointing to what part of the diaphragm pressure plate?

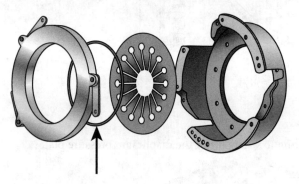

A. Drive strap
B. Pressure plate
C. Fulcrum ring
D. Cover

_____ **43.** The arrow in the image is pointing to what part of the coil-spring pressure plate?

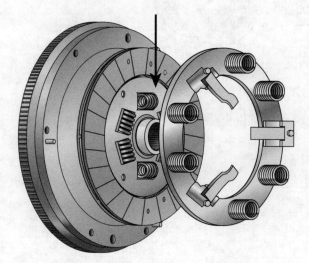

A. Flywheel
B. Clutch plate
C. Pressure plate
D. Release lever

_____ **44.** The arrow in the image is pointing to what part of the coil-spring pressure plate?

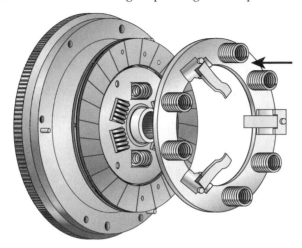

    A. Clutch plate
    B. Flywheel
    C. Coil spring
    D. Pressure plate

_____ **45.** The arrow in the image is pointing to what part of the coil-spring pressure plate?

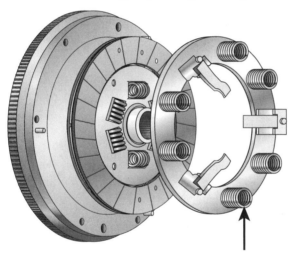

    A. Flywheel
    B. Clutch plate
    C. Release lever
    D. Coil spring

_____ **46.** The arrow in the image is pointing to what part of the clutch disc?

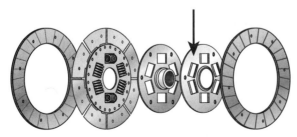

    A. Friction facing
    B. Central plate
    C. Hub
    D. Hub cover

_____ **47.** The arrow in the image is pointing to what part of the clutch disc?

    A. Waved spring
    B. Central plate
    C. Hub
    D. Torsional spring

_____ **48.** The arrows in the image are pointing to what part of the clutch disc?

    A. Friction facings
    B. Waved springs
    C. Central plates
    D. Torsional springs

_____ **49.** The arrow in the image is pointing to what part of the waved spring?

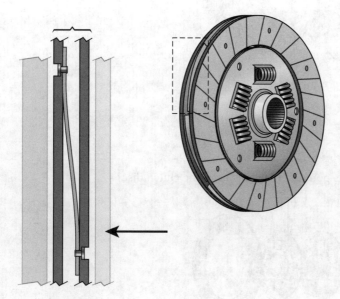

    A. Flywheel
    B. Rivet
    C. Pressure plate
    D. Friction facing

_____ **50.** The arrow in the image is pointing to what part of the waved spring?

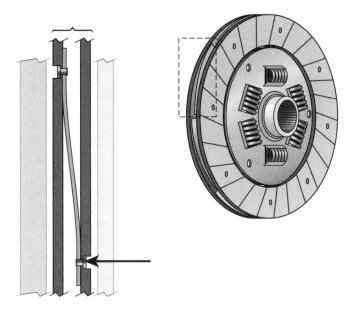

A. Flywheel
B. Rivet
C. Friction facing
D. Waved spring

_____ **51.** The arrow in the image is pointing to what part of the waved spring?

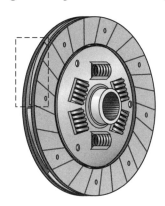

A. Flywheel
B. Rivet
C. Pressure plate
D. Friction facing

_____ **52.** The arrow in the image is pointing to what part of the cable-operated clutch?

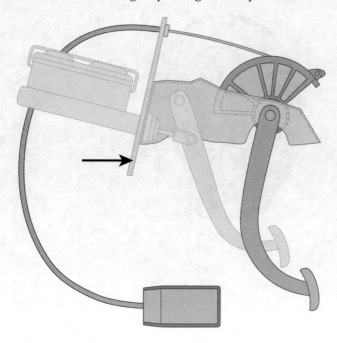

    A. Firewall
    B. Pawl
    C. Brake pedal
    D. Clutch pedal

_____ **53.** The arrow in the image is pointing to what part of the cable-operated clutch?

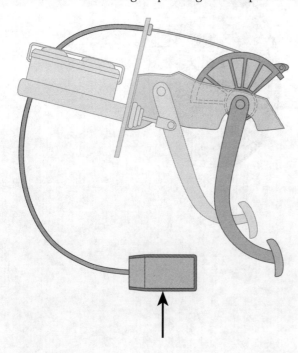

    A. Master cylinder
    B. Pedal support
    C. Dust cover
    D. Brake pedal

_____ **54.** The arrow in the image is pointing to what part of the cable-operated clutch?

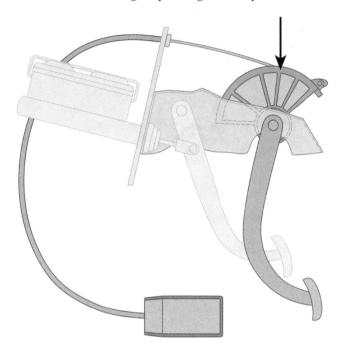

    A. Dust cover
    B. Firewall
    C. Clutch cable
    D. Quadrant gear

_____ **55.** The arrow in the image is pointing to what part of the hydraulic clutch control?

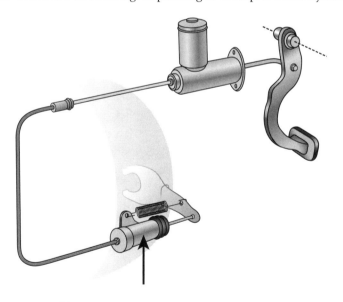

    A. Metal line
    B. Master cylinder
    C. Reservoir
    D. Slave cylinder

_____ **56.** The arrow in the image is pointing to what part of the hydraulic clutch control?

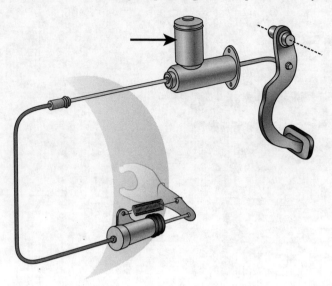

   A. Master cylinder
   B. Reservoir
   C. Clutch housing
   D. Slave cylinder

_____ **57.** The arrow in the image is pointing to what part of the hydraulic clutch control?

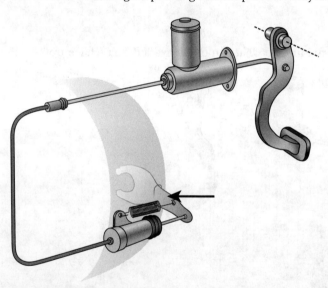

   A. Rubber hose
   B. Metal line
   C. Clutch fork
   D. Return spring

_____ **58.** The arrow in the image is pointing to what part of the linkage-operated system?

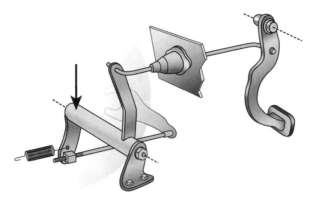

- **A.** Return spring
- **B.** Bell crank
- **C.** Clutch push rod
- **D.** Release rod

_____ **59.** The arrow in the image is pointing to what part of the linkage-operated system?

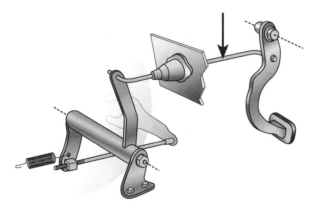

- **A.** Return spring
- **B.** Bell crank
- **C.** Clutch pushrod
- **D.** Release rod

_____ **60.** The arrow in the image is pointing to what part of the linkage-operated system?

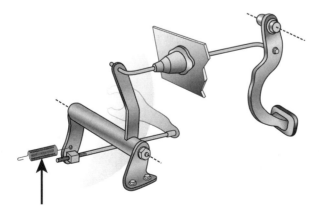

- **A.** Return spring
- **B.** Bell crank
- **C.** Firewall
- **D.** Clutch pedal

_____ **61.** Tech A says that the pressure plate friction surface rides on the flywheel friction surface to transmit torque. Tech B says that the flywheel can either be flat or stepped. Who is correct?
   A. Tech A
   B. Tech B
   C. Both Tech A and Tech B
   D. Neither Tech A nor Tech B

_____ **62.** Tech A says that insufficient clutch pedal clearance (free play) can cause gear clashing when shifting. Tech B says that when the engine is idling and the clutch pedal is released, the friction disc should stop rotating. Who is correct?
   A. Tech A
   B. Tech B
   C. Both Tech A and Tech B
   D. Neither Tech A nor Tech B

_____ **63.** Tech A says a dual mass flywheel improves the engine's fuel economy by smoothing out the power pulses and focusing them in the direction of engine rotation. Tech B says there are two types of dual mass flywheel. Who is correct?
   A. Tech A
   B. Tech B
   C. Both Tech A and Tech B
   D. Neither Tech A nor Tech B

_____ **64.** Tech A says that on larger vehicles, such as trucks, a ball bearing type may be used for a pilot bearing. Tech B says that in some applications, pilot bearings may be placed or pressed into the center of the flywheel. Who is correct?
   A. Tech A
   B. Tech B
   C. Both Tech A and Tech B
   D. Neither Tech A nor Tech B

_____ **65.** Tech A says that there are three main types of mechanical systems used for clutch operation. Tech B says that a hydraulically operated system is also used in some applications. Who is correct?
   A. Tech A
   B. Tech B
   C. Both Tech A and Tech B
   D. Neither Tech A nor Tech B

_____ **66.** Tech A says that it is important to check the clutch linkage mechanism for proper operation and correct the adjustment (free play) periodically. Tech B says that if the clutch pedal has too little free play, the throw-out bearing could remain in contact with the pressure plate levers. Who is correct?
   A. Tech A
   B. Tech B
   C. Both Tech A and Tech B
   D. Neither Tech A nor Tech B

_____ **67.** Tech A says that the pilot bearing can be a needle-style bearing. Tech B says that the pilot bearing can be a brass bushing style. Who is correct?
   A. Tech A
   B. Tech B
   C. Both Tech A and Tech B
   D. Neither Tech A nor Tech B

_____ **68.** Tech A says that when bleeding a clutch hydraulic system, one method uses pressure or vacuum to push or pull fluid and air from the system. Tech B says that this method does not require special bleeding tools or equipment. Who is correct?
   A. Tech A
   B. Tech B
   C. Both Tech A and Tech B
   D. Neither Tech A nor Tech B

_____ **69.** Tech A says that in a mechanical system, if the clutch pedal adjustment has too much free play, there may not be enough travel to fully release the pressure plate. Tech B says that this condition can cause the gears to clash (grind) when shifting and result in heavy synchronizer wear. Who is correct?
   A. Tech A
   B. Tech B
   C. Both Tech A and Tech B
   D. Neither Tech A nor Tech B

_____ **70.** Tech A says that diaphragm pressure plates require more pedal effort than coil-spring pressure plates. Tech B says that the two types of diaphragm pressure plates require different free pedal adjustment procedures. Who is correct?
   A. Tech A
   B. Tech B
   C. Both Tech A and Tech B
   D. Neither Tech A nor Tech B

Driveshafts, Axles, and Final Drives

At the start of each chapter, you will find the Learning Objectives from the textbook. These are your objectives as you make your way through the exercises in this workbook and the chapter in your textbook. The following activities have been designed to help you refresh your knowledge of the material in this chapter.

Learning Objectives

After reading this chapter, you will be able to:

- LO 23-01 Describe drivetrain layout.
- LO 23-02 Inspect and service driveshafts.
- LO 23-03 Describe the operation and maintenance of final drives and differentials.
- LO 23-04 Service axles.
- LO 23-05 Service universal (U-) joints and constant-velocity (CV) joints.

Matching

Match the following terms with the correct description or example.

A. Backlash
B. Companion flange
C. CV joint
D. Dead axle
E. Full-floating axle
F. Hypoid bevel gear
G. Live axle
H. Longitudinal
I. Power take-off (PTO)

J. Rzeppa joint
K. Side gear
L. Speedy sleeve
M. Torque steer
N. Transfer case
O. Transverse
P. Tulip/tripod joint
Q. Universal joint (U-joint)

_____ **1.** An aftermarket repair kit that consists of a thin metal sleeve that fits tightly over the seal surface of the axle, providing a new, undamaged surface for the seal to ride against.

_____ **2.** A CV joint that has three equally spaced fingers shaped like a star. This configuration allows in-and-out movement of the shaft while allowing flexing.

_____ **3.** A device attached to the transmission that is gear driven and can be used to run accessories such as winches and towing equipment. It can also refer to the gears that send power to the rear axle in a predominantly front-wheel drive vehicle.

_____ **4.** An axle that does not support any weight; if removed, the vehicle will still roll on its wheels.

_____ **5.** A term used to describe the side-to-side engine orientation when mounted in the engine compartment.

_____ **6.** A flexible cross-shaped joint used to transmit torque.

_____ **7.** A special design of spiral bevel gears where the centerline of the pinion is below the centerline of the ring gear.

_____ **8.** An axle that does not have the capability to drive the vehicle. It is usually found on the rear of front-wheel drive vehicles.

_____ **9.** The orientation of the engine in which the front of the engine is facing the front of the vehicle. It is most commonly found in rear-wheel drive vehicles.

_____ **10.** A type of fixed CV joint that has an inner race, six steel ball bearings, a bearing cage, and an outer race.

_____ **11.** A splined flange attached to a vehicle component, such as a drive axle pinion shaft, that bolts to a flange.

_____ **12.** A gear that is splined to the axle shaft and meshes with the spider gears and that allows the axles to rotate at their own speeds when cornering and turning.

_____ **13.** The required clearance between two meshing gears.

_____ **14.** An axle that is powered and can move the vehicle. It is usually found on the rear of rear-wheel drive vehicles.

_____ **15.** A condition in which the vehicle pulls to one side during hard acceleration.

_____ **16.** A joint used to transmit torque through wider angles and without the change of velocity that occurs in U-joints.

_____ **17.** A component that is bolted to the back of the transmission and connects the front and rear axles via the driveshafts.

Match the following steps with the correct sequence for diagnosing CV joint issues.

A. Step 1   **C.** Step 3
B. Step 2   **D.** Step 4

_____ **1.** Thoroughly road test the vehicle to verify the customer complaint. Safely raise the vehicle on an approved lift, and make sure it is secure.

_____ **2.** Look for any type of axle damage, such as a bent axle from a recent accident.

_____ **3.** Visually inspect all four CV joints. Look for broken or ruptured boots. Look for cracked or dry-rotted boots, as they are subject to all types of weather.

_____ **4.** Manually move the axle shaft up and down, looking for unnecessary play or bad bearings.

Match the following steps with the correct sequence for inspecting half-shaft components.

A. Step 1   **C.** Step 3
B. Step 2   **D.** Step 4

_____ **1.** Inspect the old joint to gain an accurate assessment of the failure to prevent a reoccurrence of this failure. Reinstall the new CV joint onto the shaft splines as required.

_____ **2.** Wipe out as much grease as possible to be able to access the retaining ring from the CV joint itself. If there is a retaining ring present, remove the retaining ring with the appropriate tool. When the retaining ring is removed, remove the CV joint from the half shaft, and inspect the splines on the end of the half shaft. This also applies to the other end of the half shaft.

_____ **3.** Clamp the entire half shaft into a soft-jawed vise, and make sure it is secure. Remove the retaining clamps from the CV boot. Slide the boot down the shaft, paying attention to the condition of the boot.

_____ **4.** Apply the lubrication grease that comes with the new CV joint. Tighten the boot clamp to ensure that grease will not be lost. This applies to both sides of the half shaft. Reinstall the half-shaft, following the manufacturer's guidelines.

Match the following steps with the correct sequence for inspecting fluid leakage.

A. Step 1   **D.** Step 4
B. Step 2   **E.** Step 5
C. Step 3   **F.** Step 6

_____ **1.** If necessary, remove the rear wheels and install a dial indicator on the axle flange to check for any distortion.

_____ **2.** Put the vehicle on an approved lift, and make sure it is secure. Visually inspect around the housing where the axle seats for any seepage.

_____ **3.** Check the differential rear cover for a leaking gasket, if so equipped, and tighten if loose or replace as necessary.

_____ **4.** Check the fluid level for lack of fluid or overfilling of the differential, as either one can indicate that a leak is present.

_____ **5.** Inspect the pinion flange for any seepage.

_____ **6.** Check and clean the breather or vent for any obstructions that may cause a pressure buildup to occur.

Match the following steps with the correct sequence for removing and reinstalling sealed wheel bearings using the unitized wheel bearing hub style.

A. Step 1   **D.** Step 4

B. Step 2   **E.** Step 5

C. Step 3   **F.** Step 6

_____ **1.** If necessary, remove the rear wheels and install a dial indicator on the axle flange to check for any distortion.

_____ **2.** Put the vehicle on an approved lift, and make sure it is secure. Visually inspect around the housing where the axle seats for any seepage.

_____ **3.** Check the differential rear cover for a leaking gasket, if so equipped, and tighten if loose or replace as necessary.

_____ **4.** Check the fluid level for lack of fluid or overfilling of the differential, as either one can indicate that a leak is present.

_____ **5.** Inspect the pinion flange for any seepage.

_____ **6.** Check and clean the breather or vent for any obstructions that may cause a pressure buildup to occur.

Multiple Choice

Read each item carefully, and then select the best response.

_____ **1.** The _____ slides in and out on the output shaft, allowing the length changes needed as the suspension moves up and down.
   A. Slip yoke
   B. U-joint
   C. Half shaft
   D. Speedy sleeve

_____ **2.** A two-piece driveshaft, joined in the middle with splines that can slide on itself to increase and decrease in length, is called a _____.
   A. Half shaft
   B. Full-floating driveshaft
   C. Sliding spline driveshaft
   D. Rzeppa joint

_____ **3.** The _____ is a device that transfers torque from one component to another, such as from the transmission output shaft to the final drive assembly.
   A. Half shaft
   B. Driveshaft
   C. Drive axle
   D. CV joint

_____ **4.** The _____ does not provide any drive capabilities and is used on the rear of front-wheel drive vehicles, and the front of rear-wheel drive vehicles.
   A. Dead axle
   B. Half shaft
   C. Drive axle
   D. Driveshaft

_____ **5.** What type of rear-wheel drive solid axle has a single roller bearing between the hub and the outside of the axle housing?
   A. Dead axle
   B. Semi-floating axle
   C. Three-quarter floating axle
   D. Full-floating axle

_____ **6.** A(n) _____ can be of the single-lip or double-lip design depending on the application.
   A. Flange
   B. Yoke
   C. Speedy sleeve
   D. Axle seal

_____ **7.** The use of different-sized axle shafts can cause a problem known as _____.
   A. Misalignment
   B. Torque steer
   C. Slip differential
   D. Angularity

_____ **8.** What type of CV joint does not slide to allow for shaft lengthening or shortening; it simply allows for angle changes as the suspension moves?
   A. Fixed-type joint
   B. Rzeppa joint
   C. Plunge-type joint
   D. Tulip tripod joint

_____ **9.** The _____ differential is a gear-style limited slip differential design that multiplies the torque available from the wheel that is losing traction and turns it over to the slower-turning wheel with better traction.
   A. Clutch pack
   B. Rzeppa
   C. Torsen style
   D. Viscous coupling

_____ **10.** Longitudinally mounted means that the front of the engine is facing the side of the vehicle.
   A. True
   B. False

_____ **11.** On some all-wheel drive vehicles, only two of the wheels are normally powered, typically the front wheels.
   A. True
   B. False

_____ **12.** The purpose of the drive axle is to transfer the torque that comes from the engine, transmission, driveshaft, and differential to the wheels and tires, propelling the vehicle forward or backward.
   A. True
   B. False

_____ **13.** If a vehicle with a semi-floating axle were to hit a curb and break or snap the axle shaft, the wheel would come off the vehicle.
   A. True
   B. False

_____ **14.** Full-floating axles can be removed without removing the wheels.
   A. True
   B. False

_____ **15.** A double Cardan joint greatly increases the change in velocity compared with a single Cardan joint.
   A. True
   B. False

_____ **16.** U-joints are larger than CV joints.
   A. True
   B. False

_____ **17.** If you hold the ring gear and turn the pinion, you will have a slight clearance back and forth.
   A. True
   B. False

_____ **18.** On part-time four-wheel drive vehicles, differentials are fitted to both the front and rear axle assemblies.
   A. True
   B. False

_____ **19.** A transfer case is used on vehicles that are primarily front-wheel drive; a PTO is used on vehicles that are primarily rear-wheel drive.
   A. True
   B. False

_____ **20.** _____ joints allow for smoother transfer of power and allow for the vehicle to turn more tightly without the joint binding.
  A. Fixed-type
  B. Double Cardan
  C. CV
  D. Universal

_____ **21.** The _____ can be made up of as many as four separate segments.
  A. Companion flange
  B. Driveshaft
  C. Side gear
  D. Output shaft

_____ **22.** On a(n) _____ axle, the axle shafts are splined to the differential side gears, and the outer bearing is between the outer end of the axle shaft and the inside of the axle housing.
  A. Full-floating
  B. Semi-floating
  C. Three-quarter floating
  D. Universal

_____ **23.** Lug studs are pressed into the axle _____ and used to secure the wheel onto the vehicle.
  A. Joint
  B. Shaft
  C. Case
  D. Flange

_____ **24.** In front-wheel drive vehicles and all-wheel drive vehicles, the driveshafts transfer the drive directly from the _____ inside the transaxle to the front wheels.
  A. Final drive
  B. Half shaft
  C. Transfer case
  D. Slip yoke

_____ **25.** The _____ is a short section of shaft that typically has a bearing pressed onto it, used to make both half-shafts the same length from left to right.
  A. Output shaft
  B. Crankshaft
  C. Intermediate shaft
  D. Pinion shaft

_____ **26.** The _____ joint consists of a steel cross with four hardened bearing journals, mounted on needle rollers in hardened caps, which locate the cross in the eyes of the yokes.
  A. Double Cardan
  B. CV
  C. Plunge-type
  D. Hooke's

_____ **27.** Often, a sliding spline or a _____ joint is used as the inner half-shaft joint to accommodate for changes in shaft length when traveling on different types of terrain.
  A. Tulip/tripod
  B. Plunge-type
  C. Fixed-type
  D. Rzeppa

_____ **28.** A CV _____ is used to keep the grease inside the joint and to keep dirt and debris out.
  A. Boot
  B. Shoe
  C. Glove
  D. Cover

_____ **29.** The speed reduction gears in the final drive are called the _____ and _____ gears.
   A. Spider, ring
   B. Differential, side
   C. Ring, pinion
   D. Hypoid bevel, side

_____ **30.** A _____ limited slip differential responds very quickly to changes in traction. It also does not bind from friction in turns and does not lose its effectiveness because there are no clutches, like a clutch-style unit.
   A. Hypoid-geared
   B. Helical-geared
   C. Pinion-geared
   D. Clutch-style

_____ **31.** The transfer case is typically bolted to the rear of the _____.
   A. Transmission
   B. Companion flange
   C. Driveshaft
   D. Transaxle

_____ **32.** What type of drive axle is shown in the image?

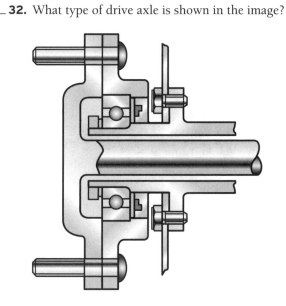

   A. Full-floating axle
   B. Half-floating axle
   C. Semi-floating axle
   D. Three-quarter floating axle

_____ **33.** What type of drive axle is shown in the image?

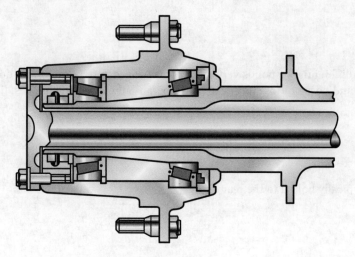

    A. Half-floating axle
    B. Three-quarter floating axle
    C. Full-floating axle
    D. Semi-floating axle

_____ **34.** What type of CV joint is shown in the image?

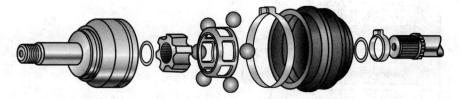

    A. Tulip joint
    B. Plunge-type joint
    C. Rzeppa joint
    D. Tripod joint

_____ **35.** What type of CV joint is shown in the image?

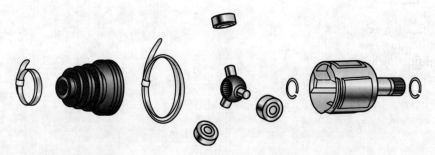

    A. Tulip/tripod joint
    B. Fixed-type joint
    C. Rzeppa joint
    D. Round-end joint

_____ **36.** Tech A says that full-floating axles use a single wheel bearing that sits between the axle and the axle housing. Tech B says that a differential allows the wheels on a common axle to rotate at different speeds. Who is correct?
 A. Tech A
 B. Tech B
 C. Both Tech A and Tech B
 D. Neither Tech A nor Tech B

_____ **37.** Tech A says that CV joints are used on some four-wheel drive vehicles. Tech B says that a slip joint is part of a driveshaft. Who is correct?
 A. Tech A
 B. Tech B
 C. Both Tech A and Tech B
 D. Neither Tech A nor Tech B

_____ **38.** Tech A says that a limited slip differential causes spring-loaded clutch plates to provide torque to both wheels. Tech B says that wrong tire sizes can affect the transfer case operation. Who is correct?
 A. Tech A
 B. Tech B
 C. Both Tech A and Tech B
 D. Neither Tech A nor Tech B

_____ **39.** Tech A says that some transfer cases have a chain inside them to transfer torque to the wheels. Tech B says that a rear-wheel drive axle housing can leak due to being overfilled with fluid. Who is correct?
 A. Tech A
 B. Tech B
 C. Both Tech A and Tech B
 D. Neither Tech A nor Tech B

_____ **40.** Tech A says that CV boots should be patched if they have cracks or splits. Tech B says that CV joints may use a special type of grease, possibly containing lead. Who is correct?
 A. Tech A
 B. Tech B
 C. Both Tech A and Tech B
 D. Neither Tech A nor Tech B

Wheels and Tires Theory

At the start of each chapter, you will find the Learning Objectives from the textbook. These are your objectives as you make your way through the exercises in this workbook and the chapter in your textbook. The following activities have been designed to help you refresh your knowledge of the material in this chapter.

Learning Objectives

After reading this chapter, you will be able to:

- LO 24-01 Describe tire and wheel physics.
- LO 24-02 Describe wheel construction.
- LO 24-03 Describe tire construction.
- LO 24-04 Interpret tire markings.
- LO 24-05 Describe tire safety features.

Matching

Match the following terms with the correct description or example.

A. Casing plies	**J.** Slip angle
B. Drop center	**K.** Tapered seat
C. EH2 rim	**L.** Traction grade
D. Negative offset	**M.** Tread wear grade
E. Neutral steer	**N.** Valve core
F. Ply rating	**O.** Valve stem
G. Rim flanges	**P.** Wheel flange
H. Schrader valve	**Q.** Wheel studs
I. Side force	**R.** Zero offset

_____ **1.** A condition in which the plane of the hub mounting surface is positioned toward the brake side or back of the wheel centerline.

_____ **2.** Part that allows air to be added or removed from a tire.

_____ **3.** A one-way valve used in a valve stem.

_____ **4.** A standardized grading system that indicates how well a tire will maintain contact with the road surface when wet.

_____ **5.** A rating system that denotes the number of belt layers or plies that make up the tire carcass. In radial tires, it denotes the relative strength of the plies, not the actual number of plies.

_____ **6.** The part of the wheel containing the holes for the lug studs.

_____ **7.** The specialized rim design that is used with some run-flat tires.

_____ **8.** The number imprinted on the sidewall of a tire by the manufacturer as required by the National Highway Traffic Safety Administration that indicates the tread life of a tire's tread.

_____ **9.** A condition in which both the front and the rear tires of a vehicle are experiencing the same slip angle.

_____ **10.** A condition in which the plane of the hub mounting surface is even with the centerline of the wheel.

_____ **11.** The pressure on the wheel that pushes it toward the outside or inside of the rim as the vehicle makes a turn.

_____ **12.** A wheel design with part of the center section of the wheel having a smaller diameter than the rest. It is used for mounting and demounting the tire.

_____ **13.** The angle between which the tire is pointing and the vehicle is moving.

_____ **14.** Threaded fasteners that are pressed into the wheel hub flange and used to bolt the wheel onto the vehicle.

_____ **15.** A type of lug nut with a tapered end toward the rim that helps center the wheel on the wheel studs.

_____ **16.** A network of cords that give the tire shape and strength.

_____ **17.** The one-way spring-loaded valve that screws into the valve stem; it allows air to be pumped into a tire and prevents it from flowing out.

_____ **18.** The outside edges of the wheel that help keep the tire from popping off the wheel.

Multiple Choice

Read each item carefully, and then select the best response.

_____ **1.** The force that acts between the tread and the road surface as the vehicle turns is called _____.
   A. Side force
   B. Road force
   C. Cornering force
   D. Slip angle

_____ **2.** When the front slip angles are larger than the rear slip angles, the vehicle is said to be in a(n) _____ condition, which is referred to as the vehicle "pushing" in the corners.
   A. Understeer
   B. Oversteer
   C. Neutral steer
   D. Lateral steer

_____ **3.** The distance across the rim flanges at the bead seat is called the _____.
   A. Drop center
   B. Flange width
   C. Contact area
   D. Rim width

_____ **4.** What type of rim is designed to hold the tire bead in place on the rim in the event of inadequate tire pressure?
   A. Split rims
   B. Drop well rims
   C. Safety rims
   D. Steel-disc type

_____ **5.** All of the following are standard types of wheel retaining studs or nuts, EXCEPT _____.
   A. Tapered seat
   B. Tapered seat with washer
   C. Flat seat without washer
   D. Flat seat with washer

_____ **6.** When the plane of the hub mounting surface is shifted from the centerline toward the outside or front side of the wheel, it has a _____.
   A. Positive offset
   B. Negative offset
   C. Positive caster
   D. Negative camber

_____ **7.** The measure of the innate strength of a material is called _____.
   A. Ply rating
   B. Tensile strength
   C. Flexibility
   D. Elasticity

_____ **8.** The first two letters following the letters "DOT" on a tire identify the _____ and the _____.
  A. Tire manufacturer, manufacturing plant
  B. Tire's size, type of tire construction
  C. Tire manufacturer, manufacturer-specified characteristics
  D. Tire's size, traction grade

_____ **9.** The standardized system designed to provide tire buyers with a comparative measure of a tire's tread life, traction, and temperature characteristics, is known as the _____.
  A. Uniform Tire Quality Grading System
  B. Department of Transportation compliance code
  C. International Organization for Standardization Tire Class
  D. Manufacturer's load index

_____ **10.** If a tire has a four-digit Department of Transportation code, when was it manufactured?
  A. 1970s
  B. 1980s
  C. 1990s
  D. After 2000

_____ **11.** What type of monitoring system utilizes the wheel speed antilock brake system to measure the difference in the rotational speed of the four wheels?
  A. Direct tire pressure monitoring system (TPMS)
  B. Centrifugal TPMS
  C. Indirect TPMS
  D. Automatic TPMS

_____ **12.** What type of tires are designed to get the driver to a service center in order to have the regular tire fixed or purchase a new one?
  A. Space-saver
  B. Self-sealing
  C. Run-flat
  D. Zero-pressure

_____ **13.** During cornering, centrifugal force puts more weight on the outside wheels.
  A. True
  B. False

_____ **14.** The bead seat is the edge of the rim that creates a seal between the tire bead and the wheel.
  A. True
  B. False

_____ **15.** Vehicles manufactured in the mid-1970s through the 1980s were typically built with positive offset wheels.
  A. True
  B. False

_____ **16.** Most lug nuts and studs are left-hand threaded, which means they tighten when turned clockwise.
  A. True
  B. False

_____ **17.** Pick-up trucks and large sport utility vehicles (SUVs) can have as many as 6, 8, or 10 studs.
  A. True
  B. False

_____ **18.** The primary function of the Schrader valve is to assist in keeping debris out of the valve stem.
  A. True
  B. False

_____ **19.** Bias-ply tires have much more flexible sidewalls because they use two or more layers of casing plies.
  A. True
  B. False

_____ **20.** A properly inflated radial tire runs cooler than a comparable bias-ply tire.
  A. True
  B. False

_____ **21.** Asymmetric tread patterns are designed in such a way that the tire can be mounted on the wheel for any direction of rotation.
   A. True
   B. False

_____ **22.** Radial tires are marked with the section width in millimeters, but with the rim diameter in inches.
   A. True
   B. False

_____ **23.** The lower a tire's aspect ratio, the wider the tire is in relation to its height.
   A. True
   B. False

_____ **24.** The DOT tire date manufacturing code is a six-digit code.
   A. True
   B. False

_____ **25.** It is illegal to sell a tire intended for use on a public road within the United States without a DOT stamp.
   A. True
   B. False

_____ **26.** The use of a centrifugal switch in the indirect TPMS sensor allows the sensor to go to sleep when the vehicle stops.
   A. True
   B. False

_____ **27.** Self-sealing tires feature standard tire construction with the addition of a flexible and malleable lining inside the tire in the tread area.
   A. True
   B. False

_____ **28.** The proper tire pressure for the vehicle is referred to as the recommended cold inflation pressure.
   A. True
   B. False

_____ **29.** Nitrogen-filled tires can be identified by the green valve stem caps that are placed on the valve stem when the tire is inflated.
   A. True
   B. False

_____ **30.** _____ is when the rear slip angle is larger than the front slip angle; it is referred to as the vehicle being "loose" in the corners.
   A. Neutral steer
   B. Understeer
   C. Oversteer
   D. Pushing

_____ **31.** The _____ of a wheel is the outer circular lip of the metal on which the inside edge of the tire is mounted.
   A. Seat
   B. Rim
   C. Axle
   D. Drop center

_____ **32.** In _____ wheels, the drop center is closer to the rear of the wheel.
   A. Split rim
   B. Drop well
   C. Safety rim
   D. Deep dish

_____ **33.** The _____ of a wheel is the distance from its hub mounting surface to the centerline of the wheel.
   A. Offset
   B. Tread
   C. Torque
   D. Side force

_____ **34.** The _____ refers to the number and spacing of the lug nuts or wheel studs on the wheel hub on the wheel rim.
  A. Bolt pattern
  B. Traction grade
  C. Tread pattern
  D. Aspect ratio

_____ **35.** A _____ is a specially designed opening that allows a tire to be inflated and then automatically closes to prevent air from escaping.
  A. Tire cap
  B. Tire stem
  C. Tire core
  D. Tire valve

_____ **36.** A _____ tire is the older form of tire and is still in use on some trailers and off-road vehicles, primarily because of a slightly lower cost and their more durable construction.
  A. Casing
  B. Bias-ply
  C. Radial
  D. Schrader

_____ **37.** Directional _____ are designed to provide a range of attributes during particular driving conditions.
  A. Radial patterns
  B. Bolt patterns
  C. Tread patterns
  D. Traction patterns

_____ **38.** _____ tread patterns have the same tread pattern on both sides of the tire and are usually nondirectional tires.
  A. Directional
  B. Nondirectional
  C. Asymmetric
  D. Symmetric

_____ **39.** _____ is a representation of a tire's ability to resist and dissipate heat.
  A. Traction grade
  B. Six-ply rating
  C. Temperature grade
  D. Tread wear grade

_____ **40.** Tires with a speed rating designation of _____ may be driven up to 124 mph (200 kph).
  A. Q
  B. S
  C. U
  D. W

_____ **41.** The _____ of a tire is the ratio of its height to its width. It is usually given as a percentage.
  A. Tensile strength
  B. Aspect ratio
  C. Tread wear grade
  D. Temperature grade

_____ **42.** As part of the U.S. _____ regulations, there must be a tire manufacture date code stamped on the sidewall of every tire.
  A. Highway Traffic Safety Administration
  B. Uniform Tire Quality Grading
  C. Tire Safety Administration
  D. Department of Transportation

_____ **43.** A _____ system monitors the tires for low air pressure and alerts the driver when one or more tires are lower than (or in some cases, higher than) the designated thresholds.

    A. Tire quality grading

    B. Transportation compliance

    C. Tire pressure monitoring

    D. Traction monitoring

_____ **44.** The major safety benefit of _____ technology is that it enables a driver to maintain vehicle control if a tire suffers a rapid pressure loss when in motion.

    A. TPMS

    B. Spare tire

    C. Tire tread

    D. Run-flat

_____ **45.** The arrow in the image is pointing to what part of the tire?

    A. Bead

    B. Sidewall

    C. Cap ply

    D. Inner liner

_____ **46.** The arrows in the image are pointing to what part of the tire?

    A. Grooves

    B. Belts

    C. Sidewalls

    D. Shoulders

_____ **47.** The arrow in the image is pointing to what part of the tire?

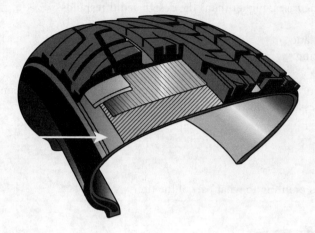

A. Carcass
B. Tread
C. Bead
D. Groove

_____ **48.** The arrows in the image are pointing to what part of the radial tire?

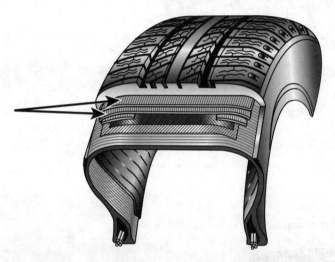

A. Radial plies
B. Cap plies
C. Steel belts
D. Bead chaffers

_____ **49.** The arrow in the image is pointing to what part of the radial tire?

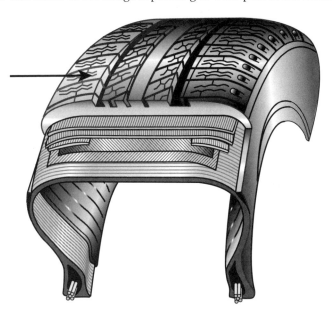

A. Tread block
B. Cap plies
C. Grooves
D. Sipes

_____ **50.** The arrow in the image is pointing to what part of the radial tire?

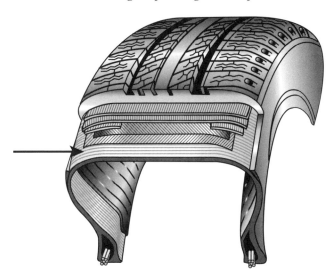

A. Radial plies
B. Steel belts
C. Grooves
D. Ribs

_____ **51.** Tech A says that when turning a corner, both wheels being steered remain parallel to each other as the wheels are steered. Tech B says that on some vehicles the rear wheels also can be steered. Who is correct?
   A. Tech A
   B. Tech B
   C. Both Tech A and Tech B
   D. Neither Tech A nor Tech B

_____ **52.** Tech A says that most wheels have a drop center or deep well that is used in installing a tire on the wheel. Tech B says that the drop center or deep well is used to prevent the tire from coming off the wheel in the case of low tire pressure. Who is correct?
   A. Tech A
   B. Tech B
   C. Both Tech A and Tech B
   D. Neither Tech A nor Tech B

_____ **53.** Tech A says that all wheels must be torqued to prevent wheels from loosening up and falling off. Tech B says that all wheels must be torqued to prevent overtightening, which can weaken lug studs and warp brake rotors. Who is correct?
   A. Tech A
   B. Tech B
   C. Both Tech A and Tech B
   D. Neither Tech A nor Tech B

_____ **54.** Tech A says that underinflated tires reduce fuel economy. Tech B says that underinflated tires are unsafe and cause accelerated tire wear. Who is correct?
   A. Tech A
   B. Tech B
   C. Both Tech A and Tech B
   D. Neither Tech A nor Tech B

_____ **55.** Tech A says that tires are marked with a date code indicating the date the tires should be discarded. Tech B says that any tires with a three-digit date code should be discarded. Who is correct?
   A. Tech A
   B. Tech B
   C. Both Tech A and Tech B
   D. Neither Tech A nor Tech B

_____ **56.** Tech A says that a TPMS system can save fuel over time. Tech B says that a TPMS will help prevent blowouts. Who is correct?
   A. Tech A
   B. Tech B
   C. Both Tech A and Tech B
   D. Neither Tech A nor Tech B

Servicing Wheels and Tires

At the start of each chapter, you will find the Learning Objectives from the textbook. These are your objectives as you make your way through the exercises in this workbook and the chapter in your textbook. The following activities have been designed to help you refresh your knowledge of the material in this chapter.

Learning Objectives

After reading this chapter, you will be able to:

- LO 25-01 Describe tire maintenance preliminaries.
- LO 25-02 Properly check and adjust tire pressure.
- LO 25-03 Identify tire wear patterns and perform tire rotation.
- LO 25-04 Perform tire balance.
- LO 25-05 Dismount and mount a tire without a tire pressure monitoring system (TPMS).
- LO 25-06 Dismount and remount a tire with TPMS.
- LO 25-07 Replace tire valve stems.
- LO 25-08 Perform TPMS service.
- LO 25-09 Perform tire diagnosis and repair.
- LO 25-10 Measure wheel, tire, axle flange, and hub runout.

Multiple Choice

Read each item carefully, and then select the best response.

_____ 1. The amount of air pressure in the tire that provides it with load-carrying capacity and affects the overall performance of the vehicle is called the _____.
   A. Maximum inflation pressure
   B. Pressure index
   C. Tire inflation pressure
   D. International Organization for Standards (ISO) pressure

_____ 2. What type of rotation pattern should you NOT use for a four-tire rotation with nondirectional tires?
   A. Forward-cross
   B. Rearward-cross
   C. X pattern
   D. Side to side

_____ 3. What type of balancing does NOT take into consideration that the tire has width?
   A. Dynamic
   B. Road force
   C. Static
   D. Horizontal

_____ 4. What does a green valve stem cap on a newer tire signify?
   A. The tire is constructed with a green design.
   B. The tire is filled with nitrogen.
   C. The tire is filled with helium.
   D. The tire is filled with hydrogen.

_____ **5.** Which of the following statements is correct regarding tire inflation?
  A. Tire inflation pressure is stated in ounces per square inch (OPSI).
  B. Foreign/import vehicles may specify tire pressure in knots per square centimeters (KPSC).
  C. Proper tire inflation pressure allows a tire to carry up to 75% of its maximum load limit.
  D. Proper tire pressure allows a vehicle to carry weight up to its designed load limit.

_____ **6.** Which of the following statements is correct regarding tire inflation?
  A. Oxygen, due to its reactivity, is harmful to rubber and other tire materials.
  B. Oxygen reacts negatively with the tire liner and creates particles of rubber that can clog valve stems.
  C. Plain compressed air consists of approximately 50% oxygen and 50% nitrogen.
  D. Nitrogen-filled tires usually have green valve stem caps.

_____ **7.** The cold inflation pressure of a tire is determined on the basis of which of the following?
  A. The type of transmission in the vehicle
  B. The type of gas in the tire
  C. The vehicle's design load limit
  D. The engine capacity of the vehicle

_____ **8.** Directional tires require keeping the tires on the same side of the vehicle during rotation and generally use a front-to-rear pattern.
  A. True
  B. False

_____ **9.** Dynamic imbalance tends to cause the tire to move purely up and down.
  A. True
  B. False

_____ **10.** A tire pressure gauge measures pressures in pounds per square inch (psi), kilopascals (kPa), or bars.
  A. True
  B. False

_____ **11.** Uniform tread wear can boost the vehicle's fuel economy and increase the vehicle's performance.
  A. True
  B. False

_____ **12.** In recent years, some manufacturers have filled their tires with pure or nearly pure oxygen in hopes of avoiding the problems associated with nitrogen-filled tires.
  A. True
  B. False

_____ **13.** While both nitrogen and oxygen can permeate rubber, nitrogen does so at a much faster rate.
  A. True
  B. False

_____ **14.** The proper tire pressure for the vehicle is referred to as the recommended warm inflation pressure.
  A. True
  B. False

_____ **15.** To get an accurate pressure reading, the tires must be checked when cold.
  A. True
  B. False

_____ **16.** When checking air pressure, the tire sidewall should always be referred to, because this is the most accurate air pressure reading.
  A. True
  B. False

_____ **17.** The TPMS must be on all new vehicles, starting with the 2008 model year.
  A. True
  B. False

_____ **18.** The wear patterns on a tire correlate with damage to particular vehicle components.
  A. True
  B. False

_____ **19.** The _____ is inspected to ensure that it is above the built-in wear indicators, which indicate that the tire is at the end of its legal life and should be replaced.
   A. Tire pressure
   B. Tread area
   C. Tread depth
   D. Sidewalls

_____ **20.** Filling the tires with _____ will increase the vehicle's fuel efficiency and tire life by reducing the effects of pressure loss due to permeation.
   A. Oxygen
   B. Nitrogen
   C. Hydrogen
   D. Carbon dioxide

_____ **21.** _____ balancing is performed by placing specific amounts of weight on each side of the rim to provide the exact counterbalance needed.
   A. Static
   B. Dynamic
   C. Road force
   D. Positive

_____ **22.** _____ psi of air pressure can generate over 30,000 lb of force on a tire.
   A. 8
   B. 16
   C. 24
   D. 32

_____ **23.** As the air within the tire warms, it expands, causing the air pressure within the tire to _____.
   A. Increase
   B. Decrease
   C. Stay the same

_____ **24.** The air that we breathe and that is typically used to inflate tires chemically consists of 78% _____, 21% _____, and 1% _____.
   A. Helium, nitrogen, oxygen
   B. Oxygen, nitrogen, other
   C. Nitrogen, oxygen, other
   D. Oxygen, helium, nitrogen

_____ **25.** Nitrogen-filled tires can be identified by the _____ valve stem caps that are placed on the valve stem when the tire is inflated.
   A. Yellow
   B. Red
   C. Blue
   D. Green

_____ **26.** _____-filling will increase the vehicle's fuel efficiency and tire life by reducing the effects of pressure loss due to permeation.
   A. Nitrogen
   B. Oxygen
   C. Helium
   D. Sealant

_____ **27.** In general, a four-tire rotation with nondirectional tires can have one of three rotation patterns: forward-cross, rearward-cross, and _____ pattern.
   A. Static
   B. Lateral-cross
   C. T
   D. X

_____ **28.** If the vehicle uses a full-sized spare tire, the rotation pattern will be one of two _____-tire rotations, which are variations of the four-tire forward-cross or rearward-cross pattern.
   A. Three
   B. Four
   C. Five
   D. Two

_____ **29.** It is generally recommended that tires be rotated on vehicles approximately every _____ miles (8,000 km) or with every oil change.
   A. 4,000
   B. 3,000
   C. 5,000
   D. 6,000

_____ **30.** Many TPMS sensors use a(n) _____-plated valve core in an aluminum valve stem.
   A. Brass
   B. Aluminum
   C. Nickel
   D. Copper

_____ **31.** TPMS sensor batteries are designed to last for _____ years without replacement.
   A. 1–3
   B. 2–4
   C. 5–10
   D. Over 10

_____ **32.** Tech A says that incorrect toe settings will cause feathered wear across the tire tread. Tech B says that underinflated tires have more wear in the center of the tread. Who is correct?
   A. Tech A
   B. Tech B
   C. Both Tech A and Tech B
   D. Neither Tech A nor Tech B

_____ **33.** Tech A says that the use of nitrogen to fill a tire will prevent the tire from blowing out. Tech B says that when mounting or dismounting a tire on a wheel with a TPMS sensor, you need to position the wheel/tire properly on the tire machine, or the TPMS sensor can be easily broken. Who is correct?
   A. Tech A
   B. Tech B
   C. Both Tech A and Tech B
   D. Neither Tech A nor Tech B

_____ **34.** Tech A says that using a tire plug to repair a hole in a tire is the best and fastest way to fix a tire. Tech B says that using a tire plug patch is the only approved method of repairing a tire in many states. Who is correct?
   A. Tech A
   B. Tech B
   C. Both Tech A and Tech B
   D. Neither Tech A nor Tech B

_____ **35.** Tech A says that directional tires cannot be rotated. Tech B says that all old wheel weights should be removed before balancing a tire. Who is correct?
   A. Tech A
   B. Tech B
   C. Both Tech A and Tech B
   D. Neither Tech A nor Tech B

_____ **36.** Tech A says that tire inflation needs to be checked when the tire is cold. Tech B says that tire pressure can be checked as long as the tire has been driven less than 1 mile in the last 3 hours. Who is correct?
   A. Tech A
   B. Tech B
   C. Both Tech A and Tech B
   D. Neither Tech A nor Tech B

_____ **37.** Tech A says that when checking a tire's air pressure, the tire sidewall information should always be referred to, and the tire should be inflated to that pressure. Tech B says that when checking air pressure, the pressure should be compared with the specifications on the tire placard, which is usually located on the driver's door, door pillar, or glove compartment lid. Who is correct?
  A. Tech A
  B. Tech B
  C. Both Tech A and Tech B
  D. Neither Tech A nor Tech B

_____ **38.** Tech A says that when checking tire pressure, the tires should be "cold." Tech B says that tires should be driven more than 3 miles before tire pressure is checked. Who is correct?
  A. Tech A
  B. Tech B
  C. Both Tech A and Tech B
  D. Neither Tech A nor Tech B

Steering Systems Theory

At the start of each chapter, you will find the Learning Objectives from the textbook. These are your objectives as you make your way through the exercises in this workbook and the chapter in your textbook. The following activities have been designed to help you refresh your knowledge of the material in this chapter.

Learning Objectives

After reading this chapter, you will be able to:

- LO 26-01 Describe steering system preliminaries.
- LO 26-02 Describe steering geometry and rack-and-pinion layout.
- LO 26-03 Describe parallelogram steering layout.
- LO 26-04 Describe steering columns and their components.
- LO 26-05 Describe rack-and-pinion steering boxes.
- LO 26-06 Describe worm gear steering boxes.
- LO 26-07 Describe hydraulic power steering system operation.
- LO 26-08 Describe electric power steering system (EPS) operation.
- LO 26-09 Describe four-wheel steering operation.

Matching

Match the following terms with the correct description or example.

A. Active control		**K.** Power unit	
B. Actuating		**L.** Rack	
C. Beam axle		**M.** Relay lever	
D. Chassis		**N.** Spline	
E. Drag link		**O.** Steering box	
F. Gear reduction		**P.** Steering linkage	
G. Inertia		**Q.** Steering system	
H. Intermediate shaft		**R.** Tie-rod assembly	
I. Knuckle		**S.** Torque sensor	
J. Power section			

_____ **1.** Steel rods that connect the steering box to the steering arms on the steering knuckle.

_____ **2.** A steel rod that transfers movement from the drag link to an idler arm.

_____ **3.** A device used to measure the load on the steering wheel.

_____ **4.** A chamber in the rack where pressurized fluid acts upon pistons that assist in steering.

_____ **5.** A steel or iron rod that transfers movement of the pitman arm to a relay lever.

_____ **6.** A device that converts the rotary motion of the steering wheel to the linear motion needed to steer the vehicle.

_____ **7.** A suspension system in which one set of wheels is connected laterally by a single beam or shaft.

_____ **8.** The part that fits between the rack and the steering arms and transfers the movement of the rack.

_____ **9.** A system of providing constant feedback from sensors in the vehicle to the control unit.

_____ **10.** A term used to describe all of the components and parts involved in steering a vehicle.

_____ **11.** The act of making something move or work.

_____ **12.** The part that contains the wheel hub or spindle and attaches to the suspension components.

_____ **13.** A gear ratio used to make large turns of the steering wheel into smaller turns of the tire to ease steering for the driver.

_____ **14.** A belt- or gear-driven pump that produces hydraulic pressure for use in the steering box or rack.

_____ **15.** The resistance to a change in motion.

_____ **16.** A ridge or tooth on a driveshaft that meshes with grooves in a mating piece and transfers torque to it, maintaining the angular correspondence between them.

_____ **17.** The frame of a vehicle, to which the suspension pieces attach.

_____ **18.** A steel rod positioned at an angle from the steering column to the steering gear that functions in transferring movement from the one to the other.

_____ **19.** A steel rod driven by the pinion with tie-rods on each end or tie-rods connected to the center of the rack.

Multiple Choice

Read each item carefully, and then select the best response.

_____ **1.** Parallelogram steering uses a _____ gearbox, which changes the direction of steering wheel rotation 90 degrees.
  A. Pitman
  B. Rack and pinion
  C. Worm
  D. Linear

_____ **2.** What type of steering system is used on the majority of front-wheel drive vehicles due to the restriction of space under the hood?
  A. Worm and sector
  B. Rack and pinion
  C. Recirculating ball
  D. Direct linkage

_____ **3.** As the driver turns the steering wheel, the forces are transferred to the _____, causing the rack to move in either direction.
  A. Worm gear
  B. Pitman arm
  C. Tie-rods
  D. Pinion

_____ **4.** The _____ is attached between the tie-rod shaft and the steering arm; it pivots as the rack is extended or retracted when the vehicle is negotiating turns.
  A. Inner tie-rod
  B. Outer tie-rod
  C. Pitman arm
  D. Idler arm

_____ **5.** The _____ connects the pitman arm to the idler arm.
  A. Center link
  B. Tie-rod
  C. Steering knuckle
  D. Adjustment sleeve

_____ **6.** The _____ connects the tie-rod to the tie-rod end and provides the adjustment point for setting toe-in or toe-out.
  A. Adjustment sleeve
  B. Idler arm
  C. Pitman arm
  D. Center link

_____ 7. The rotary electrical connector located between the steering wheel and the steering column that maintains a constant electrical connection with the wiring while the vehicle's steering wheel is being turned is called the _____.
   A. Swivel switch
   B. Airbag relay
   C. Slot switch
   D. Clock spring

_____ 8. In the rack-and-pinion steering system, the steering rack is supported at the pinion end by being sandwiched between the pinion and _____.
   A. Flexible rubber bellows
   B. A spring-loaded rack guide yoke
   C. An idler arm
   D. A pitman arm

_____ 9. The worm-and-nut steering gear is also known as a _____ steering gear.
   A. Recirculating ball
   B. Box-type
   C. Worm and sector
   D. Rack and pinion

_____ 10. When the rear wheels can be steered independently of or in conjunction with the front wheels, it is known as _____.
   A. Independent steering
   B. Four-wheel drive
   C. Four-wheel steering
   D. Worm and roller steering

_____ 11. All of the following are types of power steering, EXCEPT _____.
   A. Hydraulically assisted power steering
   B. Compressed air–assisted power steering
   C. Electrically powered hydraulic steering
   D. Fully electric power steering

_____ 12. The spring-loaded piece of steel connected to the pinion gear or worm at its bottom end and the input shaft at its top end is called a _____.
   A. Clock spring
   B. Helix
   C. Leaf spring
   D. Torsion bar

_____ 13. All power steering pumps have a(n) _____ to vary fluid flow and power steering system pressures.
   A. Pressure relief valve
   B. Spool valve
   C. Actuator
   D. Flow-control valve

_____ 14. What type of steering system still uses fluid and a pump, but the pump is driven by an electric motor to reduce power drawn from the engine?
   A. Electrically assisted steering system
   B. EPS
   C. Electrically powered hydraulic steering system
   D. Hybrid steering system

_____ 15. The steering box converts the rotary motion of the steering wheel into the linear motion needed to pivot the wheels.
   A. True
   B. False

_____ 16. The parallelogram steering system gets its name because the center link and axle, along with the pitman arm and idler arm, always move parallel to each other.
   A. True
   B. False

_____ **17.** The outer tie-rod is attached to the end of the rack and allows for suspension movement and slight changes in steering angles.
   A. True
   B. False

_____ **18.** Toe-out is a condition where the fronts of the wheels, as seen from above, are closer together than the rears of the wheels.
   A. True
   B. False

_____ **19.** The toe-setting is the symmetric angle that each wheel makes with the longitudinal axis of the vehicle.
   A. True
   B. False

_____ **20.** To reduce serious injury, all steering columns are now fitted with collapsible sections that help protect the driver.
   A. True
   B. False

_____ **21.** When a technician needs to carry out a servicing procedure on the steering column, it is good practice to disarm the triggering system for the driver's side airbag.
   A. True
   B. False

_____ **22.** Both the pinion and the rack teeth are worm gears.
   A. True
   B. False

_____ **23.** A disadvantage of the rack-and-pinion steering system is that it typically is not manually adjustable.
   A. True
   B. False

_____ **24.** Worm gear steering boxes made the process of turning the front wheels an easier task for drivers of the early automobile.
   A. True
   B. False

_____ **25.** A ball-return guide is a special passage or metal tube through which the balls move in recirculating ball steering boxes.
   A. True
   B. False

_____ **26.** Power steering fluid gets contaminated with rubber and metallic particles from internal wear in the system.
   A. True
   B. False

_____ **27.** Direct drive steering is a completely electrically powered power-assist system that eliminates all hydraulic components and fluid.
   A. True
   B. False

_____ **28.** In a pinion-assist type EPS, the power-assist unit, controller, and torque sensor are attached to the steering column.
   A. True
   B. False

_____ **29.** In an EPS system, sensor inputs are compared to determine how much power assistance is required according to the forces capability map data stored in the Engine Control Unit's (ECU's) memory.
   A. True
   B. False

_____ **30.** The _____ transmits the driver's steering effort from the steering wheel down to the steering box.
   A. Steering arm
   B. Steering column
   C. Steering damper
   D. Steering linkage

_____ **31.** The steering linkage transfers the linear steering effort to the wheels by connecting the steering box to the _____ on each of the steering knuckles, which pivot on the ball joints, allowing the wheels to steer the vehicle.
   A. Steering arm
   B. Steering column
   C. Steering damper
   D. Steering sensor

_____ **32.** The steering knuckle pivots on one or two _____, depending on the type of suspension.
   A. Beam axles
   B. Clock springs
   C. Drag links
   D. Ball joints

_____ **33.** The rack slides in the housing and is moved by the action of the _____ pinion.
   A. Preloaded
   B. Linear
   C. Meshed
   D. Spring-loaded

_____ **34.** The _____ protects the inner joints from dirt and contaminants and retains the grease lubricant inside the rack-and-pinion housing.
   A. CV boot
   B. Adjustable bushing
   C. Beam axle
   D. Rubber bellows

_____ **35.** The _____ transfers movement from the steering box to the center link.
   A. Pitman arm
   B. Idler arm
   C. Control arm
   D. Spline

_____ **36.** The _____ is attached to the chassis (the frame of the vehicle) and is positioned parallel to the pitman arm.
   A. Chassis arm
   B. Sway bar
   C. Control arm
   D. Idler arm

_____ **37.** _____ is the undesired condition produced when, upon hitting a bump, the vehicle darts to one side as the steering linkage is pushed or pulled as a result of the travel of the suspension.
   A. Oversteer
   B. Understeer
   C. Bump steer
   D. Constant steer

_____ **38.** To compensate for variations in driving positions, many manufacturers have included a steering column _____ and/or _____ mechanism to their vehicles.
   A. Tilting, telescoping
   B. Actuating, moving
   C. Tilting, actuating
   D. Dragging, moving

_____ **39.** The rack is typically supported at both ends of the _____ or tube by a nylon bushing.
   A. Spline
   B. Control unit
   C. Rack housing
   D. Power section

_____ **40.** The pinion is supported by two bearings in the rack housing that must be _____ to ensure the pinion is in the correct position, relative to the rack, and to eliminate free play.

A. Unloaded

B. Preloaded

C. Connected

D. Adjusted

_____ **41.** The steering arm, the stub-axle knuckle, and the stub-axle carrier can be forged as one piece and can be referred to as a _____.

A. Steering column

B. Steering linkage

C. Steering system

D. Steering knuckle

_____ **42.** The _____ can read both torque and rotation from the steering wheel and convert them into voltage signals for the ECU to monitor.

A. Steering sensor

B. Steering box

C. Steering linkage

D. Steering arm

_____ **43.** The arrow in the image is pointing to what part of the basic steering system?

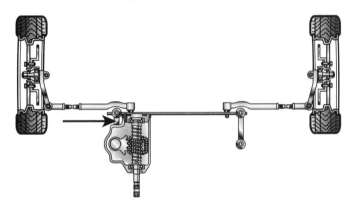

A. Pitman arm

B. Steering arm

C. Idler arm

D. Tie-rod

_____ **44.** The arrow in the image is pointing to what part of the basic steering system?

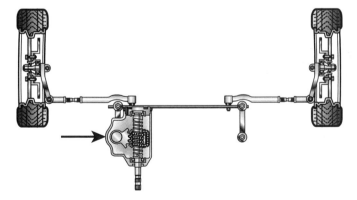

A. Worm shaft

B. Sector shaft

C. Steering box

D. Ball joint

_____ **45.** The arrow in the image is pointing to what part of the basic steering system?

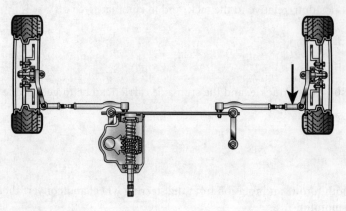

    A. Track rod
    B. Inner tie-rod
    C. Outer tie-rod end
    D. Idler arm

_____ **46.** The arrow in the image is pointing to what part of the rack-and-pinion steering system?

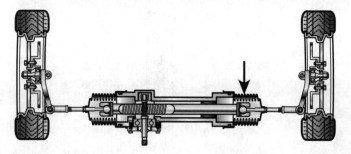

    A. Steering arm
    B. Rack boot
    C. Toe angle adjuster
    D. End bushing

_____ **47.** The arrow in the image is pointing to what part of the rack-and-pinion steering system?

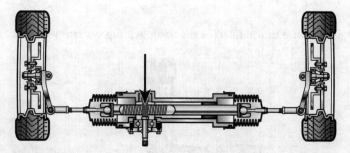

    A. Steering arm
    B. Rack
    C. Ball joint
    D. Pinion

_____ **48.** The arrow in the image is pointing to what part of the rack-and-pinion steering system?

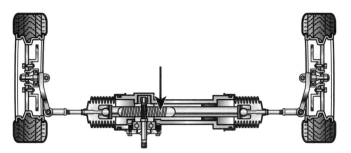

   A. Steering arm
   B. Rack
   C. Pinion
   D. End bushing

_____ **49.** The arrow in the image is pointing to what part of the recirculating ball steering system?

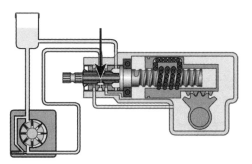

   A. Power steering fluid reservoir
   B. Pinion and rotary control valve
   C. Power piston and recirculating ball assembly
   D. Pressure relief valve

_____ **50.** The arrow in the image is pointing to what part of the recirculating ball steering system?

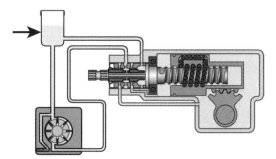

   A. Pump
   B. Power piston and recirculating ball assembly
   C. Power steering fluid reservoir
   D. Sector shaft

_____ **51.** The arrow in the image is pointing to what part of the recirculating ball steering system?

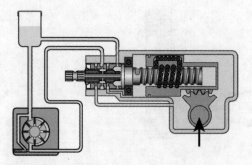

   A. Power steering fluid reservoir
   B. Sector shaft
   C. Power piston and recirculating ball assembly
   D. Pump

_____ **52.** The arrow in the image is pointing to what part of the power-assisted rack-and-pinion system?

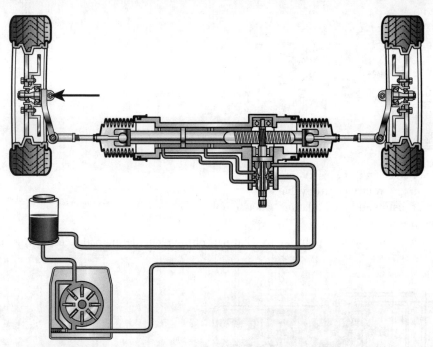

   A. Ball joint
   B. Tie-rod end
   C. Toe angle adjuster
   D. End bushing

_____ **53.** The arrows in the image are pointing to what part of the power-assisted rack-and-pinion system?

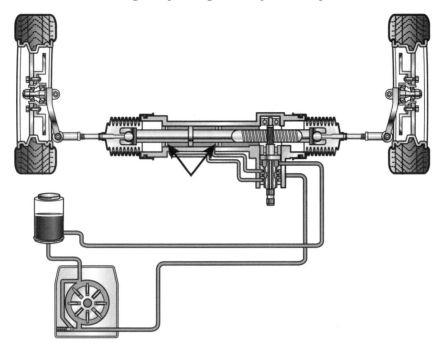

A. Wheel bearings
B. Pinion and rotary control valve
C. Pressure seats
D. Pressure relief valves

_____ **54.** The arrow in the image is pointing to what part of the power-assisted rack-and-pinion system?

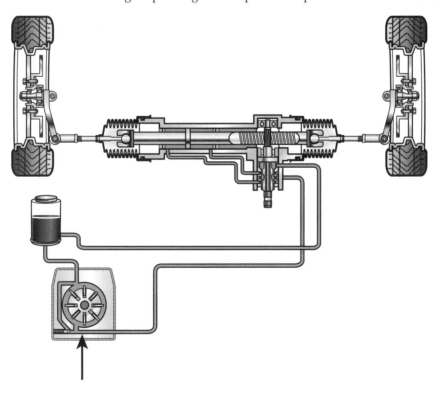

A. Power piston
B. Housing
C. Power steering fluid reservoir
D. Pump

_____ **55.** The arrow in the image is pointing to what part of the power-assisted recirculating ball gearbox?

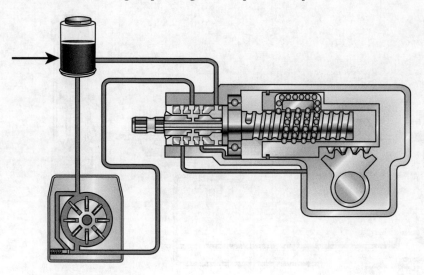

A. Power steering fluid reservoir

B. Power piston and recirculating ball assembly

C. Pump

D. Sector shaft

_____ **56.** The arrow in the image is pointing to what part of the power-assisted recirculating ball gearbox?

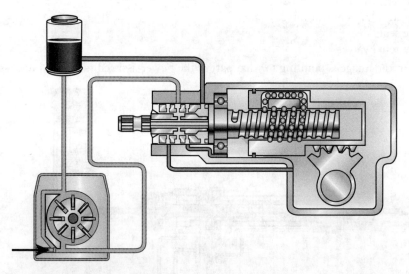

A. Power steering fluid reservoir

B. Pinion and rotary control valve

C. Power piston and recirculating ball assembly

D. Pressure relief valve

_____ **57.** The arrow in the image is pointing to what part of the power-assisted recirculating ball gearbox?

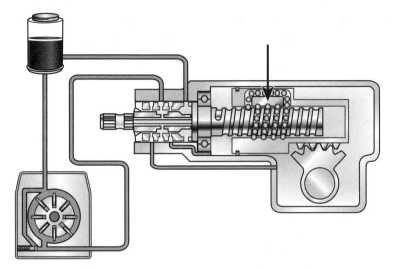

   A. Sector shaft
   B. Power steering fluid reservoir
   C. Power piston and recirculating ball assembly
   D. Pump

_____ **58.** Tech A says that the steering column uses one or more flexible joints to connect to the steering gearbox. Tech B says that the pitman arm is bolted to the frame and relays the steering linkage movement to the opposite wheel from the steering gearbox. Who is correct?
   A. Tech A
   B. Tech B
   C. Both Tech A and Tech B
   D. Neither Tech A nor Tech B

_____ **59.** Tech A says that the clock spring assists the turning of the steering wheel on vehicles with EPS. Tech B says that the clock spring is used to transmit an electrical signal to the driver's side airbag. Who is correct?
   A. Tech A
   B. Tech B
   C. Both Tech A and Tech B
   D. Neither Tech A nor Tech B

_____ **60.** Tech A says that power steering fluid is universal and can be used in virtually any vehicle's power steering system. Tech B says that in a power steering system, the force needed to turn the wheels is created by the hydraulic pump or electric motor. Who is correct?
   A. Tech A
   B. Tech B
   C. Both Tech A and Tech B
   D. Neither Tech A nor Tech B

_____ **61.** Tech A says that rack-and-pinion steering systems generally do not use power steering due to their lighter duty construction. Tech B says that rack-and-pinion steering systems use a worm gear arrangement to move the rack. Who is correct?
   A. Tech A
   B. Tech B
   C. Both Tech A and Tech B
   D. Neither Tech A nor Tech B

_____ **62.** Tech A says that to properly check a tie-rod end, the technician should twist the tie-rod end, and any rotational movement means the joint is bad. Tech B says that to properly check tie-rod ends, the vehicle's weight should be on the wheels, and as an assistant wiggles the steering wheel, the technician can check for movement in the tie-rod ends. Who is correct?
  A. Tech A
  B. Tech B
  C. Both Tech A and Tech B
  D. Neither Tech A nor Tech B

_____ **63.** Tech A says that worn rack-and-pinion mount bushings can cause excessive play in the steering system. Tech B says that a worn idler arm can cause excessive play in the steering system. Who is correct?
  A. Tech A
  B. Tech B
  C. Both Tech A and Tech B
  D. Neither Tech A nor Tech B

_____ **64.** Tech A says that when the vehicle is being driven straight ahead, a belt-driven power steering pump will pump fluid continuously, placing a minimal load on the engine. Tech B says that a belt-driven power steering pump is activated by an electromagnetic clutch, so it only pumps fluid when the wheels are being steered. Who is correct?
  A. Tech A
  B. Tech B
  C. Both Tech A and Tech B
  D. Neither Tech A nor Tech B

_____ **65.** Tech A says that the pressure relief valve maintains a preset minimum pressure in the system. Tech B says that the pressure relief valve prevents excessive pressure. Who is correct?
  A. Tech A
  B. Tech B
  C. Both Tech A and Tech B
  D. Neither Tech A nor Tech B

_____ **66.** Tech A says that worn tie-rod ends can cause a steering wandering complaint. Tech B says that a hard steering complaint could be caused by a worn power steering pump. Who is correct?
  A. Tech A
  B. Tech B
  C. Both Tech A and Tech B
  D. Neither Tech A nor Tech B

_____ **67.** Tech A says that a pickle fork is used to hold a tie-rod while it is being tightened. Tech B says that the pitman arm is usually threaded so that the front-wheel toe can be adjusted. Who is correct?
  A. Tech A
  B. Tech B
  C. Both Tech A and Tech B
  D. Neither Tech A nor Tech B

Servicing Steering Systems

At the start of each chapter, you will find the Learning Objectives from the textbook. These are your objectives as you make your way through the exercises in this workbook and the chapter in your textbook. The following activities have been designed to help you refresh your knowledge of the material in this chapter.

Learning Objectives

After reading this chapter, you will be able to:

- LO 27-01 Describe steering system service preliminaries.
- LO 27-02 Describe the steering system diagnosis procedure.
- LO 27-03 Perform power steering fluid maintenance.
- LO 27-04 Perform rack-and-pinion service.
- LO 27-05 Perform parallelogram steering linkage service.
- LO 27-06 Inspect electric power steering and identify high-voltage electrical circuits.
- LO 27-07 Disable supplemental restraint system (SRS) and service clock spring.

Multiple Choice

Read each item carefully, and then select the best response.

_____ 1. Safety concerns encountered during power steering service should include all of the following, EXCEPT _____.
   A. Power steering fluid can be very hot
   B. Belts can move when the engine is running
   C. Fluid pressure can be as high as 1,400 psi
   D. Power steering fluid contact can cause skin cancer

_____ 2. What is the first step to flushing the power steering system?
   A. Add fluid.
   B. Start the engine.
   C. Raise the vehicle.
   D. Plug the return fitting.

_____ 3. What is the first step in removing and replacing the power steering pump?
   A. Drain the pump.
   B. Disconnect the hoses.
   C. Disconnect the drive belts.
   D. Research the proper procedure.

_____ 4. Which of the following statements regarding power steering pump replacement is correct?
   A. A special puller is used to remove the pulley from the old pump.
   B. A soft-faced hammer is used to drive the pulley from the old pump onto the shaft of the new pump.
   C. A ball peen hammer is used to tap the pulley from the old pump onto the shaft.
   D. The pulley from the old pump is a slip-fit and needs no force to install it.

_____ 5. What may occur if a vehicle's steering linkage damper fails?
   A. Excessive toe change
   B. Wandering conditions
   C. Extreme positive camber
   D. Shimmy in the steering wheel after hitting a bump in the road

_____ **6.** The _____ connect the pitman arm to the idler arm.
 A. Center link and drag link
 B. Center link and tie rod
 C. Control arm and tie rod
 D. Drag link and control arm

_____ **7.** Which of the following systems should be disabled while working on or around the steering column, any of the airbags, or other pyrotechnic devices?
 A. Electric power steering system (EPS)
 B. Antilock brake system (ABS)
 C. SRS
 D. TRW

_____ **8.** If removing a tie rod from an aluminum steering knuckle, a pickle fork should be used; otherwise, the soft aluminum will be damaged.
 A. True
 B. False

_____ **9.** It is important to know that most airbags are inflated by igniting a solid fuel similar to rocket fuel.
 A. True
 B. False

_____ **10.** When disabling a vehicle's SRS, it is critical to use a memory minder or auxiliary power supply on the vehicle to retain settings and presets as well as module memory.
 A. True
 B. False

_____ **11.** The clock spring coils and uncoils as needed during turning of the steering wheel to ensure that the airbag is ready to be deployed in the event of a collision.
 A. True
 B. False

_____ **12.** The steering column has a sliding shaft that will slide into itself in the event of an accident, and the column housing mounting points are designed to allow the column to slip downward as the driver makes contact with the steering wheel.
 A. True
 B. False

_____ **13.** The _____ connects the center link to the steering arms.
 A. Tie rod
 B. Control arm
 C. Pitman arm
 D. Idler arm

_____ **14.** The _____ is a flexible ribbon that is coiled inside of a plastic holder.
 A. Steering linkage
 B. Timing coil
 C. Clock spring
 D. Timing spring

_____ **15.** Typically, a power steering system flush is performed with a _____ that is connected in line with the power steering system pump to remove all old dirty fluid and install new clean fluid without introducing air.
 A. Power steering return hose
 B. Suction gun
 C. Flushing machine
 D. Steering gear

_____ **16.** Some power steering pump pulleys are now made of _____. These pulleys pose an additional challenge, especially if they become brittle.
 A. Aluminum alloy
 B. Copper
 C. Plastic
 D. Cast iron

_____ **17.** A(n) _____ may be spliced into the hose between the steering gear and the pump, or it may fit inside the return hose where the return hose connects onto the pump.
   A. Inline filter
   B. In-reservoir filter
   C. Reservoir
   D. Cooling grid

_____ **18.** The _____ is parallel to the pitman arm and provides the pivoting support for the chassis end of the steering linkage.
   A. Center link
   B. Tie rod
   C. Steering gear
   D. Idler arm

_____ **19.** The _____ keeps the shock forces from being transmitted through the steering linkage and back to the steering wheel.
   A. Steering shock
   B. Steering angle sensor
   C. Steering boot
   D. Steering damper

_____ **20.** The _____ keeps the steering wheel from shaking when the driver hits a pothole or other road irregularity.
   A. Steering shock
   B. Steering angle sensor
   C. Steering boot
   D. Steering damper

_____ **21.** The _____ transfer(s) movement from the steering box to the center link.
   A. Inner tie rods
   B. Pitman arm
   C. Ball joints
   D. Control arm

_____ **22.** The _____ is used to measure the runout or movement of different parts of the steering system, such as play in tie-rod ends.
   A. Power steering analyzer
   B. Dial indicator
   C. Digital volt-ohmmeter
   D. Circuit tester

_____ **23.** The _____ connects the tie rod to the tie-rod end.
   A. Pitman arm
   B. Idler arm
   C. Center link
   D. Adjustment sleeve

_____ **24.** Replacement of _____ requires an alignment to be performed to prevent rapid tire wear from occurring.
   A. Tie-rod ends
   B. The tie-rod cotter pin
   C. Clamps
   D. Grease fitting

_____ **25.** Because the clock spring constantly winds and unwinds, the wires in the clock spring can _____ over time, illuminating the SRS light and disabling the SRS system.
   A. Become brittle
   B. Tangle
   C. Break
   D. Become loose

_____ **26.** Tech A says that when replacing the clock spring, the spring should be turned all the way to either end and then installed in the steering column. Tech B says that clock springs are used to return the steering wheel to its centered position. Who is correct?
   A. Tech A
   B. Tech B
   C. Both Tech A and Tech B
   D. Neither Tech A nor Tech B

_____ **27.** Tech A says that when disabling the SRS system, the key should be turned on to confirm that the SRS light remains lit for at least 30 seconds; this verifies that the correct fuse was removed. Tech B says that after enabling the SRS system, the system should be tested by turning the key to the run position while sitting in the driver's seat. Who is correct?
   A. Tech A
   B. Tech B
   C. Both Tech A and Tech B
   D. Neither Tech A nor Tech B

_____ **28.** Tech A says that paying attention to tires and wheels can assist in identifying problems in the steering system. Tech B says that paying attention to tires and wheels can assist in identifying problems in the suspension system. Who is correct?
   A. Tech A
   B. Tech B
   C. Both Tech A and Tech B
   D. Neither Tech A nor Tech B

_____ **29.** Tech A says that the main problems that occur in the power steering system are play and poor fuel economy. Tech B says that the main problems that occur in the power steering system are play and unusual noises. Who is correct?
   A. Tech A
   B. Tech B
   C. Both Tech A and Tech B
   D. Neither Tech A nor Tech B

_____ **30.** Tech A says power steering pressure hose leaks can be catastrophic because the fluid can spray all over the hot engine and exhaust system, which can result in fire. Tech B says low fluid level will not result in noise or warning and the steering will feel fine. Who is correct?
   A. Tech A
   B. Tech B
   C. Both Tech A and Tech B
   D. Neither Tech A nor Tech B

_____ **31.** Tech A says that a steering wandering complaint can be caused by worn tie-rod ends. Tech B says that a hard-steering complaint can be caused by a worn power steering pump. Who is correct?
   A. Tech A
   B. Tech B
   C. Both Tech A and Tech B
   D. Neither Tech A nor Tech B

_____ **32.** Tech A says the power steering system should be flushed whenever the manufacturer specifies a fluid change or whenever the fluid appears to be contaminated or dirty. Tech B says the power steering system should be flushed whenever the power steering belt is replaced. Who is correct?
   A. Tech A
   B. Tech B
   C. Both Tech A and Tech B
   D. Neither Tech A nor Tech B

_____ **33.** Tech A says that to properly check a tie-rod end, the tie-rod end should be twisted and that any rotational movement means the joint is bad. Tech B says that to properly check a tie-rod end, the vehicle's weight should be on the wheels, and as the steering wheel is wiggled, movement in the tie-rod ends can be checked. Who is correct?

    A. Tech A

    B. Tech B

    C. Both Tech A and Tech B

    D. Neither Tech A nor Tech B

_____ **34.** Tech A says that when checking a tie-rod end, any rotational movement means the joint is bad. Tech B says that when checking a tie rod, any side-to-side movement in the joint means the tie rod is bad. Who is correct?

    A. Tech A

    B. Tech B

    C. Both Tech A and Tech B

    D. Neither Tech A nor Tech B

_____ **35.** Tech A says the SRS system should be disabled whenever someone is working on or around the steering column, airbags, other pyrotechnics, and any of the sensors. Tech B says a memory minder should be used to save any important data while the airbag system is disabled. Who is correct?

    A. Tech A

    B. Tech B

    C. Both Tech A and Tech B

    D. Neither Tech A nor Tech B

Suspension Systems Theory

At the start of each chapter, you will find the Learning Objectives from the textbook. These are your objectives as you make your way through the exercises in this workbook and the chapter in your textbook. The following activities have been designed to help you refresh your knowledge of the material in this chapter.

Learning Objectives

After reading this chapter, you will be able to:

- LO 28-01 Describe suspension system principles.
- LO 28-02 Describe suspension system spring components.
- LO 28-03 Describe fixed shock absorbers and struts.
- LO 28-04 Describe manually and automatically adjustable shocks.
- LO 28-05 Describe suspension system components.
- LO 28-06 Describe the main types of suspensions.
- LO 28-07 Describe the main types of front suspension systems.
- LO 28-08 Describe the main types of rear suspension systems.
- LO 28-09 Describe active and adaptive suspension systems.

Matching

Match the following terms with the correct description or example.

A. Axle
B. Caster
C. Deflecting force
D. Elasticity
E. Independent suspension
F. Negative camber
G. Overshoot
H. Rebound clip

I. Shroud
J. Sintering
K. Suspension system
L. Torsion bar
M. Trailing arm suspension
N. Uniform pitch
O. Yaw

_____ **1.** A system within a vehicle designed to isolate the vehicle body from road bumps and vibrations.

_____ **2.** A steel or plastic cover placed over the shock rod.

_____ **3.** A spring whose pitch (the distance from the center of one coil to the center of the adjacent coil) is the same distance throughout.

_____ **4.** The angle formed through the wheel pivot points when viewed from the side in comparison with a vertical line through the wheel.

_____ **5.** Movement of a vehicle around its z-axis (vertical axis) felt when the vehicle deviates from its straight path, as when skidding sideways, and the rear comes around.

_____ **6.** The shaft of the suspension system to which the tires and wheels are attached; they are used to drive or support the wheels.

_____ **7.** A metal strap that is warped around the leaf spring to prevent excessive flexing of the main leaf during rebound.

_____ **8.** The ability to deform and reform into the same shape.

_____ **9.** The process of using pressure and heat to bond metal particles.

_____ **10.** The amount a spring extends (springs back) past its original length following compression.

_____ **11.** Tilt of the top of the tire toward the centerline of the vehicle.

_____ **12.** A force that moves an object in a different direction or into a different shape.

_____ **13.** A bar made of a steel alloy that is fixed rigidly to the chassis at one end and the suspension control arm at the other to support the weight of a vehicle.

_____ **14.** A type of suspension system that uses upper and lower control arms.

_____ **15.** A system for allowing the up and down movement of one tire without affecting the other tire on that axle.

Multiple Choice

Read each item carefully, and then select the best response.

_____ **1.** A force that moves an object in a different direction or into a different shape is known as a(n) _____.
- **A.** Applied force
- **B.** Reaction force
- **C.** Deflecting force
- **D.** Suspension force

_____ **2.** Parts of a vehicle that are not supported by the suspension system, including the wheels, tires, brakes, axles, and steering and suspension parts not supported by springs, are considered _____.
- **A.** Unsprung mass
- **B.** Free weight
- **C.** Sprung weight
- **D.** Uniform weight

_____ **3.** The force transferred from the tire contact patch through the axle housing is called _____.
- **A.** Cornering force
- **B.** Driving thrust
- **C.** Reaction force
- **D.** Roll

_____ **4.** Movement around the vehicle's y-axis, commonly felt during hard braking or fast acceleration, when the front of the vehicle noses down or rises up slightly, is called _____.
- **A.** Yaw
- **B.** Pitch
- **C.** Roll
- **D.** Oscillation

_____ **5.** When a spring deflects easily under a light load, but its resistance increases as the load increases, it is said to have a(n) _____.
- **A.** Progressive rate of deflection
- **B.** Constant rate of deflection
- **C.** Uniform pitch
- **D.** Progressive pitch

_____ **6.** The rear of the multileaf is connected to the frame by a(n) _____, which provides a link between the spring eye and a bracket on the frame.
- **A.** Rigid spring hanger
- **B.** Intermediate hook
- **C.** Rebound clip
- **D.** Swinging shackle

_____ **7.** The twisting force that is applied by anchoring one end of an object and then applying a twisting force to the other end is called _____.
- **A.** Torsional load
- **B.** Compression
- **C.** Pitch
- **D.** Deflection

_____ **8.** A(n) _____ consists of a flexible rubber bladder, which seals the outside of the upper and lower halves of the shock absorber.
  A. Strut-type shock absorber
  B. Air spring
  C. Gas-pressurized shock absorber
  D. Suspension strut

_____ **9.** The newest style of electronic adjustable shock uses a special type of fluid called _____ fluid that has the unique characteristic of changing viscosity when exposed to a magnetic field.
  A. Anisotropic
  B. Magnetic hydraulic
  C. Magneto-rheological
  D. Frequency

_____ **10.** In control arm applications, particularly at the rear of a vehicle, a rubber bushing may be molded with a voided section; this bushing is called a _____.
  A. Rubber-bonded bushing
  B. Spring shackle bushing
  C. Gap-tooth bushing
  D. Compliance bushing

_____ **11.** What is another name for a rigid-axle coil-spring suspension that uses two bars similar to a Panhard rod and a pivot point on the axle to keep the axle from moving in turns?
  A. Watt's linkage
  B. Rigid non-drive axle suspension
  C. Trailing arm suspension
  D. Rear-wheel independent suspension

_____ **12.** Which of the following is NOT an example of a computerized suspension system?
  A. Stepper motor actuated
  B. Solenoid valve actuated
  C. Electromagnetic rheological
  D. Selector knob shock absorbers

_____ **13.** The distance between the ground and a specified part of the vehicle such as the fender well, upper control arm, or rocker panel is called _____.
  A. Clearance
  B. Setback
  C. Ride height
  D. Camber

_____ **14.** The side-to-side vertical tilt of the wheel, viewed from the front of the vehicle and measured in degrees, is called _____.
  A. Caster
  B. Camber
  C. Alignment
  D. Steering axis

_____ **15.** Preventing or reducing oscillations is called dampening.
  A. True
  B. False

_____ **16.** Pitch is vehicular movement along its x-axis. It is the rolling motion you feel when making a sharp corner and is generally what causes rollovers.
  A. True
  B. False

_____ **17.** Shock absorbers dampen spring oscillations by forcing water through small holes in a piston.
  A. True
  B. False

_____ **18.** The pitch of a spring is the distance from the center of one coil to the center of the adjacent coil; if they are evenly spaced, it is called uniform pitch.
A. True
B. False

_____ **19.** The longest leaf is called the main leaf, and it is rolled at both ends to form spring eyes; they are used to mount the spring to the frame of the vehicle.
A. True
B. False

_____ **20.** Leaf springs can be used across the chassis frame in a trailing arm suspension, or as part of the connecting link between two axle assemblies on a semi-rigid axle beam.
A. True
B. False

_____ **21.** The most widely used hydraulic shock absorber is the direct-acting telescopic type.
A. True
B. False

_____ **22.** Electronic adjustable-rate shock absorbers are also called self-leveling shock absorbers.
A. True
B. False

_____ **23.** Typically, straight pieces of steel used to either transfer motion or prevent motion within a vehicle's suspension system are called control arms.
A. True
B. False

_____ **24.** Rubber-bonded bushings are normally used for the front eye of the spring at the fixed shackle point and also in control arm applications.
A. True
B. False

_____ **25.** A solid axle is a nonindependent suspension because the wheels on both sides of the axle are connected together.
A. True
B. False

_____ **26.** MacPherson strut suspension can be used on either the front or the rear of the vehicle.
A. True
B. False

_____ **27.** The primary reason that the short-/long-suspension system was designed was to ensure correct wheel alignment as the vehicle corners.
A. True
B. False

_____ **28.** Most twin I-beam systems use a coil spring to support the weight of the vehicle.
A. True
B. False

_____ **29.** Rear-wheel suspension systems can be of the independent or nonindependent design.
A. True
B. False

_____ **30.** A wheel that leans away from the center of the vehicle at the top is said to have negative camber.
A. True
B. False

_____ **31.** When a tire hits an obstruction, there is a(n) _____, meaning the tire will move in response to the force applied by the obstruction.
A. Reaction force
B. Deflecting force
C. Applied force
D. Cornering force

_____ 32. _____ refers to the fluctuating of an object between two states, basically meaning that the spring compresses and rebounds over and over again.
  A. Braking torque
  B. Driving thrust
  C. Oscillation
  D. Elasticity

_____ 33. _____ force refers to the lateral movement of the axle housing during turning.
  A. Applied
  B. Cornering
  C. Deflection
  D. Reaction

_____ 34. _____ are swivel connections mounted in the outer ends of the control arm.
  A. Air springs
  B. Bushings
  C. Rebound clips
  D. Ball joints

_____ 35. A _____ is made from a single length of special wire, which is heated and wound on a former to produce the required shape.
  A. Coil spring
  B. Swinging shackle
  C. Sway bar
  D. Strut

_____ 36. A bar similar to the torsion bar is the _____, or antiroll bar.
  A. Rigid spring bar
  B. Panhard bar
  C. Sway bar
  D. Coil bar

_____ 37. If the suspension reaches its limit of travel, rubber _____ prevent direct metal-to-metal contact, thereby reducing jarring of the suspension components.
  A. Rods
  B. Bonds
  C. Arms
  D. Stops

_____ 38. A manual adjustable-rate shock absorber has a manual, external _____ adjustment.
  A. Damper rate
  B. Progressive rate
  C. Deflection rate
  D. Resistance rate

_____ 39. The A-arm style control arm is sometimes referred to as a _____ control arm, a relatively flat triangular part that mounts to the frame or subframe at each leg of the A.
  A. Suspension
  B. Wishbone
  C. Trailing
  D. Load-bearing

_____ 40. _____ bushings can be molded to form two halves, to fit into each side of the spring eye on the swinging shackle, which is located on the vehicle frame.
  A. Compliance
  B. Strut rod
  C. Rubber-bonded
  D. Spring shackle

_____ **41.** In strut suspension systems, the _____ is contained inside the strut.
   A. Spring
   B. Rod
   C. Shock absorber
   D. Control arm

_____ **42.** Also referred to as a track bar, a _____ sits parallel with the axle. One end connects to the frame of the vehicle, and the other end connects to the axle.
   A. Tie rod
   B. Tension rod
   C. Strut rod
   D. Panhard rod

_____ **43.** On rear-wheel drive vehicles with independent rear suspension, drive is transmitted to each wheel by external _____.
   A. Control arms
   B. Drive shafts
   C. Ball joints
   D. Shock absorbers

_____ **44.** _____ suspension is an electronically controlled air suspension system at all four wheels with a continuously adaptive dampening system.
   A. Adaptive air
   B. Independent
   C. Active
   D. Strut

_____ **45.** The arrow in the image is pointing to what part of the suspension system?

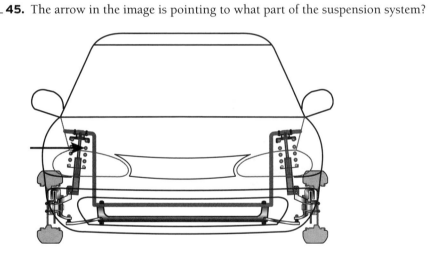

   A. Top cap (strut insulator)
   B. Coil spring
   C. Ball joint
   D. Steering knuckle

_____ **46.** The arrow in the image is pointing to what part of the suspension system?

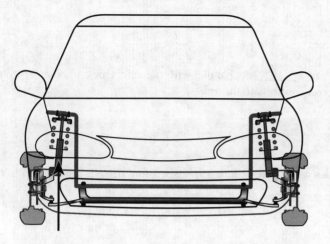

A. MacPherson strut assembly (shock absorber)
B. Lower control arm
C. Subframe
D. Wheel hub

_____ **47.** The arrow in the image is pointing to what part of the suspension system?

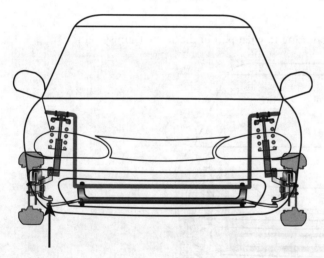

A. Top cap (strut insulator)
B. Ball joint
C. Lower control arm
D. Wheel rim

_____ **48.** The arrow in the image is pointing to what vehicular movement?

    A. Roll
    B. Yaw
    C. Pitch
    D. Coil

_____ **49.** The arrow in the image is pointing to what vehicular movement?

    A. Spring
    B. Roll
    C. Pitch
    D. Yaw

_____ **50.** The arrow in the image is pointing to what vehicular movement?

    A. Brake
    B. Roll
    C. Yaw
    D. Pitch

_____ **51.** The arrow in the image is pointing to what part of the gas-pressurized shock absorber?

    A. Rock shield
    B. Rod
    C. Oil seal
    D. Bushing

_____ **52.** The arrow in the image is pointing to what part of the gas-pressurized shock absorber?

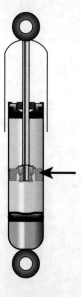

    A. Oil seal
    B. Mono tube
    C. Piston valve
    D. Free-floating separation piston

_____ **53.** The arrow in the image is pointing to what part of the gas-pressurized shock absorber?

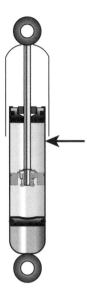

    A. Rock shield
    B. Rod
    C. Mono tube
    D. Piston valve

_____ **54.** The arrows in the image are pointing to what part of the electromagnetic shock absorber?

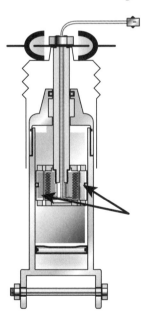

    A. Nitrogen gas
    B. Fluid paths
    C. High magnetic flux zones
    D. Free-floating separation pistons

_____ **55.** The arrow in the image is pointing to what part of the electromagnetic shock absorber?

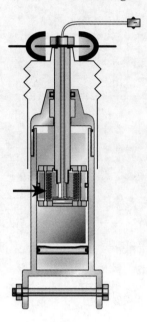

   A. Electromagnet
   B. Electrical connection
   C. Upper mount
   D. Free-floating separation piston

_____ **56.** The arrow in the image is pointing to what part of the electromagnetic shock absorber?

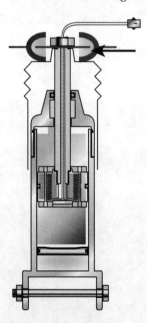

   A. Electromagnet
   B. Upper mount
   C. Free-floating separation piston
   D. Lower mount

_____ **57.** The arrow in the image is pointing to what part of the automatic load-adjustable suspension system?

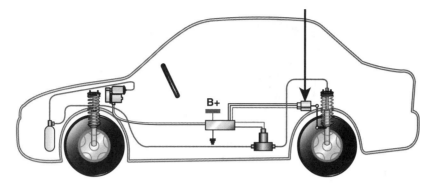

    A. Air compressor
    B. Control unit
    C. Rear height sensor
    D. 3-way solenoid valve

_____ **58.** The arrow in the image is pointing to what part of the automatic load-adjustable suspension system?

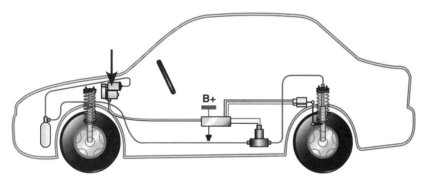

    A. Air compressor
    B. Reservoir
    C. Control unit
    D. Air-assisted strut

_____ **59.** The arrow in the image is pointing to what part of the automatic load-adjustable suspension system?

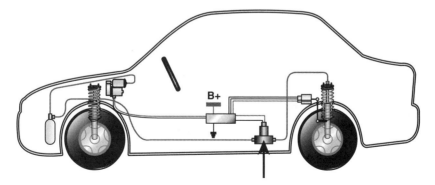

    A. Reservoir
    B. Rear height sensor
    C. Air-assisted strut
    D. 3-way solenoid valve

_____ **60.** The arrow in the image is pointing to what part of the strut suspension?

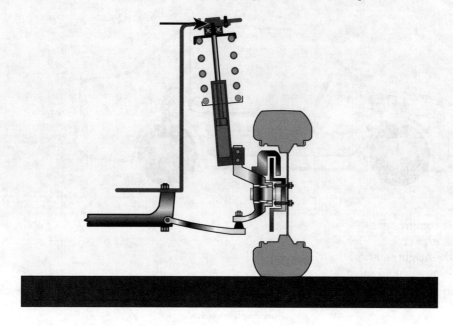

     A. Top cap (strut insulator)
     B. Coil spring
     C. Steering knuckle
     D. Wheel hub

_____ **61.** The arrow in the image is pointing to what part of the strut suspension?

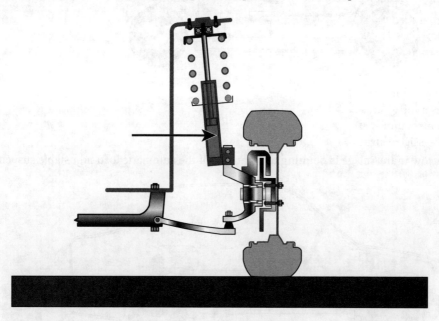

     A. Ball joint
     B. Lower control arm
     C. Body
     D. McPherson strut assembly (shock absorber)

_____ **62.** The arrow in the image is pointing to what part of the strut suspension?

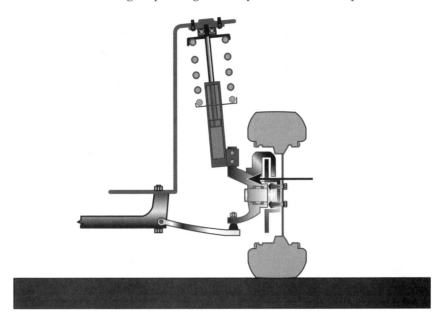

    A. Lower control arm
    B. Coil spring
    C. Steering knuckle
    D. Wheel hub

_____ **63.** The arrow in the image is pointing to what part of the adaptive air suspension?

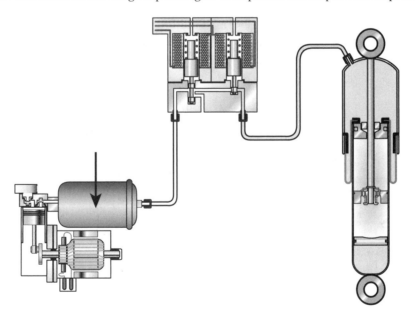

    A. Compressor
    B. Air drier
    C. Isolation solenoid valve
    D. Air shock

_____ **64.** The arrow in the image is pointing to what part of the adaptive air suspension?

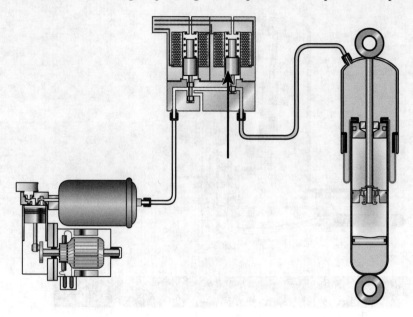

    A. Isolation solenoid valve
    B. Air shock
    C. Compressor
    D. Vent solenoid valve

_____ **65.** The arrow in the image is pointing to what part of the adaptive air suspension?

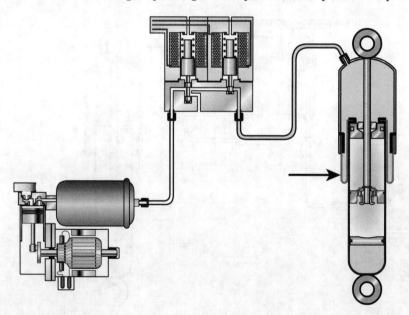

    A. Compressor
    B. Air drier
    C. Vent solenoid valve
    D. Air shock

_____ **66.** Tech A says that the wheels and tires are examples of unsprung weight. Tech B says that the exhaust system is an example of sprung weight. Who is correct?
    A. Tech A
    B. Tech B
    C. Both Tech A and Tech B
    D. Neither Tech A nor Tech B

_____ **67.** Tech A says that yaw is when a vehicle deviates from its straight path. Tech B says that roll is felt during hard braking. Who is correct?
   A. Tech A
   B. Tech B
   C. Both Tech A and Tech B
   D. Neither Tech A nor Tech B

_____ **68.** Tech A says that a vehicle will tend to pull toward the side with the most positive camber. Tech B says that a progressive-rate spring offers a soft ride but can also carry a heavier load. Who is correct?
   A. Tech A
   B. Tech B
   C. Both Tech A and Tech B
   D. Neither Tech A nor Tech B

_____ **69.** Tech A says that a shock dampens movement of the suspension in both upward and downward movement. Tech B says that a shock dampens only upward movement. Who is correct?
   A. Tech A
   B. Tech B
   C. Both Tech A and Tech B
   D. Neither Tech A nor Tech B

_____ **70.** Tech A says that ball joints must be unloaded when checking them for wear. Tech B says that ball joints must be loaded when checking them for wear. Who is correct?
   A. Tech A
   B. Tech B
   C. Both Tech A and Tech B
   D. Neither Tech A nor Tech B

_____ **71.** Tech A says that a dead axle is designed to carry the weight of the vehicle with no drive capability. Tech B says that checking toe-out on turns is a prealignment check. Who is correct?
   A. Tech A
   B. Tech B
   C. Both Tech A and Tech B
   D. Neither Tech A nor Tech B

_____ **72.** Tech A says that loose ball joints can cause the vehicle to wander. Tech B says that loose control arm bushings can affect alignment angles. Who is correct?
   A. Tech A
   B. Tech B
   C. Both Tech A and Tech B
   D. Neither Tech A nor Tech B

29 Servicing Suspension Systems

At the start of each chapter, you will find the Learning Objectives from the textbook. These are your objectives as you make your way through the exercises in this workbook and the chapter in your textbook. The following activities have been designed to help you refresh your knowledge of the material in this chapter.

Learning Objectives

After reading this chapter, you will be able to:

- LO 29-01 Describe suspension system service preliminaries.
- LO 29-02 Describe suspension system diagnosis.
- LO 29-03 Measure ride height and test shock absorbers.
- LO 29-04 Unload a suspension and measure ball joint play.
- LO 29-05 Replace stabilizer components and shock absorbers.
- LO 29-06 Remove coil springs and steering knuckles.
- LO 29-07 Remove control arms and ball joints.
- LO 29-08 Install and lubricate short-/long-arm (SLA) components.
- LO 29-09 Inspect and service strut assembly.
- LO 29-10 Inspect strut rods and bushings, leaf springs, and torsion bars.

Matching

Match the following steps with the correct sequence for performing ride height diagnosis.

A. Step 1   **D.** Step 4
B. Step 2   **E.** Step 5
C. Step 3

_____ **1.** Refer to the manufacturer's service information for correct measurement points and specifications.
_____ **2.** Check the vehicle for any nonstandard loads in the trunk or luggage area. Remove them temporarily while measuring ride height.
_____ **3.** Measure from points specified, such as from frame to ground on all four corners of the vehicle. Compare measurements to specifications.
_____ **4.** Check for properly sized, matching, and inflated tires. Correct any issues found.
_____ **5.** Inspect for bent components or a weak or broken spring if any measurements are not correct. If you are working with a torsion bar suspension, you may be able to adjust ride height to correct the condition.

Match the following steps with the correct sequence for replacing a shock absorber.

A. Step 1   **C.** Step 3
B. Step 2

_____ **1.** Pull the shock out by hand. Replace with new shock. Repeat on the other side.
_____ **2.** Raise the vehicle on a lift, and support the axle/control arm with a jack stand. Remove the upper bolts holding the shocks in place.
_____ **3.** Remove the lower bolts holding the shock in place.

Match the following steps with the correct sequence for lubricating a suspension and steering system.

A. Step 1   **D.** Step 4
B. Step 2   **E.** Step 5
C. Step 3   **F.** Step 6

_____ **1.** Push the grease gun nozzle fully over the fitting. It should snap into place. Add enough grease to see the seal or rubber boot rise slightly. Do not overfill a lubricated joint with grease.

_____ **2.** Attach a static cling sticker to the windshield, or reset the maintenance reminder system. Lower the vehicle and remove it from the lifting device.

_____ **3.** Clean each of the lubrication fittings and the grease gun nozzle by wiping all of them with a clean rag. You may need to remove a component's plugs and temporarily install a lubrication fitting. After the component has been lubricated, reinstall the original plug.

_____ **4.** If the fitting is clean and will not take grease, remove the grease zerk, and check for blockage. If found, the fitting must be replaced with a new fitting of the same size and angle and the joint relubricated.

_____ **5.** Check the service information to determine where the grease points are and the type of grease required. Also, look for any aftermarket grease fittings that may be installed. If so, grease them.

_____ **6.** Remove the nozzle from the fitting, and wipe away any excess grease from it. Repeat the procedure until all the appropriate joints have been lubricated.

Match the following steps with the correct sequence for inspecting leaf springs.

A. Step 1   **C.** Step 3
B. Step 2

_____ **1.** Test the security of the spring center bolt, and make sure the U-bolts are tight.

_____ **2.** Check the condition of the bushings and the spring shackles by placing a lever between the frame and the eye of the spring and prying against the spring.

_____ **3.** Safely raise and support the vehicle on a lift. Support the rear axle with tall screw jack stands. Check whether any of the leaves are cracked or broken and ensure that the noise deadening inserts are positioned correctly between the leaves.

Multiple Choice

Read each item carefully, and then select the best response.

_____ **1.** A customer has concerns of a bouncy ride, a reduction in steering control, and unusual noises when going over bumps. This would indicate problems in what system?
   A. Fuel
   B. Ignition
   C. Exhaust
   D. Suspension

_____ **2.** Which of the following tools can be used to diagnose the source of noise?
   A. Strut servicing kit
   B. Air chisel
   C. Chassis ear
   D. Scan tool

_____ **3.** Vehicle wander may be caused by all of the following conditions, EXCEPT _____.
   A. Worn joints and bushings
   B. Looseness in steering and suspension components
   C. Displaced caster angle
   D. Bent steering components

_____ **4.** You can check for noises by performing all of the following steps, EXCEPT _____.
   A. Driving over bumps
   B. Using chassis ears
   C. Racing the engine up and listening for noise
   D. Using an assistant to bounce the car up and down while you check for parts

_____ **5.** All of the following statements regarding suspension systems are correct, EXCEPT _____.
    A. An electronic stethoscope is the best tool for diagnosing caster angle problems
    B. If one of the front tires is low, the vehicle may pull to one side
    C. A four-wheel alignment is needed to verify the alignment angles
    D. Torque steer is a pull that occurs during heavy acceleration

_____ **6.** What is the best method for testing the lower ball joint on a MacPherson strut suspension?
    A. Raise the vehicle by the frame and allow the suspension to hang free.
    B. Test it at the floor level.
    C. Place a floor jack under the lower control arm and raise the wheel off the ground.
    D. Fit a wooden block between the arm and the frame.

_____ **7.** When the driver feels the steering wheel pull to one side, which of the following should be checked first?
    A. Looseness in the suspension system
    B. Problems in the braking system
    C. Tire pressure
    D. Four-wheel alignment

_____ **8.** Incorrect angles in the steering linkage due to a component being bent can result in what condition?
    A. Noise
    B. Bump steer
    C. Pull
    D. Hard steering

_____ **9.** Which of the following conditions should be inspected when the ride height does not match the manufacturer's specifications?
    A. Shock absorbers
    B. Suspension springs
    C. Upper ball joints
    D. Lower ball joints

_____ **10.** Any amount of play in a fixed part is problematic and frequently a result of part failure.
    A. True
    B. False

_____ **11.** If the measured ride height is greater or less than specified, the vehicle is not in correct alignment and may be causing or contributing to the customer's concern.
    A. True
    B. False

_____ **12.** Diagnosis of a problem must always start with a good interview of the technician.
    A. True
    B. False

_____ **13.** A vehicle with camber pull will pull in the direction of the highest negative camber.
    A. True
    B. False

_____ **14.** When checking shocks, the car should be bounced.
    A. True
    B. False

_____ **15.** If the lower control arm of a vehicle is replaced, the vehicle will likely need a wheel alignment to make sure that the _____.
    A. Engine is balanced
    B. MacPherson struts are properly installed
    C. Suspension system is adjusted properly
    D. Ball joints have proper play

_____ **16.** If a vehicle suspension is being serviced, then the vehicle will require a(n) _____ after the service is completed.
    A. Wheel alignment
    B. Steering column adjustment
    C. Tie-rod replacement
    D. Strut removal

_____ **17.** It is sometimes desirable to learn to perform wheel alignments before suspension system part replacement so students will understand _____.
   A. Steering concerns
   B. Body sway
   C. Alignment angles
   D. Steering alignment

_____ **18.** A(n) _____ is an air-operated hammer that can accept a variety of bits, including chisels and punches.
   A. Pickle fork
   B. Air chisel
   C. Strut compressor
   D. Strut servicing kit

_____ **19.** A _____ is used to remove and replace press-fit ball joints.
   A. Coil spring compressor
   B. Strut servicing kit
   C. Universal strut nut wrench kit
   D. Ball joint press tool

_____ **20.** A(n) _____ is used to separate ball joints, but ruins the grease seal.
   A. Pickle fork
   B. Air chisel
   C. Coil spring compressor
   D. Strut compressor

_____ **21.** _____ occurs when the vehicle's driver is holding the steering wheel steady and straight but the vehicle slowly begins to move to either the right or the left.
   A. Pull
   B. Hard steering
   C. Drift
   D. Torque steer

_____ **22.** _____ is a pull that occurs during heavy acceleration.
   A. Hard steering
   B. Torque steer
   C. Drift
   D. Vehicle wander

_____ **23.** _____ is a condition that occurs when the vehicle turns itself on hitting a bump.
   A. Bump steer
   B. Drift
   C. Pull
   D. Torque steer

_____ **24.** Ride height should typically not vary by more than _____ from side to side.
   A. Half a foot
   B. A foot
   C. An inch
   D. Half an inch

_____ **25.** A vehicle will pull to the side with the _____ camber.
   A. Most even
   B. Most negative
   C. Least positive
   D. Most positive

_____ **26.** Unwanted _____ in the suspension system can be extremely dangerous and should be corrected as soon as possible.
   A. Fit
   B. Tightness
   C. Looseness
   D. Tolerances

_____ **27.** Tech A states that a wheel alignment should be performed after any suspension system components are replaced. Tech B states that a wheel alignment does not need to be performed after suspension system component replacement. Who is correct?
   A. Tech A
   B. Tech B
   C. Both Tech A and Tech B
   D. Neither Tech A nor Tech B

_____ **28.** Tech A states that if the coil springs of a vehicle are damaged and need replacement, the vehicle will require a wheel alignment. Tech B states that replacement of coil springs does not require a wheel alignment. Who is correct?
   A. Tech A
   B. Tech B
   C. Both Tech A and Tech B
   D. Neither Tech A nor Tech B

_____ **29.** Major suspension components have been replaced on the vehicle in question. Tech A states that the vehicle will need a wheel alignment. Tech B states that the wheels need to be balanced after the replacement of the components. Who is correct?
   A. Tech A
   B. Tech B
   C. Both Tech A and Tech B
   D. Neither Tech A nor Tech B

_____ **30.** Tech A says that worn suspension components can oftentimes be located during a visual inspection and can be confirmed as faulty with a measuring tool if necessary. Tech B says that damaged components may require an alignment machine to be identified. Who is correct?
   A. Tech A
   B. Tech B
   C. Both Tech A and Tech B
   D. Neither Tech A nor Tech B

_____ **31.** Tech A says that when checking some suspension parts, a hydraulic jack needs to be placed under the lower control arm. Tech B says that when checking certain suspension parts, a long pry bar should be used to compress the coil springs, allowing you to test the ball joints. Who is correct?
   A. Tech A
   B. Tech B
   C. Both Tech A and Tech B
   D. Neither Tech A nor Tech B

_____ **32.** Tech A says torque steer is caused by low tire pressure. Tech B says steering return concerns may be caused by binding components. Who is correct?
   A. Tech A
   B. Tech B
   C. Both Tech A and Tech B
   D. Neither Tech A nor Tech B

_____ **33.** Tech A says pull is felt when the driver feels the steering wheel wanting to go to one side. Tech B says drift is when the driver is holding the steering wheel steady and straight but the vehicle slowly begins to move either right or left. Who is correct?
   A. Tech A
   B. Tech B
   C. Both Tech A and Tech B
   D. Neither Tech A nor Tech B

_____ **34.** Tech A says steering return problems describe the condition that occurs when the steering wheel does not return to straight ahead after a turn, also known as memory steer. Tech B says steering return can be caused by binding at pivot points or can be the result of insufficient positive caster. Who is correct?

A. Tech A

B. Tech B

C. Both Tech A and Tech B

D. Neither Tech A nor Tech B

_____ **35.** Tech A says that worn sway bar bushings, brackets, and link bushings can cause excessive body sway. Tech B says that unmatched tires can cause ride height to be out of specs. Who is correct?

A. Tech A

B. Tech B

C. Both Tech A and Tech B

D. Neither Tech A nor Tech B

Wheel Alignment

At the start of each chapter, you will find the Learning Objectives from the textbook. These are your objectives as you make your way through the exercises in this workbook and the chapter in your textbook. The following activities have been designed to help you refresh your knowledge of the material in this chapter.

Learning Objectives

After reading this chapter, you will be able to:

- LO 30-01 Define camber, caster, and toe.
- LO 30-02 Describe toe-out on turns and turning radius.
- LO 30-03 Describe steering axis inclination, included angle, and scrub radius.
- LO 30-04 Describe thrust angle, centerline, setback, and ride height.
- LO 30-05 Describe types of wheel alignment.
- LO 30-06 Describe wheel alignment preliminaries and adjustment methods.
- LO 30-07 Prepare a vehicle for alignment.
- LO 30-08 Adjust caster, camber, and toe.
- LO 30-09 Measure secondary alignment angles.

Matching

Match the following terms with the correct description or example.

A. Camber
B. Caster
C. Included angle
D. Ride height

E. Scrub radius
F. Static toe
G. Thrust line
H. Toe-out on turns

_____ **1.** The tilt of the steering axis from the vertical, as viewed from the side of the vehicle.

_____ **2.** A setting designed to compensate for slight wear in steering components that may cause the wheels to turn outward or inward while the vehicle is in motion.

_____ **3.** The imaginary line drawn perpendicular to the center of the rear axle.

_____ **4.** The distance between two imaginary points on the road surface—the point of center contact between the road surface and the tire, and the intersecting point where the steering axis centerline and the tire centerline contact the road surface.

_____ **5.** The side-to-side vertical tilt of the wheel as viewed from the front of the vehicle.

_____ **6.** The distance from the ground to a certain specified part of the vehicle.

_____ **7.** The toe setting of the front wheels when the vehicle is turning.

_____ **8.** The angle that is formed between the steering axis inclination and the camber line.

Match the following steps with the correct sequence for preparing a vehicle for a wheel alignment.

A. Step 1   **F.** Step 6
B. Step 2   **G.** Step 7
C. Step 3   **H.** Step 8
D. Step 4   **I.** Step 9
E. Step 5

_____ **1.** With the vehicle raised, inspect all suspension and steering components, including the wheel bearings. Repair or replace all damaged or worn suspension components.

_____ **2.** Measure the vehicle's ride height.

_____ **3.** Remove any heavy items from the trunk and passenger compartments.

_____ **4.** Check the size and condition of all four tires. Adjust the air pressure to specifications.

_____ **5.** Bounce each corner of the vehicle to check the correct functioning of the shock absorbers.

_____ **6.** Attach the wheel units of the wheel alignment machine.

_____ **7.** Position the vehicle on the wheel alignment ramp, making sure the front tires are positioned correctly on the turntables.

_____ **8.** Check the play of the steering wheel. Correct any excess play before undertaking the wheel alignment.

_____ **9.** Position the rear wheels on the slip plates or rear turntables.

Multiple Choice

Read each item carefully, and then select the best response.

_____ **1.** The distance between two imaginary points on the road surface is called the _____.
   A. Scrub radius
   B. Ackermann angle
   C. Camber
   D. Thrust line

_____ **2.** If the camber line is outside of the steering axis inclination line, then it has _____.
   A. Negative offset
   B. Zero scrub radius
   C. Positive scrub radius
   D. Toe-out on turns

_____ **3.** The condition in which the fronts of the wheels are closer together than the rears of the wheels is called _____.
   A. Toe-in
   B. Toe-out
   C. Static toe
   D. Toe-out on turns

_____ **4.** The _____ refers to the relationship between the centerline of the vehicle and the angle of the rear tires.
   A. Thrust line
   B. Thrust angle
   C. Turning radius
   D. Ackermann angle

_____ **5.** What type of alignment is outdated and almost never performed on modern vehicles?
   A. Front-end two-wheel alignment
   B. Thrust-angle alignment
   C. Four-wheel alignment
   D. Toe-in alignment

_____ **6.** If the camber line is outside of the steering axis inclination (SAI) line, then it has positive offset or positive scrub radius.
   A. True
   B. False

_____ **7.** The Ackermann angle is the angle the steering arms make with the steering axis, projected toward the center of the rear axle.
   A. True
   B. False

_____ **8.** Setback is an alignment angle that is not adjustable but that allows the technician to diagnose the vehicle.
   A. True
   B. False

_____ **9.** Wheels are positioned on the suspension at certain angles to provide for easy driving of the vehicle. These angles, taken together, determine the vehicle's _____.
   A. Ackermann angle
   B. SAI
   C. Wheel alignment
   D. Negative scrub radius

_____ **10.** The angle formed between the SAI and the camber line is called the _____.
   A. Thrust angle
   B. Included angle
   C. Ackermann angle
   D. Static angle

_____ **11.** _____ on turns is the relative toe setting of the front wheels as the vehicle turns.
   A. Toe-in
   B. Toe-out
   C. Toe-right
   D. Toe-left

_____ **12.** The _____ is a geometric alignment of linkages in a vehicle's steering such that the wheels on the inside of a turn are able to move in a different circle radius than the wheels on the outside.
   A. Ackermann principle
   B. Scrub radius
   C. Included angle
   D. Thrust line

_____ **13.** _____ is a measure of how small a circle the vehicle can turn in when the steering wheel is turned to the limit.
   A. Scrub radius
   B. Zero scrub radius
   C. Steering radius
   D. Turning radius

_____ **14.** What alignment angle is shown in the image?

   A. Positive camber
   B. Positive caster
   C. Negative camber
   D. Negative caster

_____ **15.** What alignment angle is shown in the image?

    A. Positive camber
    B. Positive caster
    C. Negative camber
    D. Negative caster

_____ **16.** What alignment angle is shown in the image?

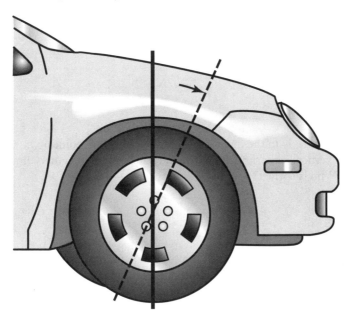

    A. Positive camber
    B. Positive caster
    C. Negative camber
    D. Negative caster

_____ **17.** What alignment angle is shown in the image?

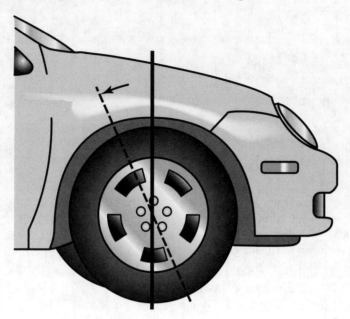

A. Positive camber
B. Positive caster
C. Negative camber
D. Negative caster

_____ **18.** Tech A says that thrust angle refers to the direction the front wheels are pointing in. Tech B says that scrub radius refers to the vertical centerline of the tire in relation to an imaginary line through the steering knuckle pivots. Who is correct?
A. Tech A
B. Tech B
C. Both Tech A and Tech B
D. Neither Tech A nor Tech B

_____ **19.** Tech A says that positive toe is when the fronts of the tires are farther apart than the rears of the tires. Tech B says that a ball joint in a MacPherson strut suspension is a follower joint (not loaded). Who is correct?
A. Tech A
B. Tech B
C. Both Tech A and Tech B
D. Neither Tech A nor Tech B

_____ **20.** Tech A says that the front wheels should be aligned before the rear wheels. Tech B says that front wheel toe should be adjusted after front wheel camber and caster. Who is correct?
A. Tech A
B. Tech B
C. Both Tech A and Tech B
D. Neither Tech A nor Tech B

CHAPTER

31

Principles of Braking

At the start of each chapter, you will find the Learning Objectives from the textbook. These are your objectives as you make your way through the exercises in this workbook and the chapter in your textbook. The following activities have been designed to help you refresh your knowledge of the material in this chapter.

Learning Objectives

After reading this chapter, you will be able to:

- LO 31-01 Describe the history of brake development.
- LO 31-02 Describe braking fundamentals.
- LO 31-03 Describe the physics of braking.
- LO 31-04 Describe friction, heat transfer, and brake fade.
- LO 31-05 Describe rotational force, weight transfer, and levers.
- LO 31-06 Describe the common types of automotive brakes.

Matching

Match the following terms with the correct description or example.

A. Band brake	**K.** Kinetic energy
B. Brake assist	**L.** Master cylinder
C. Brake fade	**M.** Mechanical advantage
D. Brakes	**N.** Mechanical disadvantage
E. Conservation of energy	**O.** Parking brake
F. Disc brakes	**P.** Rotational force
G. Drum brakes	**Q.** Scrub brakes
H. Friction	**R.** Top hat parking brake
I. Fulcrum	**S.** Weight transfer
J. Heat fade	

_____ **1.** A type of brake system that forces stationary brake pads against the outside of a rotating brake rotor.

_____ **2.** A brake system that uses leverage to force a friction block against one or more wheels.

_____ **3.** When the load distance on a lever is greater than the effort distance, which means the effort required to move the load is greater than the load itself.

_____ **4.** A physical law that states that energy cannot be created or destroyed.

_____ **5.** A drum brake that is located inside a disc brake rotor in order to act as a parking brake.

_____ **6.** A type of brake system that forces brake shoes against the inside of a brake drum.

_____ **7.** An enhanced safety system built into some antilock brake systems (ABS) that anticipates a panic stop and applies the maximum braking force to slow the vehicle as quickly as possible.

_____ **8.** The force created by the rotating wheel when the brakes are applied; it causes the brake components to twist the brake support and ultimately the vehicle in the direction of wheel rotation.

_____ **9.** The reduction in stopping power caused by a change in the brake system such as overheating, water, or overheated brake fluid.

_____ **10.** The ratio of load and effort for any simple machine such as a lever.

_____ **11.** A type of brake that utilizes a steel band lined with friction material that wraps around a brake drum to slow the drum.

_____ **12.** The half-round bearing that the rocker moves on as a bearing surface.

_____ **13.** Weight moving from one set of wheels to the other set of wheels during braking, acceleration, or cornering.

_____ **14.** Brake fade caused by the buildup of heat in braking surfaces, which get so hot they cannot create any additional heat, leading to a loss of friction.

_____ **15.** The relative resistance to motion between any two bodies in contact with each other.

_____ **16.** Converts the brake pedal force into hydraulic pressure, which is then transmitted via brake lines and hoses to one or more pistons at each wheel brake unit.

_____ **17.** A brake system used for holding the vehicle when it is stationary.

_____ **18.** A system made up of hydraulic and mechanical components designed to slow or stop a vehicle.

_____ **19.** The energy of an object in motion; it increases by the square of the speed.

Multiple Choice

Read each item carefully, and then select the best response.

_____ **1.** The old _____ braking systems caused the vehicle to veer dangerously to one side when braking and required frequent adjustment.
  A. Scrub
  B. Band
  C. Disc
  D. Drum

_____ **2.** What type of system uses a computer to monitor each wheel's speed and either hold, decrease, or apply hydraulic pressure to each wheel to prevent wheel lockup and maintain the maximum amount of braking power just short of brake lockup?
  A. Brake assist
  B. ABS
  C. Brake-by-wire
  D. Brake pedal emulator

_____ **3.** In a brake-by-wire system, a _____ tells the computer how firmly the driver intends to brake, which then sends control signals to the appropriate brake actuators.
  A. Brake pedal emulator
  B. Service brake
  C. Hydraulic actuator
  D. Kinetic sensor

_____ **4.** What law states that an object will stay at rest or at uniform speed unless it is acted on by an outside force?
  A. Euler's first law of motion
  B. The law of conservation of energy
  C. Newton's first law of motion
  D. Kirchhoff's law

_____ **5.** The amount of friction between two moving surfaces in contact with each other is expressed as a ratio and is called the _____.
  A. Law of conservation of energy
  B. Coefficient of friction
  C. Rate of heat transfer
  D. Second law of motion

_____ **6.** When brakes are operated on a moving vehicle, a _____ force is generated.
  A. Cornering
  B. Directional
  C. Deflecting
  D. Rotational

_____ **7.** Brake pedals are usually a _____. They pivot at the top end (fulcrum), the foot pressure (effort) is applied to the bottom end, and the master cylinder (load) is applied between the two.
   A. Lever of the first order
   B. Lever of the second order
   C. Lever of the third order
   D. Lever of the fourth order

_____ **8.** What type of parking brake uses a small drum brake to prevent the drive shaft from turning?
   A. Drum-style parking brake
   B. Transmission-mounted parking brake
   C. Top hat parking brake
   D. Electric parking brake

_____ **9.** The band braking system was used for more than 2,000 years with virtually no change.
   A. True
   B. False

_____ **10.** Giving greater control of the braking system to the computer increases driving safety.
   A. True
   B. False

_____ **11.** Aggressive driving causes tires to become hot and possibly overheated, thus reducing their ability to obtain maximum traction.
   A. True
   B. False

_____ **12.** The service brake is usually operated by hand, but some vehicles use a foot-activated pedal.
   A. True
   B. False

_____ **13.** Faster-moving objects have more kinetic energy than slower-moving objects of the same weight.
   A. True
   B. False

_____ **14.** An outside force needs to act upon a vehicle to cause it to decelerate; that force comes from the mass of the Earth.
   A. True
   B. False

_____ **15.** Static friction is resistance between moving surfaces and is present in standard brakes.
   A. True
   B. False

_____ **16.** Most of the heat generated by the braking process radiates into the atmosphere.
   A. True
   B. False

_____ **17.** Water fade is caused by water-soaked brake linings acting like a lubricant and lowering the coefficient of friction between the braking surfaces.
   A. True
   B. False

_____ **18.** Modern drum and disc brake systems are regularly fitted with an ABS that monitors the speed of each wheel and prevents wheel lockup or skidding, no matter how hard brakes are applied or how slippery the road surface is.
   A. True
   B. False

_____ **19.** Some vehicles come equipped with adjustable pedal assemblies that allow the driver to raise or lower the brake and throttle pedal assembly for personal comfort.
   A. True
   B. False

_____ **20.** A(n) _____ system does away with the hydraulic portion of the brake system and replaces it with sensors, wires, an electronic control unit, and electrically actuated motors to apply individual brake units at each wheel.
   A. Antilock brake
   B. Regenerative brake
   C. Brake-by-wire
   D. Traction control

_____ **21.** In a _____ braking system, the amount of stopping power is controlled by how much electricity is being generated.
   A. Regenerative
   B. Scrub
   C. Brake-by-wire
   D. Band

_____ **22.** There are two brake systems on all vehicles—a _____ brake and a _____ brake.
   A. Service, parking
   B. Traction, parking
   C. Regenerative, service
   D. Traction, service

_____ **23.** By using _____ of different sizes, hydraulic forces can be increased or reduced, allowing designers to obtain the desired braking force for each wheel.
   A. Pads
   B. Tires
   C. Cylinders
   D. Drums

_____ **24.** Heavier objects have more _____ energy than lighter objects moving at the same speed.
   A. Kinetic
   B. Thermal
   C. Potential
   D. Elastic

_____ **25.** In an automobile, _____, or an increase in kinetic energy, is caused by the power from the engine.
   A. Braking
   B. Acceleration
   C. Deceleration
   D. Idling

_____ **26.** _____ is caused by the buildup of heat in the braking surfaces, which get so hot they cannot create any additional heat, leading to a loss of friction.
   A. Brake fade
   B. Hydraulic fade
   C. Heat fade
   D. Water fade

_____ **27.** _____ fade is caused by the brake fluid becoming so hot that it boils.
   A. Water
   B. Hydraulic
   C. Heat
   D. Brake

_____ **28.** Brake systems use _____ and mechanical advantage to apply service and parking brakes.
   A. Levers
   B. Pedals
   C. Discs
   D. Fulcrums

_____ **29.** On drum brakes, a drum-style parking brake _____ applies the brake shoes against the drum.
   A. Manually
   B. Mechanically
   C. Hydraulically
   D. Pneumatically

_____ **30.** What type of lever is shown in the image?

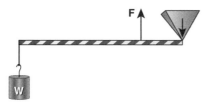

    A. Lever of the first order
    B. Lever of the second order
    C. Lever of the third order
    D. Lever of the fourth order

_____ **31.** What type of lever is shown in the image?

    A. Lever of the first order
    B. Lever of the second order
    C. Lever of the third order
    D. Lever of the fourth order

_____ **32.** What type of lever is shown in the image?

    A. Lever of the first order
    B. Lever of the second order
    C. Lever of the third order
    D. Lever of the fourth order

_____ **33.** The arrow in the image is pointing to what part of the hydraulic brake system?

    A. Wheel brake units
    B. Lines
    C. Pedals
    D. Hoses

_____ **34.** The arrow in the image is pointing to what part of the hydraulic brake system?

 A. Lines
 B. Wheel brake units
 C. Master cylinder
 D. Hoses

_____ **35.** The arrow in the image is pointing to what part of the hydraulic brake system?

 A. Pedals
 B. Wheel brake units
 C. Hoses
 D. Master cylinder

_____ **36.** The arrow in the image is pointing to what part of the power brake booster?

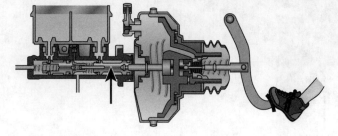

 A. Booster
 B. Pedal
 C. Secondary piston
 D. Primary piston

_____ **37.** The arrow in the image is pointing to what part of the power brake booster?

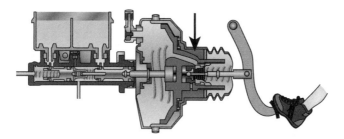

    A. Pedal
    B. Master cylinder
    C. Booster
    D. Secondary piston

_____ **38.** The arrow in the image is pointing to what part of the power brake booster?

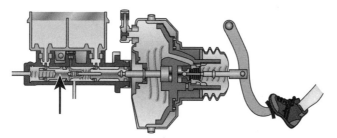

    A. Primary piston
    B. Master cylinder
    C. Pedal
    D. Secondary piston

_____ **39.** The arrow in the image is pointing to what part of the top hat parking brake design?

    A. Return spring
    B. Adjuster
    C. Spring
    D. Parking brake shoe

_____ **40.** The arrow in the image is pointing to what part of the top hat parking brake design?

A. Adjuster
B. Parking brake shoe
C. Return spring
D. Spring

_____ **41.** The arrow in the image is pointing to what part of the top hat parking brake design?

A. Parking brake shoe
B. Adjuster
C. Spring
D. Return spring

_____ **42.** The arrow in the image is pointing to what part of the drum-style design parking brake?

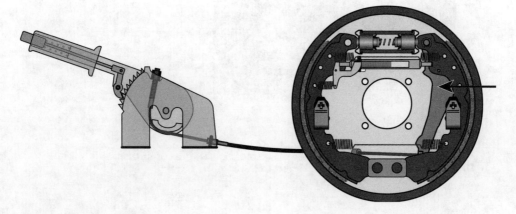

A. Parking brake strut
B. Parking brake cable
C. Parking brake lever
D. Parking brake pivot

_____ **43.** The arrow in the image is pointing to what part of the drum-style design parking brake?

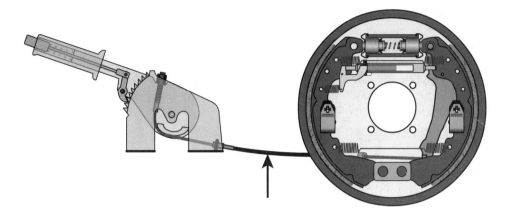

A. Parking brake cable
B. Parking brake lever
C. Parking brake pivot
D. Parking brake strut

_____ **44.** The arrow in the image is pointing to what part of the drum-style design parking brake?

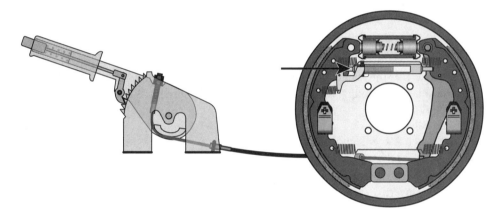

A. Parking brake pivot
B. Parking brake strut
C. Parking brake lever
D. Parking brake cable

_____ **45.** Tech A says that regenerative braking converts brake heat into electricity. Tech B says that regenerative braking converts electrical energy into braking energy. Who is correct?
A. Tech A
B. Tech B
C. Both Tech A and Tech B
D. Neither Tech A nor Tech B

_____ **46.** Tech A says that an ABS allows the front wheels to be steered during a panic stop. Tech B says that ABS sensor inputs control wheel speed during brake events. Who is correct?
A. Tech A
B. Tech B
C. Both Tech A and Tech B
D. Neither Tech A nor Tech B

_____ **47.** Tech A says that water-soaked brake shoes will cause brake fade. Tech B says that disc brakes dissipate heat faster than drum brakes. Who is correct?
  A. Tech A
  B. Tech B
  C. Both Tech A and Tech B
  D. Neither Tech A nor Tech B

_____ **48.** Tech A says that a brake-by-wire system can start to apply brakes before the driver can step on the brake pedal. Tech B says that brake-by-wire systems use heavy cables to transmit the brake pedal force to the wheel brake units. Who is correct?
  A. Tech A
  B. Tech B
  C. Both Tech A and Tech B
  D. Neither Tech A nor Tech B

_____ **49.** Tech A says that tire pressure does not affect braking. Tech B says that heavy vehicle loads increase stopping distance. Who is correct?
  A. Tech A
  B. Tech B
  C. Both Tech A and Tech B
  D. Neither Tech A nor Tech B

_____ **50.** Tech A says that light-duty service brakes are applied hydraulically. Tech B says that light-duty parking brakes are applied hydraulically. Who is correct?
  A. Tech A
  B. Tech B
  C. Both Tech A and Tech B
  D. Neither Tech A nor Tech B

_____ **51.** Tech A says that the heavier the vehicle, the more stopping power is needed. Tech B says that the faster a vehicle is moving, the more braking power is needed. Who is correct?
  A. Tech A
  B. Tech B
  C. Both Tech A and Tech B
  D. Neither Tech A nor Tech B

_____ **52.** Tech A says that kinetic energy is created during braking to stop the vehicle. Tech B says that kinetic energy is converted to heat energy during braking. Who is correct?
  A. Tech A
  B. Tech B
  C. Both Tech A and Tech B
  D. Neither Tech A nor Tech B

_____ **53.** Tech A says that the brake pedal uses leverage to multiply foot pressure. Tech B says that when braking hard while moving forward, the vehicle's weight transfers to the rear wheels, increasing their traction. Who is correct?
  A. Tech A
  B. Tech B
  C. Both Tech A and Tech B
  D. Neither Tech A nor Tech B

_____ **54.** Tech A says that friction brakes can fade due to overheating of the brake lining. Tech B says that friction brakes can fade due to overheating of the brake fluid. Who is correct?
  A. Tech A
  B. Tech B
  C. Both Tech A and Tech B
  D. Neither Tech A nor Tech B

Hydraulic and Power Brakes Theory

At the start of each chapter, you will find the Learning Objectives from the textbook. These are your objectives as you make your way through the exercises in this workbook and the chapter in your textbook. The following activities have been designed to help you refresh your knowledge of the material in this chapter.

Learning Objectives

After reading this chapter, you will be able to:

- LO 32-01 Describe hydraulic principles.
- LO 32-02 Describe brake fluid types and characteristics.
- LO 32-03 Describe master cylinder construction and operation.
- LO 32-04 Describe quick take-up and antilock brake system (ABS) master cylinders.
- LO 32-05 Describe brake pedal assemblies and divided hydraulic systems.
- LO 32-06 Describe brake lines and hoses.
- LO 32-07 Describe proportioning valves and their operation.
- LO 32-08 Describe metering, pressure differential, and combination valves.
- LO 32-09 Describe brake warning light and stop light operation.
- LO 32-10 Describe the operation of vacuum brake boosters.
- LO 32-11 Describe the operation of hydraulic brake boosters.

Matching

Match the following terms with the correct description or example.

A. Bleeding
B. Brake fluid
C. Brake hose
D. Brake lines
E. Compensating port
F. Inlet port
G. Input force
H. Load transfer
I. Metering valve
J. Output force

K. Outlet port
L. Poppet valve
M. Primary cup
N. Primary piston
O. Quick take-up valve
P. Residual pressure valve
Q. Secondary cup
R. Secondary piston
S. Working pressure

_____ 1. A brake piston in the master cylinder moved directly by the pushrod or the power booster; it generates hydraulic pressure to move the secondary piston.

_____ 2. Force that equals the working pressure multiplied by the surface area of the output piston, expressed as pounds, newtons, or kilograms.

_____ 3. In drum brake systems, a valve that maintains pressure in the wheel cylinders slightly above atmospheric pressure so air does not enter the system through the seals in the wheel cylinders.

_____ 4. The process of removing air from a hydraulic braking system.

_____ 5. The force applied to the input piston, measured in either pounds or kilograms.

_____ 6. A valve used on vehicles equipped with older rear drum/front disc brakes to delay application of the front disc brakes until the rear drum brakes are applied. Located in line with the front disc brakes.

_____ **7.** Hydraulic fluid that transfers forces under pressure through the hydraulic lines to the wheel braking units.

_____ **8.** Links the cylinder to the brake lines.

_____ **9.** Made of seamless, double-walled steel and able to transmit over 1,000 psi (6,895 kPa) of hydraulic pressure through the hydraulic brake system.

_____ **10.** A piston that is moved by hydraulic pressure generated by the primary piston in the master cylinder.

_____ **11.** Connects the reservoir with the space around the piston and between the piston cups in a brake master cylinder.

_____ **12.** The pressure within a hydraulic system while the system is being operated.

_____ **13.** A valve used to release excess pressure from the larger piston in a quick take-up master cylinder once the brake pads have contacted the brake rotors.

_____ **14.** A valve that controls the flow of brake fluid at usually preset pressures.

_____ **15.** Connects the brake fluid reservoir to the master cylinder bore when the piston is fully retracted, allowing for expansion and contraction of the brake fluid.

_____ **16.** A seal that prevents loss of fluid from the rear of each piston in the master cylinder.

_____ **17.** A seal that holds pressure in the master cylinder when force is applied to the piston.

_____ **18.** A flexible section of the brake lines between the body and the suspension to allow for steering and suspension movement.

_____ **19.** Weight transfer from one set of wheels to the other set of wheels during braking, acceleration, or cornering.

Multiple Choice

Read each item carefully, and then select the best response.

_____ **1.** What law states that pressure applied to a fluid in one part of a closed system will be transmitted without loss to all other areas of the system?
  A. Newton's law
  B. Pascal's law
  C. Kirchhoff's law
  D. Ohm's law

_____ **2.** All of the following are variables to consider when talking about pressure and force in hydraulic systems, EXCEPT _____.
  A. Input force
  B. Fluid type
  C. Output force
  D. Working pressure

_____ **3.** Standard brake fluid is _____, which means it absorbs water.
  A. Hydrotropic
  B. Aerated
  C. Hygroscopic
  D. Hydraulic

_____ **4.** Brake fluids are tested to ensure they meet the standards of quality for all of the following, EXCEPT _____.
  A. Stability
  B. Resistance to oxidation
  C. pH value
  D. Freezing point

_____ **5.** What type of brake fluid is silicone based?
  A. DOT 2
  B. DOT 3
  C. DOT 4
  D. DOT 5

_____ **6.** When the brake pedal is released quickly, the use of small holes drilled in the piston so that brake fluid from the reservoir can pass through the inlet port and past the edge of the primary cup, thus preventing a vacuum from being created, is called _____.

   A. Recuperation

   B. Compensating

   C. Aeration

   D. Residual pressure release

_____ **7.** The _____ master cylinder is a tandem master cylinder used in divided systems.

   A. Single-piston

   B. Dual-piston

   C. ABS

   D. Quick take-up

_____ **8.** What type of flare is sometimes called a bubble flare?

   A. Inverted double flare

   B. International Standards Organization flare

   C. Mushroom flare

   D. Deutsches Institut für Normung (DIN) flare

_____ **9.** Which of the following components is used to modify pressures within the hydraulic braking system?

   A. Proportioning valves

   B. Brake pedal pushrod

   C. Float switches

   D. Reservoirs

_____ **10.** All of the following conditions will cause the brake warning light to illuminate, EXCEPT _____.

   A. Parking brake is engaged

   B. Brake fluid is low

   C. Brake pedal is pressed

   D. Prove out circuit

_____ **11.** In a closed system, hydraulic pressure is transmitted equally in all directions throughout the system.

   A. True

   B. False

_____ **12.** Because silicone-based fluid tends to aerate when forced at high pressure through small passages, it is not to be used in any vehicle equipped with antilock brakes.

   A. True

   B. False

_____ **13.** The outlet port adjusts for changes in the volume of the brake fluid ahead of the piston.

   A. True

   B. False

_____ **14.** Tandem systems can be split diagonally so one front wheel is paired with the rear wheel on the opposite side of the vehicle in one brake circuit, and vice versa in the other circuit.

   A. True

   B. False

_____ **15.** Master cylinder reservoirs can be built into the master cylinder housing or can be a separate unit.

   A. True

   B. False

_____ **16.** Master cylinder reservoirs should be filled all the way to the top to prevent air bubbles from entering the system.

   A. True

   B. False

_____ **17.** A front-engine, rear-wheel drive car has around 40% of its load on its rear wheels and 60% on its front wheels.

   A. True

   B. False

_____ **18.** While brake hoses are designed to be flexible, they should never be pinched, kinked, or bent tighter than a specified radius.
   A. True
   B. False

_____ **19.** During heavy braking, master cylinder pressure can reach the poppet valve's crack point.
   A. True
   B. False

_____ **20.** Adjustable proportioning valves are not recommended for most applications due to the amount of trial and error necessary to set them properly.
   A. True
   B. False

_____ **21.** The combination valve can combine the pressure differential valve, metering valve, and proportioning valve(s) in one unit.
   A. True
   B. False

_____ **22.** The brake warning light is activated by a normally closed switch located on the brake pedal assembly.
   A. True
   B. False

_____ **23.** The vacuum-assisted power booster uses the difference between engine vacuum and atmospheric pressure to increase the force that acts on the master cylinder pistons.
   A. True
   B. False

_____ **24.** When the brake pedal is released, the atmospheric valve is opened and the vacuum valve closes.
   A. True
   B. False

_____ **25.** While hydraulic brake boosters use power steering fluid to operate the booster, the master cylinder portion of the system still uses brake fluid.
   A. True
   B. False

_____ **26.** The same _____ applied over different-sized surface areas will produce different levels of force.
   A. Temperature
   B. Weight
   C. Pressure
   D. Moisture

_____ **27.** If brake fluid boils, it turns from a liquid to a _____, which is compressible.
   A. Vapor
   B. Solid
   C. Particle
   D. Plasma

_____ **28.** Brake fluids are graded against compliance standards set by the United States _____.
   A. Transportation Organization
   B. Department of Transportation
   C. Standards Organization
   D. Department of Braking Fluids

_____ **29.** _____ master cylinders have one piston with two cups: a primary cup and a secondary cup.
   A. Double-piston
   B. Quick take-up
   C. Single-piston
   D. Tandem

_____ **30.** _____ master cylinders combine two master cylinders within a common housing that share a common cylinder bore.
   A. Single-piston
   B. Double-bore
   C. Primary piston
   D. Tandem

_____ **31.** Tandem systems can be split _____ to _____ so the front brakes operate from one circuit and the rear brakes from the other.
  A. Front, rear
  B. Left, right
  C. Front, left
  D. Right, rear

_____ **32.** _____ master cylinders are used on disc brake systems that are equipped with low-drag brake calipers.
  A. Single-piston
  B. Double-piston
  C. Quick take-up
  D. Tandem

_____ **33.** The _____ uses leverage to multiply the effort from the driver's foot to the master cylinder.
  A. Booster
  B. Brake pedal
  C. Proportioning valve
  D. Drum brake

_____ **34.** An alternative to the diagonal, or X, pattern braking system split used in front-engine, front-wheel drive vehicles is a(n) _____ system.
  A. L split
  B. Front-rear
  C. Vertical
  D. U split

_____ **35.** The _____ flare is created by first flaring the end of the tube outward in a Y shape. Then, about half of the flared end is folded inside of itself, leaving a double-thick section of brake line on the flared portion of the Y.
  A. Inverted triple
  B. International Standards Organization
  C. Inverted double
  D. Inverted Y

_____ **36.** Many brake hoses use _____ fittings to connect the hose to the wheel unit.
  A. Banjo
  B. Double flare
  C. Inverted double flare
  D. Y

_____ **37.** _____ valves reduce brake pressure to the rear wheels when their load is reduced during moderate to severe braking.
  A. Metering
  B. Pressure differential
  C. Proportioning
  D. Antilock hydraulic control

_____ **38.** A _____ valve monitors any pressure difference between the two separate hydraulic brake circuits.
  A. Proportioning
  B. Metering
  C. Antilock hydraulic control
  D. Pressure differential

_____ **39.** During hydraulic braking system bleeding, the pressure differential valve may need to be _____.
  A. Moved to the left
  B. Moved to the right
  C. Centered
  D. Closed

_____ **40.** Although not as common as a conventional brake system fitted with a vacuum booster, many vehicles are now equipped with _____ assisted brake boosters.

A. Vacuum

B. Hydraulically

C. Single-diaphragm

D. Dual-diaphragm

_____ **41.** The arrow in the image is pointing to what part of the single-piston master cylinder with primary and secondary pumps?

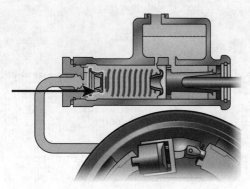

A. Outlet port

B. Master cylinder

C. Primary seal

D. Residual pressure valve

_____ **42.** The arrow in the image is pointing to what part of the single-piston master cylinder with primary and secondary pumps?

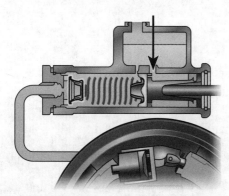

A. Outlet port

B. Compensating port

C. Inlet port

D. Piston

_____ **43.** The arrow in the image is pointing to what part of the single-piston master cylinder with primary and secondary pumps?

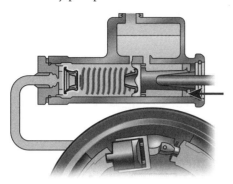

A. Master cylinder
B. Primary seal
C. Piston
D. Secondary cup

_____ **44.** The arrow in the image is pointing to what part of the single-piston master cylinder with small holes in the piston?

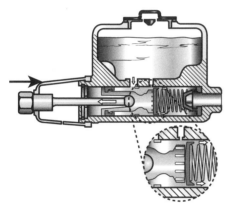

A. Piston
B. Boot
C. Reservoir
D. Filler cap

_____ **45.** The arrow in the image is pointing to what part of the single-piston master cylinder with small holes in the piston?

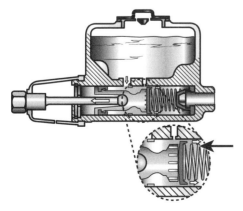

A. Piston
B. Reservoir
C. Housing
D. Return spring

_____ **46.** The arrow in the image is pointing to what part of the single-piston master cylinder with small holes in the piston?

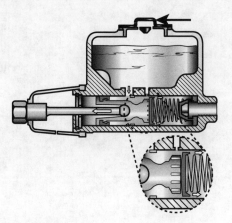

   A. Return spring
   B. Reservoir
   C. Boot
   D. Filler cap

_____ **47.** What divided hydraulic system is shown in the image?

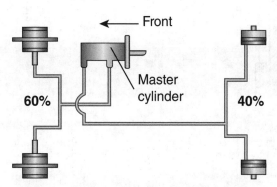

   A. Vertical or front-rear split hydraulic system
   B. Diagonal hydraulic system
   C. X pattern hydraulic system
   D. L-split hydraulic system

_____ **48.** What divided hydraulic system is shown in the image?

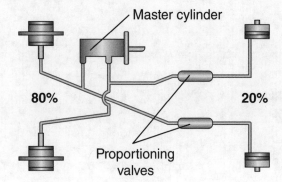

   A. Vertical hydraulic system
   B. Diagonal hydraulic system
   C. L-split hydraulic system
   D. Front-rear split hydraulic system

_____ **49.** What divided hydraulic system is shown in the image?

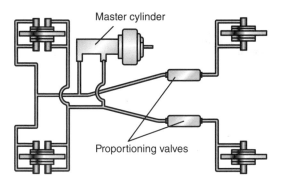

A. L-split hydraulic system
B. Vertical split hydraulic system
C. Front-rear split hydraulic system
D. Diagonal or X pattern hydraulic system

_____ **50.** The arrow in the image is pointing to what part of the flexible brake hose construction?

A. Teflon inner core
B. Protective layer
C. Kevlar braid
D. Stainless steel braid

_____ **51.** The arrow in the image is pointing to what part of the flexible brake hose construction?

A. Stainless steel braid
B. Kevlar braid
C. Teflon inner core
D. Protective layer

_____ **52.** The arrow in the image is pointing to what part of the flexible brake hose construction?

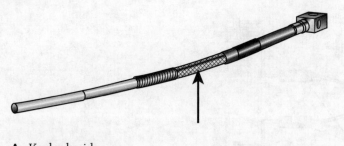

    A. Kevlar braid

    B. Protective layer

    C. Stainless steel braid

    D. Teflon inner core

_____ **53.** The arrow in the image is pointing to what part of the pressure differential valve?

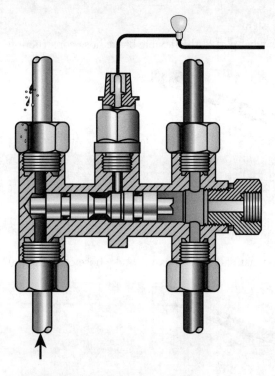

    A. Pressure in from master cylinder

    B. Piston

    C. Low pressure to front wheels

    D. High pressure to rear wheels

_____ **54.** The arrow in the image is pointing to what part of the pressure differential valve?

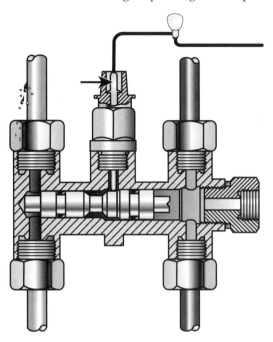

A. Brake warning lamp switch
B. Brake warning lamp
C. Ignition switch
D. Washer

_____ **55.** The arrow in the image is pointing to what part of the pressure differential valve?

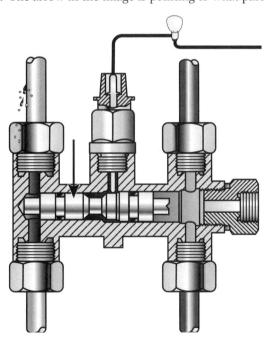

A. Ignition switch
B. Piston
C. Washer
D. Seal

_____ **56.** The arrow in the image is pointing to what part of the combination valve?

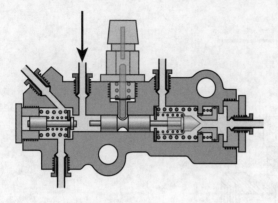

A. OUT Front circuit
B. IN Front circuit
C. IN Rear circuit
D. OUT Rear circuit

_____ **57.** The arrow in the image is pointing to what part of the combination valve?

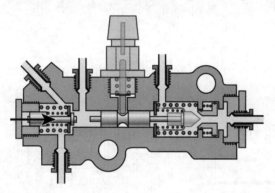

A. Warning lamp switch
B. Proportioning valve
C. Metering valve
D. Pressure differential valve

_____ **58.** The arrow in the image is pointing to what part of the combination valve?

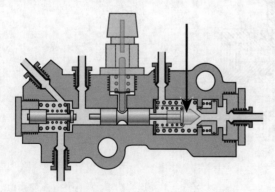

A. Metering valve
B. Proportioning valve
C. Pressure differential valve
D. Warning lamp switch

_____ **59.** Tech A says that hydraulic pressure is applied equally in all directions throughout a closed system. Tech B says that air in the hydraulic system will cause the brake pedal to be spongy. Who is correct?
  A. Tech A
  B. Tech B
  C. Both Tech A and Tech B
  D. Neither Tech A nor Tech B

_____ **60.** Tech A says that a brake pedal that does not return all the way will cause the brake warning light on the instrument panel to stay on. Tech B says that when bleeding brakes, you should normally start at the wheel that is closest to the master cylinder. Who is correct?
  A. Tech A
  B. Tech B
  C. Both Tech A and Tech B
  D. Neither Tech A nor Tech B

_____ **61.** Tech A says that brake fluid should periodically be checked for excessive moisture content. Tech B says that brake fluid should be replaced every 12,000 miles. Who is correct?
  A. Tech A
  B. Tech B
  C. Both Tech A and Tech B
  D. Neither Tech A nor Tech B

_____ **62.** Tech A says that DOT 5 brake fluid should be used in all vehicles today because it is silicone based and will not absorb water. Tech B says that mixing DOT 4 and DOT 3 is not recommended because of different boiling points, but both are hygroscopic. Who is correct?
  A. Tech A
  B. Tech B
  C. Both Tech A and Tech B
  D. Neither Tech A nor Tech B

_____ **63.** Tech A says that the secondary piston in a master cylinder is operated by hydraulic force. Tech B says that the brake pedal return spring returns the master cylinder pistons to their original position. Who is correct?
  A. Tech A
  B. Tech B
  C. Both Tech A and Tech B
  D. Neither Tech A nor Tech B

_____ **64.** Tech A says that the low-level brake fluid switch on a master cylinder will turn on the brake warning light when the system is low on fluid. Tech B says that the low-level switch also monitors the condition of the fluid and will activate the warning light when the brake fluid needs to be replaced. Who is correct?
  A. Tech A
  B. Tech B
  C. Both Tech A and Tech B
  D. Neither Tech A nor Tech B

_____ **65.** Tech A says that the metering valve controls pressure to the rear brakes. Tech B says that the proportioning valve controls pressure to the front brakes. Who is correct?
  A. Tech A
  B. Tech B
  C. Both Tech A and Tech B
  D. Neither Tech A nor Tech B

_____ **66.** Tech A says that the vacuum booster uses vacuum and atmospheric pressure to multiply the driver's foot pressure applied to the master cylinder pushrod. Tech B says that the vacuum booster increases the vacuum in the brake system. Who is correct?

A. Tech A

B. Tech B

C. Both Tech A and Tech B

D. Neither Tech A nor Tech B

_____ **67.** Tech A says that a faulty vacuum booster can affect engine operation. Tech B says that steel brake line can be replaced with a copper line, since it is easier to bend into shape. Who is correct?

A. Tech A

B. Tech B

C. Both Tech A and Tech B

D. Neither Tech A nor Tech B

Servicing Hydraulic Systems and Power Brakes

At the start of each chapter, you will find the Learning Objectives from the textbook. These are your objectives as you make your way through the exercises in this workbook and the chapter in your textbook. The following activities have been designed to help you refresh your knowledge of the material in this chapter.

Learning Objectives

After reading this chapter, you will be able to:

- LO 33-01 Describe brake repair liability and brake tools.
- LO 33-02 Inspect and test brake fluid.
- LO 33-03 Bleed and flush brake systems.
- LO 33-04 Measure brake pedal height, free play, and travel.
- LO 33-05 Describe general brake hydraulic system diagnosis.
- LO 33-06 Perform master cylinder service.
- LO 33-07 Test vacuum-style brake boosters.
- LO 33-08 Service brake lines and hoses.
- LO 33-09 Test the operation of the brake warning lamp and stop lights.

Matching

Match the following steps with the correct sequence for selecting, handling, storing, and filling brake fluid.

   A. Step 1　　　　　　　　　　**B.** Step 2　　　　　　　　　　**C.** Step 3

_____ **1.** Research the specified type of brake fluid in the appropriate service information. Wipe around the master cylinder reservoir cover to prevent any dirt from entering the system. Remove the reservoir cover.

_____ **2.** Only add the manufacturer's recommended brake fluid to bring the level to the Full mark once all brake faults have been resolved. Replace the cover, and check that it is properly seated. Check for any leaks around the master cylinder. Dilute with fresh, clean water if any brake fluid spilled.

_____ **3.** Check the fluid level in the reservoir. The fluid should be between the Full mark and the Min mark on the side of the cylinder or within half an inch of the top of each chamber if there are no marks.

Match the following steps with the correct sequence for performing pressure bleeding.

   A. Step 1　　　　　　　　　　**B.** Step 2　　　　　　　　　　**C.** Step 3

_____ **1.** Install a clear hose on the farthest bleeder screw, and open it one-quarter to one-half turn. Observe any old brake fluid and air bubbles coming out.

_____ **2.** Prepare the pressure bleeder, and install it on the master cylinder reservoir.

_____ **3.** Close off the bleeder screw when the brake fluid is clear and has no bubbles. Tighten it to the manufacturer's specifications. Repeat this procedure, moving closer to the master cylinder, one wheel at a time.

Match the following steps with the correct sequence for testing pedal free travel and performance testing the vacuum booster.

A. Step 1   **B.** Step 2   **C.** Step 3

_____ **1.** Performance test the booster by beginning with the vehicle engine off. Apply and release the brake pedal five or six times. Hold the brake pedal down with moderately firm pressure (20–30 lb [9.1–13.6 kg]).

_____ **2.** Start the engine, and observe the brake pedal.

_____ **3.** With the engine off, depress the brake pedal several times to remove any vacuum or hydraulic pressure from the power booster. Measure the distance of the brake pedal free travel by depressing the brake pedal by hand until you just feel all of the slack taken up. Compare this measurement with specifications.

Match the following steps with the correct sequence for checking vacuum supply to a vacuum-type power booster.

A. Step 1   **B.** Step 2   **C.** Step 3

_____ **1.** With the engine off, remove the inlet hose from the vacuum-type booster.

_____ **2.** Connect a vacuum gauge to the vacuum supply end of the hose.

_____ **3.** Start the engine, and read the vacuum supply available to the vacuum-type booster.

Match the following steps with the correct sequence for inspecting brake lines, brake hoses, and associated hardware.

A. Step 1   **B.** Step 2   **C.** Step 3

_____ **1.** Check for any loose fittings and supports.

_____ **2.** Inspect all flexible brake hoses for cracks, bulging, and wear.

_____ **3.** Safely raise the vehicle on a hoist. Trace all brake lines from the master cylinder to each wheel's brake assembly. Inspect the steel brake lines for leaks, dents, kinks, rust, and cracks.

Multiple Choice

Read each item carefully, and then select the best response.

_____ **1.** Which of the following statements regarding brake repair is true?
    A. Brake repair is a very easy process.
    B. Improperly repaired brakes can function reasonably well even under panic situations.
    C. Brake repair is equal to steering and suspension repair on the liability scale.
    D. A technician does not need to research service information for brake repair.

_____ **2.** All of the following areas should be researched before performing a brake task, EXCEPT _____.
    A. Service information
    B. Precautions
    C. PCM wiring diagrams
    D. Technical service bulletins

_____ **3.** Repeated exposure to particles of which of the following substances can lead to asbestosis and lung cancer?
    A. Aerosol
    B. Asbestos
    C. Ceramic
    D. Brake fluid

_____ **4.** When _____ is used in brake materials, small needle-like fibers can break off and embed themselves deep within lung tissue, causing scarring.
    A. Asbestos
    B. Rubber
    C. Natural fibers
    D. Carbon

_____ **5.** Processes that help ensure that the steps of a brake repair job are not forgotten help _____.
   A. Increase the maintenance interval
   B. Reduce liability
   C. Diagnose associated systems
   D. Reduce repair time

_____ **6.** _____ procedures should always be followed when braking systems are serviced.
   A. Manufacturer
   B. Service manager
   C. Service consultant
   D. Shop foreman

_____ **7.** Although asbestos has been removed from most brake and clutch materials, it is still present in some replacement and old components.
   A. True
   B. False

_____ **8.** Repeated exposure to asbestos can lead to asbestosis and lung cancer.
   A. True
   B. False

_____ **9.** Technicians can be found criminally negligent if they are determined to have acted maliciously.
   A. True
   B. False

_____ **10.** Tech A says that bench bleeding a master cylinder will prevent having to bleed air from the brake lines during replacement. Tech B says that bench bleeding the master cylinder makes bleeding the brakes on the vehicle easier. Who is correct?
   A. Tech A
   B. Tech B
   C. Both Tech A and Tech B
   D. Neither Tech A nor Tech B

_____ **11.** Tech A says that a technician cannot be held liable if he or she performs a faulty brake job on a vehicle. Tech B says that a technician can be held liable for improperly repaired brakes. Who is correct?
   A. Tech A
   B. Tech B
   C. Both Tech A and Tech B
   D. Neither Tech A nor Tech B

_____ **12.** Tech A says that if a technician is determined to have acted maliciously, he or she can be found criminally negligent. Tech B says that technicians have been successfully sued for improperly repaired brakes. Who is correct?
   A. Tech A
   B. Tech B
   C. Both Tech A and Tech B
   D. Neither Tech A nor Tech B

_____ **13.** Tech A says that brake repair and steering and suspension repair are equal on the liability scale. Tech B says that technicians are not liable for improper brake repair. Who is correct?
   A. Tech A
   B. Tech B
   C. Both Tech A and Tech B
   D. Neither Tech A nor Tech B

_____ **14.** Tech A says that improperly repaired brakes can function reasonably well under normal driving situations. Tech B says that improperly repaired brakes can fail during panic situations. Who is correct?
   A. Tech A
   B. Tech B
   C. Both Tech A and Tech B
   D. Neither Tech A nor Tech B

_____ **15.** Tech A says that shops and technicians have been successfully sued for improper brake repairs. Tech B says that this litigation can result in large cash settlements. Who is correct?
   A. Tech A
   B. Tech B
   C. Both Tech A and Tech B
   D. Neither Tech A nor Tech B

_____ **16.** Tech A says that forgetting to properly tighten components can lead to liability. Tech B says that shortcuts should be taken to save time while performing brake repair. Who is correct?
   A. Tech A
   B. Tech B
   C. Both Tech A and Tech B
   D. Neither Tech A nor Tech B

_____ **17.** Tech A says that that forgetting to tighten the lug nuts on a wheel can lead to liability. Tech B says that improperly tightened lug nuts can function very well in panic situations. Who is correct?
   A. Tech A
   B. Tech B
   C. Both Tech A and Tech B
   D. Neither Tech A nor Tech B

_____ **18.** Tech A says that technical service bulletins (TSBs) need to be researched before a brake repair is performed. Tech B states that TSBs do not need to be researched if the brake has been recently replaced. Who is correct?
   A. Tech A
   B. Tech B
   C. Both Tech A and Tech B
   D. Neither Tech A nor Tech B

_____ **19.** Tech A says that modern vehicles use asbestos as brake material. Tech B says that asbestos is no longer used in brakes. Who is correct?
   A. Tech A
   B. Tech B
   C. Both Tech A and Tech B
   D. Neither Tech A nor Tech B

_____ **20.** Tech A says that technicians have been successfully sued for improper brake repairs, resulting in large cash settlements. Tech B says that good technicians create processes to help ensure job steps are not forgotten in order to reduce liability. Who is correct?
   A. Tech A
   B. Tech B
   C. Both Tech A and Tech B
   D. Neither Tech A nor Tech B

Disc Brake Systems Theory

At the start of each chapter, you will find the Learning Objectives from the textbook. These are your objectives as you make your way through the exercises in this workbook and the chapter in your textbook. The following activities have been designed to help you refresh your knowledge of the material in this chapter.

Learning Objectives

After reading this chapter, you will be able to:

- LO 34-01 Describe disc brake fundamentals.
- LO 34-02 Describe disc brake caliper operation.
- LO 34-03 Describe the brake pad assembly.
- LO 34-04 Describe brake rotor construction.
- LO 34-05 Describe parking brakes on disc brakes.

Matching

Match the following terms with the correct description or example.

A. Backing plate

B. Bonded linings

C. Brake booster

D. Caliper

E. Independent rear suspension

F. Lateral runout

G. Low-drag caliper

H. Parallelism

I. Pushrod

J. Riveted linings

K. Rotor

L. Sliding or floating caliper

M. Steering knuckle

N. Ventilated rotor

_____ **1.** Also called warpage, the side-to-side movement of the rotor surfaces as the rotor turns.

_____ **2.** A type of brake caliper that only has piston(s) on the inboard side of the rotor. The caliper is free to slide or float, thus pulling the outboard brake pad into the rotor when braking force is applied.

_____ **3.** A caliper designed to maintain a larger brake pad-to-rotor clearance by retracting the pistons farther than normal.

_____ **4.** Brake linings riveted to the brake pad backing plate with metal rivets and used on heavier-duty or high-performance vehicles.

_____ **5.** Brake linings that are essentially glued to the brake pad backing plate; more common on light-duty vehicles.

_____ **6.** A type of brake rotor with passages between the rotor surfaces that are used to improve heat transfer to the atmosphere.

_____ **7.** A hydraulic device that uses pressure from the master cylinder to apply the brake pads against the rotor.

_____ **8.** A vacuum or hydraulically operated device that increases the driver's braking effort.

_____ **9.** A type of suspension system where each rear wheel is capable of moving independently of the other.

_____ **10.** A metal plate to which the brake lining is fixed.

_____ **11.** A device that connects the front wheel to the suspension; it pivots on the top and bottom, thus allowing the front wheels to turn.

_____ **12.** A mechanism used to transmit force from the brake pedal to the master cylinder.

_____ **13.** The main rotating part of a disc brake system.

_____ **14.** Also called thickness variation; both surfaces of the rotor should be perfectly parallel to each other so brake pulsations do not occur.

Multiple Choice

Read each item carefully, and then select the best response.

_____ **1.** In high-performance vehicles, the _____ are made from composite materials, ceramics, or carbon fiber; otherwise, they are usually made of cast iron.
   A. Brake pads
   B. Calipers
   C. Rotors
   D. Pushrods

_____ **2.** Disc brake caliper assemblies are bolted to the _____.
   A. Axle housing
   B. Wheel hub
   C. Splash shield
   D. Parking brake anchor plate

_____ **3.** Disc brake caliper pistons are sealed by a stationary square section sealing ring, also called a _____.
   A. Square cut O-ring
   B. Caliper gasket
   C. Square-to-round gasket
   D. Square oil seal

_____ **4.** Manufacturers have dealt with the corrosion issue by making caliper pistons out of _____, which does not corrode or rust.
   A. Aluminum
   B. Carbon fiber
   C. Phenolic resin
   D. Rubber

_____ **5.** The backing plate has _____ that correctly position the pad in the caliper assembly and help the backing plate maintain the proper position to the rotor.
   A. Rivets
   B. Tabs
   C. Bolts
   D. Lugs

_____ **6.** The amount of friction between two surfaces is expressed as a ratio and is called the _____.
   A. Sliding resistance
   B. Coefficient of friction
   C. Friction ratio
   D. Drag factor

_____ **7.** Today, brakes are manufactured from a variety of different materials, including all of the following, EXCEPT _____.
   A. Kevlar
   B. Semimetallic materials
   C. Asbestos
   D. Ceramic materials

_____ **8.** A spring steel _____ mounted to the brake pad may be used to notify the driver that the brake pads are worn to their minimum limit.
   A. Rivet
   B. Scratcher
   C. Needle
   D. Spacer

_____ 9. Which type of rotors are less expensive and usually found on smaller vehicles?
   A. Solid
   B. Ventilated
   C. Stainless steel
   D. Phenolic resin

_____ 10. Which type of parking brake is engaged by pushing a button on the dash?
   A. Top hat
   B. Integrated mechanical
   C. Electric
   D. Automatic

_____ 11. The hub can be part of the brake rotor or a separate assembly that the rotor slips over and is bolted to by the lug nuts.
   A. True
   B. False

_____ 12. Disc brake pads require much lower application pressures to operate than drum brake shoes because they are self-energizing.
   A. True
   B. False

_____ 13. The sliding or floating caliper has brake pads located on each side of the rotor, but all of the pistons are only on one side, usually the inside of the rotor.
   A. True
   B. False

_____ 14. Bonded brake linings are less susceptible to failure under high temperatures.
   A. True
   B. False

_____ 15. The lower the edge code letter, the less friction the material has and the harder the brake pedal must be applied to achieve a given amount of stopping power.
   A. True
   B. False

_____ 16. Society of Automotive Engineers standards assure that an EE-rated lining from one manufacturer will have the same braking characteristics as an EE-rated lining from another manufacturer.
   A. True
   B. False

_____ 17. Technicians can apply a high-temperature liquid rubber compound to the back of the brake pad that stays flexible, absorbs brake pad vibrations, and helps reduce brake noise.
   A. True
   B. False

_____ 18. Lateral runout tends to move the fixed caliper pistons in the same direction as each other, so brake fluid is not pushed back to the master cylinder.
   A. True
   B. False

_____ 19. Solid rotors are used to improve heat transfer to the atmosphere.
   A. True
   B. False

_____ 20. Most disc brake rotors are stamped with the manufacturer's minimum thickness specification.
   A. True
   B. False

_____ 21. The purpose of the _____ system is to provide an effective means to slow the vehicle under a variety of conditions in an acceptable distance and manner.
   A. Rotor
   B. Disc brake
   C. Booster
   D. Suspension

_____ **22.** Calipers use hydraulic pressure from the _____ to apply the brake pads.
  A. Rotors
  B. Pushrod
  C. Brake booster
  D. Master cylinder

_____ **23.** On rear-wheel drive vehicles, the _____ is mounted onto the driving axle or hub and may be held in place by the wheel.
  A. Rotor
  B. Pushrod
  C. Calipers
  D. Bleeder screws

_____ **24.** A(n) _____ on the top of the piston bore allows for the removal of air within the disc brake system as well as helping with performing routine brake fluid changes.
  A. Guide pin
  B. O-ring
  C. Bleeder screw
  D. Backing plate

_____ **25.** _____ calipers are designed to maintain a larger brake pad–to-rotor clearance by retracting the pistons a little bit farther.
  A. Fixed
  B. Low-drag
  C. Sliding
  D. Floating

_____ **26.** Floating calipers are mounted in place by _____ and _____.
  A. Lugs, bleeder screws
  B. Lugs, bushings
  C. Guide pins, bushings
  D. O-rings, guide pins

_____ **27.** The Society of Automotive Engineers has adopted letter codes to rate brake lining materials' coefficient of friction. The rating is written on the edge of the friction linings and is called the _____.
  A. Edge code
  B. Friction code
  C. Performance code
  D. Brake lining code

_____ **28.** Adding brake pad _____ and _____ to the brake pads helps cushion the brake pad and absorbs some of the vibration.
  A. Tangs, springs
  B. Scratchers, calipers
  C. Lugs, guide pins
  D. Shims, guides

_____ **29.** Incorporating _____ tangs on the brake pad backing plate allows technicians to crimp the tangs so they are more firmly mounted in the caliper.
  A. Bendable
  B. Parallel
  C. Phenolic
  D. Independent

_____ **30.** Most _____ are mechanically applied by use of a cable and ratcheting lever assembly.
  A. Service brakes
  B. Disc brakes
  C. Parking brakes
  D. Drum brakes

_____ **31.** Tech A says that disc brakes operate on the principle of friction. Tech B says that disc brakes operate on the principle of regeneration. Who is correct?
- **A.** Tech A
- **B.** Tech B
- **C.** Both Tech A and Tech B
- **D.** Neither Tech A nor Tech B

_____ **32.** Tech A says that some vehicles use fixed calipers. Tech B says that some vehicles use sliding/floating calipers. Who is correct?
- **A.** Tech A
- **B.** Tech B
- **C.** Both Tech A and Tech B
- **D.** Neither Tech A nor Tech B

_____ **33.** Tech A says that disc brakes require higher application pressures than drum brakes. Tech B says that disc brakes are self-energizing. Who is correct?
- **A.** Tech A
- **B.** Tech B
- **C.** Both Tech A and Tech B
- **D.** Neither Tech A nor Tech B

_____ **34.** Tech A says that fixed calipers use one or more pistons only on one side of the rotor. Tech B says that sliding/floating calipers use one or more pistons on both sides of the rotor. Who is correct?
- **A.** Tech A
- **B.** Tech B
- **C.** Both Tech A and Tech B
- **D.** Neither Tech A nor Tech B

_____ **35.** Tech A says that calipers use a round section O-ring to seal each piston. Tech B says that calipers use a square section O-ring to seal each piston. Who is correct?
- **A.** Tech A
- **B.** Tech B
- **C.** Both Tech A and Tech B
- **D.** Neither Tech A nor Tech B

_____ **36.** Tech A says that pistons plated with chrome resist rust. Tech B says that pistons made of phenolic resin do not corrode and rust. Who is correct?
- **A.** Tech A
- **B.** Tech B
- **C.** Both Tech A and Tech B
- **D.** Neither Tech A nor Tech B

_____ **37.** Tech A says that some vehicles are equipped with a spring-steel brake pad wear indicator that drags on the rotor when the lining thickness is low. Tech B says that some vehicles are equipped with an electric brake pad wear sensor that activates a warning on the dash. Who is correct?
- **A.** Tech A
- **B.** Tech B
- **C.** Both Tech A and Tech B
- **D.** Neither Tech A nor Tech B

_____ **38.** Tech A says that rotors should be measured for thickness variation (parallelism). Tech B says that rotors should be measured for lateral runout. Who is correct?
- **A.** Tech A
- **B.** Tech B
- **C.** Both Tech A and Tech B
- **D.** Neither Tech A nor Tech B

Servicing Disc Brakes

At the start of each chapter, you will find the Learning Objectives from the textbook. These are your objectives as you make your way through the exercises in this workbook and the chapter in your textbook. The following activities have been designed to help you refresh your knowledge of the material in this chapter.

Learning Objectives

After reading this chapter, you will be able to:

- LO 35-01 Describe disc brake diagnosis.
- LO 35-02 Perform caliper and brake pad service.
- LO 35-03 Measure and replace disc brake rotors.
- LO 35-04 Refinish disc brake rotors.
- LO 35-05 Inspect, replace, and torque lug nuts and studs.

Matching

Match the following terms with the correct description or example.

A. Bearing races
B. Brake wash station
C. Dial indicator

D. Drawing-in method
E. Electronic control module
F. Off-car brake lathe

_____ **1.** A method for replacing wheel studs that uses the lug nut to draw the wheel stud into the hub or flange.

_____ **2.** A piece of equipment designed to safely clean brake dust from drum and disc brake components.

_____ **3.** A tool used to machine (refinish) drums and rotors after they have been removed from the vehicle.

_____ **4.** A tool used to measure the lateral runout of the rotor.

_____ **5.** Hardened metal surfaces that roller or ball bearings fit into when a bearing is properly assembled.

_____ **6.** A computer that receives signals from input sensors, compares that information with preloaded software, and sends an appropriate command signal to output devices; used to manage the antilock brake system.

Multiple Choice

Read each item carefully, and then select the best response.

_____ **1.** Brake diagnosis usually starts with understanding the customer's _____.
 A. Position
 B. Concern
 C. Own diagnosis
 D. Cause

_____ **2.** A safe way to test whether a braking problem is affecting the front or rear brakes is to drive at a low speed and _____.
 A. Lightly apply the service brakes
 B. Lightly apply the parking brake
 C. Repeatedly tap the service brakes
 D. Pull hard on the parking brake lever

_____ **3.** All of the following can cause brake system dragging, EXCEPT _____.
   A. A stuck caliper piston
   B. A misadjusted master cylinder pushrod length
   C. A restricted brake line or hose
   D. An internal master cylinder leak

_____ **4.** If a retracting tool is not available, which of the following tools can be used to retract the pistons in a caliper with nonintegrated parking brakes?
   A. A pair of vice grips
   B. Slip joint pliers
   C. C-clamp
   D. Soft-faced mallet

_____ **5.** A dial indicator may be used for measuring which rotor characteristic?
   A. Thickness
   B. Runout
   C. Hot spots
   D. Groove wear

_____ **6.** What is the first step in checking the operation of the brake pad wear indicator system?
   A. Check that the sensor or scratcher is not too close to the rotor.
   B. Research the service information to determine the type of brake pad wear indicator.
   C. Check that the sensor is grounded with a test lead.
   D. Test the system.

_____ **7.** To perform a thorough inspection of brake pads, you should do all of the following, EXCEPT _____.
   A. Inspect the pads to see if they are wearing evenly
   B. Inspect all pads, retaining hardware, and antinoise shims
   C. Measure the remaining brake pad thickness and compare the measurements with specifications
   D. Measure the pad-to-caliper piston clearance

_____ **8.** Which of the following steps is not normally performed when disassembling and inspecting a caliper?
   A. Using an impact wrench to remove the caliper piston
   B. Cleaning all of the caliper parts
   C. Checking the caliper pin bores or bushings for wear or damage
   D. Measuring the piston-to-cylinder bore clearance with a feeler gauge

_____ **9.** "Scratcher" and "sensor" are types of disc brake pad _____ indicators.
   A. Height
   B. Weight
   C. Wear
   D. Position

_____ **10.** When retracting the caliper piston on an integrated parking brake system, what should you do?
   A. Apply and release the parking brake at least 10 times to retract the piston.
   B. Use a pry bar to gently pry the piston back into the bore.
   C. Use a special C-clamp to push the piston back into the bore.
   D. Use a special tool to screw the piston back into the bore.

_____ **11.** Excessive disc brake rotor thickness variation will cause which of the following conditions?
   A. Brake pads wearing on a slant
   B. Brake pedal pulsation
   C. Caliper pistons seizing in their bores
   D. Squishy brake pedal

_____ **12.** What disc brake tool is shown in the image?

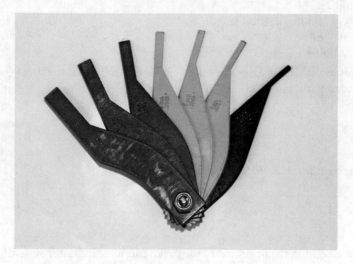

A. Brake lining thickness gauges
B. Disc brake rotor micrometer
C. Dial indicator
D. Parking brake cable tool

_____ **13.** What disc brake tool is shown in the image?

A. Off-car brake lathe
B. On-car brake lathe
C. Brake wash station
D. Dust boot seal/bushing driver set

_____ **14.** What disc brake tool is shown in the image?

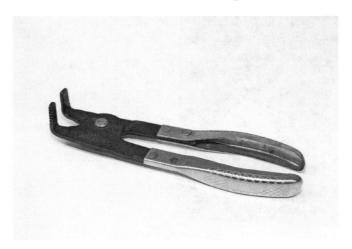

    A. Disc brake rotor micrometer
    B. Parking brake cable tool
    C. Caliper piston retracting tool
    D. Caliper piston pliers

_____ **15.** What disc brake tool is shown in the image?

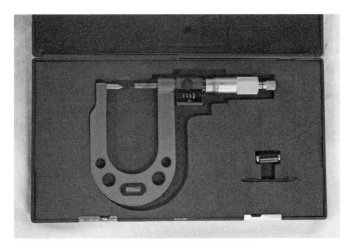

    A. Brake lining thickness gauge
    B. Disc brake rotor micrometer
    C. Dial indicator
    D. Off-car brake lathe

_____ **16.** What disc brake tool is shown in the image?

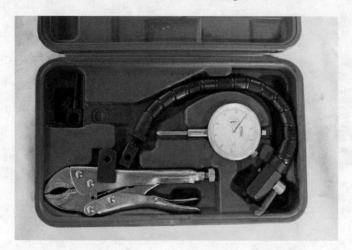

    A. Caliper piston pliers
    B. Dial indicator
    C. Parking brake cable tool
    D. On-car brake lathe

_____ **17.** What disc brake tool is shown in the image?

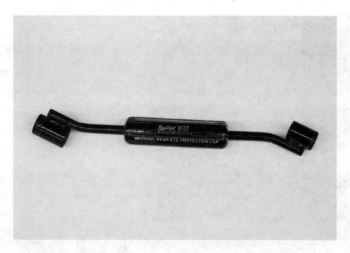

    A. Parking brake cable tool
    B. Caliper piston retracting tool
    C. Off-car brake lathe
    D. On-car brake lathe

_____ **18.** What disc brake tool is shown in the image?

    A. Caliper piston pliers
    B. Disc brake rotor micrometer
    C. Caliper piston retracting tool
    D. Dust boot seal/bushing driver set

_____ **19.** What disc brake tool is shown in the image?

    A. Caliper piston retracting tool
    B. Off-car brake lathe
    C. On-car brake lathe
    D. Dust boot seal/bushing driver set

_____ **20.** What disc brake tool is shown in the image?

    A. Brake lining thickness gauge
    B. Brake wash station
    C. Caliper piston pliers
    D. On-car brake lathe

_____ **21.** What disc brake tool is shown in the image?

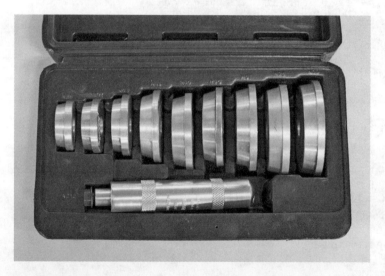

    A. Brake wash station
    B. Dial indicator
    C. Parking brake cable tool
    D. Dust boot seal/bushing driver set

_____ **22.** Tech A says that rotors that are too thin cannot handle as much heat and will experience brake fade sooner. Tech B says that brake pedal pulsation is the result of air in the hydraulic system. Who is correct?
    A. Tech A
    B. Tech B
    C. Both Tech A and Tech B
    D. Neither Tech A nor Tech B

_____ **23.** Tech A says that a micrometer is used to measure rotor thickness variation. Tech B says that a micrometer is used to measure rotor lateral runout. Who is correct?

A. Tech A

B. Tech B

C. Both Tech A and Tech B

D. Neither Tech A nor Tech B

_____ **24.** Tech A says that caliper mountings and slides/pins should be cleaned using the equipment and procedures used for dealing with asbestos or hazardous dust. Tech B says that once the dust is taken care of, a brake cleaning solvent may be needed to clean the components further. Who is correct?

A. Tech A

B. Tech B

C. Both Tech A and Tech B

D. Neither Tech A nor Tech B

_____ **25.** Tech A says that the piston on a brake caliper that integrates the parking brake will need to be screwed back in on the threaded shaft to retract it into the bore. Tech B says that a special tool that mates to slots, grooves, or holes in the outer face of the caliper piston is needed. Who is correct?

A. Tech A

B. Tech B

C. Both Tech A and Tech B

D. Neither Tech A nor Tech B

_____ **26.** Tech A says that the caliper piston on integrated parking brake calipers can best be retracted with a C-clamp. Tech B says that integrated parking brake calipers require a tool to turn the piston in the direction that causes the piston to retract. Who is correct?

A. Tech A

B. Tech B

C. Both Tech A and Tech B

D. Neither Tech A nor Tech B

_____ **27.** Tech A says that pinching off the flexible brake hoses with vice grip pliers is a recommended method of closing rubber brake lines. Tech B says that using a brake pedal holding tool to slightly apply the brakes and block off the compensating ports of the master cylinder is a recommended method of minimizing the brake fluid that drains when the brake lines or hoses are removed. Who is correct?

A. Tech A

B. Tech B

C. Both Tech A and Tech B

D. Neither Tech A nor Tech B

_____ **28.** Tech A says that when measuring a rotor's thickness variation, the rotor should be measured in a minimum of five to eight places around the face of the rotor. Tech B says that as long as the rotor faces look smooth after machining them, they do not need to be measured again. Who is correct?

A. Tech A

B. Tech B

C. Both Tech A and Tech B

D. Neither Tech A nor Tech B

_____ **29.** Tech A says that rotors should be measured for thickness variation (parallelism). Tech B says that rotors should be measured for lateral runout. Who is correct?

A. Tech A

B. Tech B

C. Both Tech A and Tech B

D. Neither Tech A nor Tech B

_____ **30.** Tech A says the thickness of the rotor should be measured in a minimum of five to eight places around the face of the rotor. Tech B says the thickness of the rotor should be measured in the lowest spot that can be eyeballed on the rotor. Who is correct?

A. Tech A

B. Tech B

C. Both Tech A and Tech B

D. Neither Tech A nor Tech B

_____ **31.** Tech A says that disc brake rotors come in a hub style. Tech B says that disc brake rotors come in a hubless style. Who is correct?

  A. Tech A

  B. Tech B

  C. Both Tech A and Tech B

  D. Neither Tech A nor Tech B

_____ **32.** Tech A says that brake rotors can be refinished on or off the vehicle. Tech B says that brake drums can be refinished on and off the vehicle. Who is correct?

  A. Tech A

  B. Tech B

  C. Both Tech A and Tech B

  D. Neither Tech A nor Tech B

Drum Brake Systems Theory

At the start of each chapter, you will find the Learning Objectives from the textbook. These are your objectives as you make your way through the exercises in this workbook and the chapter in your textbook. The following activities have been designed to help you refresh your knowledge of the material in this chapter.

Learning Objectives

After reading this chapter, you will be able to:

- LO 36-01 Describe drum brake fundamentals.
- LO 36-02 Describe the types of drum brake systems.
- LO 36-03 Describe brake drums and backing plates.
- LO 36-04 Describe wheel cylinders.
- LO 36-05 Describe brake shoes and lining.
- LO 36-06 Describe drum brake springs.
- LO 36-07 Describe drum brake self-adjusters and parking brake operation.

Matching

Match the following terms with the correct description or example.

A. Anchor pin
B. Automatic brake self-adjuster
C. Backing plate
D. Brake drum
E. Duo-servo drum brake system
F. Leading/trailing shoe drum brake system
G. Parking brake mechanism

H. Self-energizing
I. Servo action
J. Specialty springs
K. Springs and clips
L. Twin leading shoe drum brake system
M. Wheel cylinder

_____ **1.** A type of brake shoe arrangement where one shoe is positioned in a leading manner and the other shoe in a trailing manner.

_____ **2.** A drum brake design where one brake shoe, when activated, applies an increased activating force to the other brake shoe, in proportion to the initial activating force; this further enhances the self-energizing feature of some drum brakes.

_____ **3.** A mechanism that operates the brake shoes or pads to hold the vehicle stationary when the parking brake is applied.

_____ **4.** Brake shoe arrangement in which both brake shoes are self-energizing in the forward direction.

_____ **5.** A short, wide, hollow cylinder that is capped on one end and bolted to a vehicle's wheel; it has an inner friction surface that the brake shoe is forced against.

_____ **6.** A hydraulic cylinder with one or two pistons, seals, dust boots, and a bleeder screw that pushes the brake shoes into contact with the brake drum to slow or stop the vehicle.

_____ **7.** Springs used to return links and levers on the parking brake system or the self-adjuster mechanism.

_____ **8.** A system on drum brakes that automatically adjusts the brakes to maintain a specified amount of running clearance between the shoes and the drum.

_____ **9.** A system that uses servo action in both the forward and the reverse directions.

_____ **10.** Various devices that hold the brake shoes in place or return them to their proper place.

_____ **11.** A component of the backing plate that takes all of the braking force from the brake shoes.

_____ **12.** A metal plate to which the brake lining is fixed.

_____ **13.** The property of drum brakes that assists the driver in applying the brakes; when brake shoes come into contact with the moving drum, the friction tends to wedge the shoes against the drum, thus increasing the braking force.

Multiple Choice

Read each item carefully, and then select the best response.

_____ **1.** All of the following are main components of the drum brake system, EXCEPT _____.
  A. The wheel cylinder
  B. Brake shoes
  C. The rotor
  D. The parking brake mechanism

_____ **2.** In drum brake systems, when the brake pedal is depressed, a _____ transfers the force to a hydraulic master cylinder.
  A. Linkage
  B. Pushrod
  C. Cable
  D. Brake line

_____ **3.** All of the following are types of drum brake systems, EXCEPT _____.
  A. Twin leading shoe
  B. Leading/trailing shoe
  C. Twin trailing shoe
  D. Duo-servo

_____ **4.** Brake drums are usually made out of _____ owing to their ability to withstand high temperatures, absorb a lot of heat, and maintain shape.
  A. Cast iron
  B. Stainless steel
  C. Chrome
  D. Phenolic resin

_____ **5.** The _____ must be able to take all of the braking force when the brakes are applied, so it must be strong and firmly attached to the backing plate.
  A. Wheel cylinder
  B. Anchor pin
  C. Brake drum
  D. Primary shoe

_____ **6.** A reduction in the coefficient of friction capability as the heat in brake pads and linings builds up is called _____.
  A. Bonding fail
  B. Slippage
  C. Brake fade
  D. Wear

_____ **7.** Which of the following can cause drum brakes to make a groaning noise?
  A. Excessive brake dust
  B. Overheating
  C. Fluid leak
  D. Worn brake pads

_____ **8.** What type of brake springs are generally quite stiff, making them a challenge to install?
  A. Specialty springs
  B. Hold-down springs
  C. Torsion springs
  D. Return springs

_____ **9.** What type of brake springs can be of all different shapes and sizes and can be used to push or pull components into their proper position?

    A. Torsion springs

    B. Return springs

    C. Specialty springs

    D. Hold-down springs

_____ **10.** What type of self-adjuster uses two toothed pieces, held in contact with each other by spring pressure, that can slide over each other in one direction but hold in the other direction?

    A. Star-wheel type

    B. Ratcheting-style

    C. Servo-style

    D. Cable-style

_____ **11.** The master cylinder converts brake pedal force into hydraulic pressure.

    A. True

    B. False

_____ **12.** Trailing shoes are self-energizing.

    A. True

    B. False

_____ **13.** Brake drums provide the rotating friction surface that the brake lining contacts.

    A. True

    B. False

_____ **14.** Hubless-style drums have a one-piece integrated hub/drum assembly.

    A. True

    B. False

_____ **15.** Cylinder bores on aluminum wheel cylinders are usually honed to help them resist corrosion.

    A. True

    B. False

_____ **16.** Drum brakes are usually designed so that the condition of the lining can be checked only once the drum has been removed.

    A. True

    B. False

_____ **17.** In a duo-servo brake installation, the matching shoes belong on the same side of the vehicle.

    A. True

    B. False

_____ **18.** The lining on the brake shoes is much thinner than on the disc brake pads.

    A. True

    B. False

_____ **19.** If you switch a self-adjuster from one side of the vehicle to the other, it will retract the adjustment, causing the brake shoe clearance to increase as the brakes adjust.

    A. True

    B. False

_____ **20.** Drum parking brake systems mechanically apply the regular service brake shoes.

    A. True

    B. False

_____ **21.** The drum is bolted to the vehicle's axle _____ by the lug nuts.

    A. Plate

    B. Lining

    C. Flange

    D. Cylinder

_____ **22.** Each drum brake has two brake shoes with a friction material called a _____ attached.

   A. Spring

   B. Lining

   C. Clip

   D. Plate

_____ **23.** Brake drums are machined to a specific diameter from the manufacturer, which is called its _____ diameter.

   A. Inside

   B. Maximum

   C. Standard

   D. Minimum

_____ **24.** All of the brake unit components, except the _____, are mounted on a backing plate bolted to the vehicle axle housing or suspension.

   A. Anchoring pin

   B. Brake drum

   C. Brake shoes

   D. Wheel cylinder

_____ **25.** The cylinder bore, or inside diameter of the cylinder, is created by drawing a properly sized _____ through the bore.

   A. Spring

   B. Bleeder screw

   C. Pushrod

   D. Ball bearing

_____ **26.** The _____ is a hollow screw with a taper on the end that mates with a matching tapered seat in the wheel cylinder.

   A. Bleeder screw

   B. Dust boot

   C. Pushrod

   D. Lip seal

_____ **27.** Most modern vehicles use _____ wheel cylinders because they are simpler to design, install, and bleed.

   A. Single-acting

   B. Double-acting

   C. Unidirectional

   D. Bidirectional

_____ **28.** The terms "primary" and "secondary" refer to the _____ in a duo-servo brake system.

   A. Brake drums

   B. Wheel cylinders

   C. Brake shoes

   D. Backing plates

_____ **29.** Since 1968, manufacturers have incorporated a _____ mechanism into their drum brake systems that is capable of maintaining proper shoe-to-drum clearance.

   A. Self-adjusting

   B. Retracting

   C. Self-energizing

   D. Self-cleaning

_____ **30.** The arrow in the image is pointing to what component of the drum brake system?

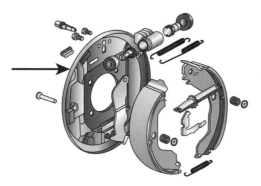

A. Hole cover
B. Backing plate
C. Wheel cylinder
D. Loading shoe

_____ **31.** The arrow in the image is pointing to what component of the drum brake system?

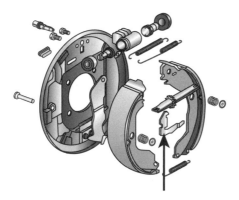

A. Shoe hold spring
B. Parking lever
C. Adjusting lever
D. Star-wheel adjusting screw

_____ **32.** The arrow in the image is pointing to what component of the drum brake system?

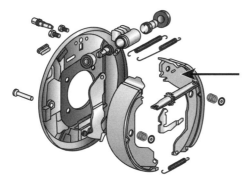

A. Wheel cylinder
B. Leading shoe
C. Anchor
D. Lining

_____ **33.** The arrow in the image is pointing to what component of the leading/trailing shoe drum brake system?

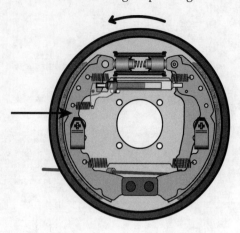

    A. Trailing shoe
    B. Anchor point
    C. Self-adjuster
    D. Leading shoe

_____ **34.** The arrow in the image is pointing to what component of the leading/trailing shoe drum brake system?

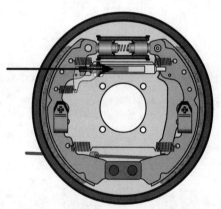

    A. Park brake pushrod
    B. Park brake cable
    C. Park brake lever
    D. Return spring

_____ **35.** The arrow in the image is pointing to what component of the leading/trailing shoe drum brake system?

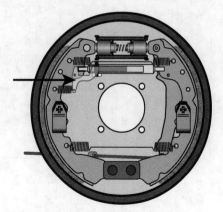

    A. Brake clearance adjuster
    B. Shoe retaining clip
    C. Self-adjuster
    D. Brake drum

_____ **36.** The arrow in the image is pointing to what component of the duo-servo drum brake system?

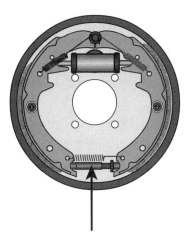

    A. Adjustable floating link
    B. Anchor pin
    C. Shoe return spring
    D. Primary shoe

_____ **37.** The arrow in the image is pointing to what component of the duo-servo drum brake system?

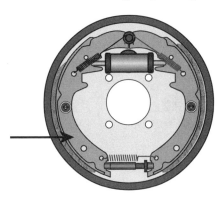

    A. Secondary shoe
    B. Brake drum
    C. Adjustable floating link
    D. Primary shoe

_____ **38.** The arrow in the image is pointing to what component of the duo-servo drum brake system?

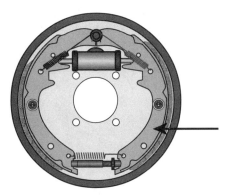

    A. Anchor pin
    B. Wheel cylinder
    C. Secondary shoe
    D. Brake drum

_____ **39.** The arrow in the image is pointing to what component of the star-wheel assembly?

   A. Washer
   B. Socket
   C. Adjuster screw
   D. Button

_____ **40.** The arrow in the image is pointing to what component of the star-wheel assembly?

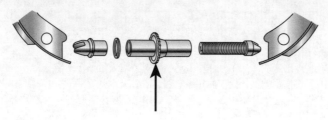

   A. Washer
   B. Button
   C. Adjuster screw
   D. Star wheel

_____ **41.** The arrow in the image is pointing to what component of the star-wheel assembly?

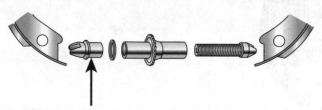

   A. Button
   B. Washer
   C. Adjuster screw
   D. Socket

_____ **42.** The arrow in the image is pointing to what component of the parking brake assembly for a drum brake?

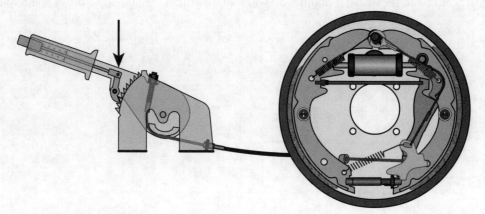

   A. Park brake lever
   B. Park brake pushrod
   C. Park brake handle
   D. Park brake cable

_____ **43.** The arrow in the image is pointing to what component of the parking brake assembly for a drum brake?

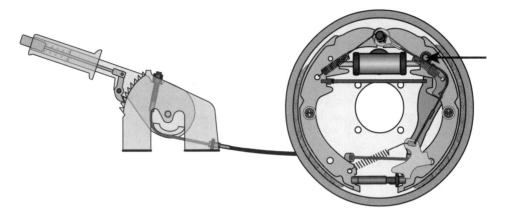

   A. Park brake cable
   B. Pivot point
   C. Adjustable floating link
   D. Park brake pushrod

_____ **44.** The arrow in the image is pointing to what component of the parking brake assembly for a drum brake?

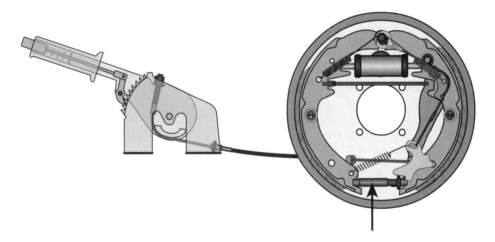

   A. Park brake cable
   B. Parking brake pushrod
   C. Park brake lever
   D. Adjustable floating link

_____ **45.** Tech A says that parking brakes on drum brake vehicles use separate brake shoes for backup in an emergency. Tech B says that parking brakes on drum brake vehicles mechanically operate the standard drum brake shoes. Who is correct?
   A. Tech A
   B. Tech B
   C. Both Tech A and Tech B
   D. Neither Tech A nor Tech B

_____ **46.** Tech A says that most brake drums are designed to be machined if minor surface issues are present. Tech B says that brake drums can be reused if they are machined over specifications, as long as the surface is smooth. Who is correct?
   A. Tech A
   B. Tech B
   C. Both Tech A and Tech B
   D. Neither Tech A nor Tech B

_____ **47.** Tech A says that duo-servo brake shoes are only anchored on the top. Tech B says that typically duo-servo brake shoes adjust automatically during normal brake applications in a forward direction. Who is correct?
  A. Tech A
  B. Tech B
  C. Both Tech A and Tech B
  D. Neither Tech A nor Tech B

_____ **48.** Tech A says that when inspecting brake shoes, if the shoes are found to be unequally worn, this could be caused by a stuck wheel cylinder piston. Tech B says that the lining on the primary shoe is typically shorter in length and is installed toward the front of the vehicle. Who is correct?
  A. Tech A
  B. Tech B
  C. Both Tech A and Tech B
  D. Neither Tech A nor Tech B

_____ **49.** Tech A says that it is almost impossible to install self-adjusters on the wrong side of the vehicle. Tech B says that grease seals can be reused. Who is correct?
  A. Tech A
  B. Tech B
  C. Both Tech A and Tech B
  D. Neither Tech A nor Tech B

_____ **50.** Tech A says to use an air hose to clean the backing plate of dust and contamination. Tech B says to use a brake cleaning solution to clean the backing plate of dust and contamination. Who is correct?
  A. Tech A
  B. Tech B
  C. Both Tech A and Tech B
  D. Neither Tech A nor Tech B

_____ **51.** Tech A says that riveted lining is usually for heavy-duty or high-performance vehicles. Tech B says that you need to identify each shoe individually to help ensure proper installation. Who is correct?
  A. Tech A
  B. Tech B
  C. Both Tech A and Tech B
  D. Neither Tech A nor Tech B

_____ **52.** Tech A says that a grinding noise in drum brakes generally requires replacing the brake shoes and resurfacing or replacing the drums. Tech B says that a click noise in drum brakes requires replacing the brake shoes and resurfacing or replacing the drums. Who is correct?
  A. Tech A
  B. Tech B
  C. Both Tech A and Tech B
  D. Neither Tech A nor Tech B

_____ **53.** Tech A says that when performing a brake job on the rear axle of an older vehicle, inspection finds brake fluid under the dust boot; this suggests wheel cylinder replacement on both rear wheels. Tech B says that when a drum brake return spring has failed, springs on both rear wheels need to be replaced. Who is correct?
  A. Tech A
  B. Tech B
  C. Both Tech A and Tech B
  D. Neither Tech A nor Tech B

Servicing Drum Brakes

At the start of each chapter, you will find the Learning Objectives from the textbook. These are your objectives as you make your way through the exercises in this workbook and the chapter in your textbook. The following activities have been designed to help you refresh your knowledge of the material in this chapter.

Learning Objectives

After reading this chapter, you will be able to:

- LO 37-01 Describe drum brake diagnosis.
- LO 37-02 Inspect and measure brake drums.
- LO 37-03 Refinish brake drums.
- LO 37-04 Service brake shoes and hardware.
- LO 37-05 Service wheel cylinders.
- LO 37-06 Install drums, wheels, and perform final checks.

Matching

Match the following terms with the correct description or example.

A. Brake shoe adjustment gauge   **F.** Hold-down spring tool
B. Brake spoon   **G.** Off-car brake lathe
C. Brake spring pliers   **H.** Parking brake cable pliers
D. Brake wash station   **I.** Parking brake cable removal tool
E. Drum brake micrometer   **J.** Wheel cylinder piston clamp

_____ **1.** A tool used for removing and installing hold-down springs.

_____ **2.** A tool used to compress the spring steel fingers of the parking brake cable so that the cable can be removed from the backing plate.

_____ **3.** A tool used to machine (refinish) drums and rotors after they have been removed from the vehicle.

_____ **4.** An adjustable tool used to preadjust the brake shoes to the diameter of the brake drum.

_____ **5.** A tool used for measuring the inside diameter of the brake drum.

_____ **6.** A piece of equipment designed to safely clean brake dust from drum and disc brake components.

_____ **7.** A tool that prevents the pistons from being pushed out of the wheel cylinders while the brake shoes are being replaced.

_____ **8.** A tool used to install parking brake cables.

_____ **9.** A tool used for removing and installing brake return springs.

_____ **10.** A tool used to adjust the brake lining-to-drum clearance when the drum is installed on the vehicle.

Match the following steps with the correct sequence for refinishing brake drums.

A. Step 1
B. Step 2
C. Step 3
D. Step 4
E. Step 5
F. Step 6

G. Step 7
H. Step 8
I. Step 9
J. Step 10
K. Step 11
L. Step 12

_____ 1. Clean any nicks, burrs, or rust from the mounting surfaces of the drum, including the centering hole, if used.

_____ 2. Make sure the cutting bits will not contact the face of the drum, and move the cutting head about 0.5″ (12.7 mm) in from the outside of the drum.

_____ 3. Once the ridge is removed, run the drum all the way in so the cutting bit is in the inner corner of the drum. Set the cutting bit to the proper depth for machining the surface of the drum, and lock it in place.

_____ 4. Install the antichatter band on the drum.

_____ 5. Move the drum well away from the cutting bit, and use sandpaper to give the drum surface a nondirectional finish.

_____ 6. Remeasure the drum diameter to determine whether the drum is above the maximum-diameter specifications. If so, discard the drum.

_____ 7. Mount the drum on the brake lathe.

_____ 8. Turn on the brake lathe, and set the depth of the cutting tool so it just touches the surface of the drum. Rotate the brake lathe's handwheel so that the drum moves outward and the cutting bit contacts the ridge. Keep turning slowly to remove the ridge.

_____ 9. Visually check to see that the drum is running true on the lathe and is not wobbling. If it wobbles, turn off the lathe and recheck for the condition that causes it to wobble.

_____ 10. If necessary, repeat this step until the worn surface areas have been removed all the way around the surface of the drum. If the brake lathe is not a single-cut machine, perform a finish cut on the drum.

_____ 11. Set the position of the cutting tool so that the brake drum is close to the brake lathe when the cutting bit is in the far corner of the drum.

_____ 12. Engage the automatic spindle feed, and set it to the proper speed; lock it in place, and watch for proper machining action.

Match the following steps with the correct sequence for removing, cleaning, inspecting, and reassembling a duo-servo brake.

A. Step 1
B. Step 2
C. Step 3
D. Step 4
E. Step 5
F. Step 6
G. Step 7
H. Step 8
I. Step 9

J. Step 10
K. Step 11
L. Step 12
M. Step 13
N. Step 14
O. Step 15
P. Step 16
Q. Step 17
R. Step 18

_____ 1. Install the cable guide and return spring in the secondary shoe. Also, align the wheel cylinder pushrod in the shoe.

_____ 2. Remove the secondary hold-down spring, retainer, pin, and secondary shoe.

_____ 3. Clean and inspect all parts according to the manufacturer's procedure.

_____ 4. Install the parking brake strut rod onto the secondary shoe, and pull the primary shoe engaged with the parking brake strut rod.

_____ **5.** Clean the brake shoes, hardware, and backing plates using equipment and procedures for dealing with asbestos and/or dust.

_____ **6.** Position the secondary shoe in place, and use brake spring pliers to install the return spring over the anchor pin.

_____ **7.** Remove the self-adjuster spring and star wheel assembly and primary shoe.

_____ **8.** Remove the return springs, cable guide (if installed), and shoe guide. Set everything aside in the order it will go back on. Remove the parking brake strut and spring.

_____ **9.** Check the fit of all springs, clips, and levers. Operate the self-adjuster cable or lever.

_____ **10.** Install the shoe guide and self-adjuster cable over the anchor pin.

_____ **11.** Reassemble the star wheel assembly, lubricate the floating end, and set aside. Also, lubricate the contact pads on the backing plate.

_____ **12.** Remove the primary shoe hold-down spring, retainer, and pin.

_____ **13.** Install the star wheel between the bottoms of the two shoes.

_____ **14.** Install both shoes to the backing plate with the hold-down spring assemblies.

_____ **15.** Disassemble the parking brake lever from the brake shoe and hardware from the backing plate, being careful not to lose any parts and remembering how they go back together.

_____ **16.** Install the return spring in the primary shoe, and use brake spring pliers to stretch the return spring over the anchor pin.

_____ **17.** Install the self-adjuster link, cable, and spring into position.

_____ **18.** Reassemble the parking brake lever on the brake shoe and parking.

Match the following steps with the correct sequence for disassembling, cleaning, inspecting, and reassembling a non-servo brake.

A. Step 1		**G.** Step 7
B. Step 2		**H.** Step 8
C. Step 3		**I.** Step 9
D. Step 4		**J.** Step 10
E. Step 5		**K.** Step 11
F. Step 6		**L.** Step 12

_____ **1.** Assemble and lube the self-adjuster/parking brake strut assembly. Lube the backing plate pads.

_____ **2.** Remove the hold-down springs, retainers, and pins.

_____ **3.** Remove the return springs.

_____ **4.** Clean and inspect all parts according to the manufacturer's procedure.

_____ **5.** Place one shoe on the backing plate, and install the hold-down spring and pin.

_____ **6.** Clean the brake shoes, hardware, and backing plates, using equipment and procedures for dealing with asbestos and/or dust.

_____ **7.** Check the fit of all springs, clips, and levers.

_____ **8.** Place the return springs on both shoes and fit the loose shoe to the backing plate, being sure to line up the wheel cylinder pushrods, the self-adjuster, and the parking brake mechanism.

_____ **9.** Spread the shoes apart, and remove the parking brake strut and self-adjuster components.

_____ **10.** Disassemble the parking brake lever from the brake shoe and hardware from the backing plate, being careful not to lose any parts and remembering how they go back together.

_____ **11.** Install the hold-down spring and pin.

_____ **12.** Place the self-adjuster/parking brake strut on the installed brake shoe.

Match the following steps with the correct sequence for removing, inspecting, and installing wheel cylinders.

A. Step 1 **D.** Step 4
B. Step 2 **E.** Step 5
C. Step 3 **F.** Step 6

_____ **1.** Install and tighten any mounting screws. After that, tighten the brake line with a flare nut or line wrench.

_____ **2.** Peel back the dust boots, and check for brake fluid behind them. Determine any necessary actions.

_____ **3.** Disassemble the wheel cylinder, and inspect each part. Determine any necessary actions.

_____ **4.** Use a flare nut or line wrench to unscrew the brake line from the wheel cylinder.

_____ **5.** Remove wheel cylinder mounting bolts and wheel cylinder from the backing plate.

_____ **6.** If the cylinder can be reused, rebuild it with new seals and dust boots. In most cases, replace it with a new wheel cylinder. Reinstall the brake line by hand.

Match the following steps with the correct sequence for preadjusting brakes and installing drums.

A. Step 1 **D.** Step 4
B. Step 2 **E.** Step 5
C. Step 3 **F.** Step 6

_____ **1.** Adjust the parking brake according to the manufacturer's procedure. If the drum has serviceable wheel bearings, repack, install, adjust, and secure them.

_____ **2.** Place the preadjustment gauge over the center of the brake shoes.

_____ **3.** Adjust the star wheel until the centers of the brake shoes just contact the preadjustment gauge.

_____ **4.** Set the preadjustment gauge to the drum diameter, and lock it in place.

_____ **5.** Make sure the brake shoes are fully up against their stops and centered on the backing plate.

_____ **6.** Test install the brake drum to verify the drum fits. Adjust the shoes as necessary.

Multiple Choice

Read each item carefully, and then select the best response.

_____ **1.** A brake drum should be removed during all of the following tasks, EXCEPT _____.
A. Inspection of the flexible brake hose
B. Inspection for wheel cylinder leaks
C. Inspection of the brake lining thickness
D. Replacement of the rear wheel bearings

_____ **2.** When removing a hubless brake drum, it may be necessary to use a(n) _____ hammer to hammer on the drum between the lug studs.
A. Plastic
B. Ball peen
C. Air
D. Carbon fiber

_____ **3.** A(n) _____ is used to machine drums and rotors that are off the vehicle.
A. Hold-down spring tool
B. Drum brake micrometer
C. Parking brake cable plier
D. Off-car brake lathe

_____ **4.** A _____ refinishes the drum friction surface by removing metal and making it perfectly round with the proper finish.
A. Brake wash station
B. Brake lathe
C. Drum friction refinisher
D. Drum brake micrometer

_____ **5.** A(n) _____ band should be installed on a brake drum before refinishing its surface.
   A. Broad
   B. Antichatter
   C. Coarse
   D. Hypoid

_____ **6.** What drum brake tool is shown in the image?

   A. Drum brake micrometer
   B. Brake wash station
   C. Brake shoe adjustment gauge
   D. Parking brake cable removal tool

_____ **7.** What drum brake tool is shown in the image?

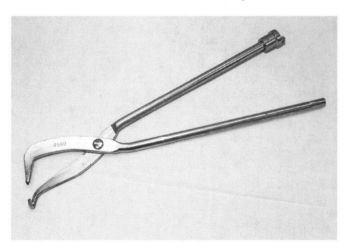

   A. Brake spring pliers
   B. Hold-down spring
   C. Wheel cylinder piston clamp
   D. Parking brake cable pliers

_____ **8.** What drum brake tool is shown in the image?

    A. Brake spring pliers
    B. Hold-down spring tool
    C. Brake spoons
    D. Parking brake cable pliers

_____ **9.** What drum brake tool is shown in the image?

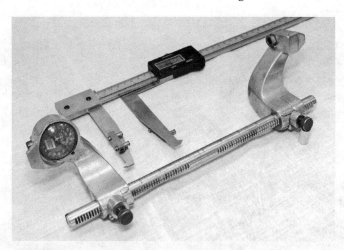

    A. Brake shoe adjustment gauge
    B. Brake spoon
    C. Drum brake micrometer
    D. Parking brake cable removal tool

_____ **10.** What drum brake tool is shown in the image?

    A. Wheel cylinder piston clamp
    B. Brake shoe adjustment gauge
    C. Off-car brake lathe
    D. Drum brake micrometer

_____ **11.** What drum brake tool is shown in the image?

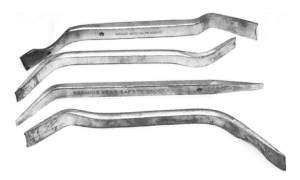

    A. Brake shoe adjustment gauges
    B. Brake spoons
    C. Wheel cylinder piston clamps
    D. Off-car brake lathes

_____ **12.** What drum brake tool is shown in the image?

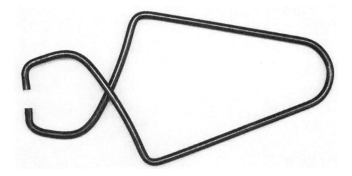

    A. Brake spring pliers
    B. Brake shoe adjustment gauge
    C. Brake spoon
    D. Wheel cylinder piston clamp

_____ **13.** What drum brake tool is shown in the image?

A. Brake wash station
B. Wheel cylinder piston clamp
C. Off-car brake lathe
D. Parking brake cable removal tool

_____ **14.** What drum brake tool is shown in the image?

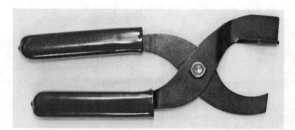

A. Brake spring pliers
B. Brake spoon
C. Wheel cylinder piston clamp
D. Parking brake cable pliers

_____ **15.** What drum brake tool is shown in the image?

A. Parking brake cable removal tool
B. Brake spoon
C. Wheel cylinder piston clamp
D. Parking brake cable pliers

_____ **16.** Tech A says that brake shoe linings saturated with brake fluid from a leaky wheel cylinder can be cleaned with brake cleaner and reused as long as they are not worn out. Tech B says that brake shoes should be replaced in axle sets. Who is correct?
  A. Tech A
  B. Tech B
  C. Both Tech A and Tech B
  D. Neither Tech A nor Tech B

_____ **17.** Tech A says that to reassemble a duo-servo brake, the brake shoes on the backing plate should first be reassembled, and then the parking brake lever and parking brake cable should be attached to the brake shoes. Tech B says that to reassemble a duo-servo brake, the parking brake lever should be reassembled on the brake shoe and parking brake cable, and then both brake shoes should be mounted on the backing plate. Who is correct?
  A. Tech A
  B. Tech B
  C. Both Tech A and Tech B
  D. Neither Tech A nor Tech B

_____ **18.** Tech A says that to reassemble a duo-servo brake, the backing plate pads should be lubricated. Tech B says that to reassemble a duo-servo brake, the star wheel assembly should be lubricated on the floating end. Who is correct?
  A. Tech A
  B. Tech B
  C. Both Tech A and Tech B
  D. Neither Tech A nor Tech B

_____ **19.** Tech A says that when removing a hubless-style drum, matching marks should be made on the drum and hub for reinstallation in the correct position. Tech B says that a center punch can be used to mark the position of the drum to hub. Who is correct?
  A. Tech A
  B. Tech B
  C. Both Tech A and Tech B
  D. Neither Tech A nor Tech B

_____ **20.** Tech A says that if a hubless-style drum has screws or speed nuts holding the drum to the hub, they should not be discarded when removed. Tech B says that if a hubless-style drum has screws or speed nuts holding the drum to the hub, the speed nuts can be discarded and are not needed upon reassembly, but the screws will be reused. Who is correct?
  A. Tech A
  B. Tech B
  C. Both Tech A and Tech B
  D. Neither Tech A nor Tech B

_____ **21.** Tech A says that brake drums should only be refinished when they are excessively grooved. Tech B says that after refinishing a brake drum, it should be remeasured to ensure that it is under the manufacturer's maximum diameter. Who is correct?
  A. Tech A
  B. Tech B
  C. Both Tech A and Tech B
  D. Neither Tech A nor Tech B

_____ **22.** Tech A says that when measuring the drum diameter, the measurement should be taken at the deepest groove or worn part of the drum. Tech B says that if it exceeds the size limit, the drum must be replaced. Who is correct?
  A. Tech A
  B. Tech B
  C. Both Tech A and Tech B
  D. Neither Tech A nor Tech B

_____ **23.** Tech A says that most brake drums are designed to be machined if minor surface issues are present. Tech B says that brake drums can be reused if they are machined over specifications and the surface is smooth. Who is correct?

    A. Tech A

    B. Tech B

    C. Both Tech A and Tech B

    D. Neither Tech A nor Tech B

_____ **24.** Tech A says that the antichatter band used for turning brake drums is not necessary and that not installing it on the brake drum will not affect the results. Tech B says the antichatter band should always be installed to reduce vibration and to get the correct finish on the brake drum. Who is correct?

    A. Tech A

    B. Tech B

    C. Both Tech A and Tech B

    D. Neither Tech A nor Tech B

Wheel Bearings

At the start of each chapter, you will find the Learning Objectives from the textbook. These are your objectives as you make your way through the exercises in this workbook and the chapter in your textbook. The following activities have been designed to help you refresh your knowledge of the material in this chapter.

Learning Objectives

After reading this chapter, you will be able to:

- LO 38-01 Describe wheel bearing fundamentals.
- LO 38-02 Describe wheel bearing types.
- LO 38-03 Describe grease seals and lubricants.
- LO 38-04 Describe wheel bearing arrangements for rear drive axles.
- LO 38-05 Describe wheel bearing diagnosis and failure analysis.
- LO 38-06 Perform maintenance tasks on serviceable wheel bearings.
- LO 38-07 Remove and reinstall sealed wheel bearings.

Matching

Match the following terms with the correct description or example.

A. Antifriction bearing
B. Ball bearings
C. Bearing packer
D. Castellated nut
E. Cylindrical roller bearing assembly
F. End play
G. Grease seal
H. Interference fit

I. Lithium soap
J. Outer race
K. Preload
L. Running clearance
M. Tapered roller bearing assembly
N. Unitized wheel bearing hub
O. Wheel bearing

_____ **1.** A component that is designed to keep grease from leaking out and contaminants from leaking in.

_____ **2.** A condition where the wheel bearing components are forced together under pressure and therefore have no end play.

_____ **3.** An adjusting nut with slots cut into the top such that it resembles a castle; used with a cotter pin to prevent the nut from turning.

_____ **4.** An assembly consisting of the hub, wheel bearing(s), and possibly the wheel flange, which is preassembled and ready to be installed on a vehicle.

_____ **5.** The outside component of a wheel bearing that has a smooth, hardened surface for rollers or balls to ride on.

_____ **6.** A component that allows the wheels to rotate freely while supporting the weight of the vehicle, made up of an inner race, outer race, rollers or balls, and a cage.

_____ **7.** Wheel bearing assemblies that use surfaces that are in rolling contact with each other to greatly reduce friction compared with surfaces in sliding contact.

_____ **8.** A condition when two parts are held together by friction because the outside diameter of the inner component is slightly larger than the inside diameter of the outer component.

_____ **9.** A tool that forces grease into the spaces between the bearing rollers.

_____ **10.** A type of wheel bearing with races and rollers that are tapered in such a manner that all of the tapered angles meet at a common point, which allows them to roll freely and yet control thrust.

_____ **11.** A thickening agent for grease to give it the proper consistency.

_____ **12.** The in-and-out movement of the hub caused by clearance within the wheel bearing assembly.

_____ **13.** The amount of space between wheel bearing components while in operation.

_____ **14.** The rolling components of a wheel bearing consisting of hardened balls that roll in matching grooves in the inner and outer races.

_____ **15.** A type of wheel bearing with races and rollers that are cylindrical in shape and roll between inner and outer races, which are parallel to each other.

Multiple Choice

Read each item carefully, and then select the best response.

_____ **1.** What type of bearings must be serviced periodically by disassembling, cleaning, inspecting, and repacking them with the specified lubricant?
   A. Sealed bearings
   B. Serviceable bearings
   C. Ball bearings
   D. Unitized bearings

_____ **2.** What type of wheel bearing assembly is used when heavier loads need to be supported and the wheel bearings are put under a side load condition?
   A. Tapered roller bearing assemblies
   B. Unitized bearings
   C. Cylindrical roller bearing assemblies
   D. Ball bearings

_____ **3.** Lubrication of serviceable tapered wheel bearing assemblies is usually accomplished by using _____ or _____.
   A. Bearing grease, gear lube
   B. Bearing grease, lithium soap
   C. Gear lube, molybdenum grease
   D. Lithium soap, molybdenum grease

_____ **4.** Many seals use a(n) _____ to help hold the lips of the seal in contact with the shaft it is sealing.
   A. O-ring
   B. Garter spring
   C. Rubber gasket
   D. Sealing bushing

_____ **5.** In a _____, the weight of the vehicle is fully carried by a pair of tapered roller bearing assemblies, which ride between the hub and the axle tube.
   A. Semifloating axle
   B. ¾ floating axle
   C. Full-floating axle
   D. ½ floating axle

_____ **6.** In a(n) _____, the wheel flange is part of the axle, which is supported by a single bearing assembly near the flange end of the axle.
   A. ¾ floating axle
   B. Semifloating axle
   C. Full-floating axle
   D. Solid axle

_____ **7.** In a _____ design, there is a single bearing assembly between the outside of the axle tube and the hub.
   A. ¾ floating axle
   B. Full-floating axle
   C. Semifloating axle
   D. Splined axle

_____ **8.** The level of gear lube should normally be within _____ of the bottom of the fill plug hole.
   A. 0.25″
   B. 0.50″
   C. 0.75″
   D. 1″

_____ **9.** In adjustable wheel bearings, the proper clearance must be set using the _____.
   A. Keyed washer
   B. Adjusting nut
   C. Lock cage
   D. Lock nut

_____ **10.** Sealed bearings are designed so they cannot be disassembled or adjusted.
   A. True
   B. False

_____ **11.** The outer bearing assembly is typically larger than the inner bearing assembly.
   A. True
   B. False

_____ **12.** Packing is best performed with a bearing packing tool, but it can be successfully performed by hand.
   A. True
   B. False

_____ **13.** Roller bearings roll easier than ball bearings since they have a smaller contact area; thus, they provide a small increase in vehicle efficiency.
   A. True
   B. False

_____ **14.** Sealed wheel bearings need to be adjusted for proper running clearance after installation.
   A. True
   B. False

_____ **15.** On some vehicles equipped with an antilock braking system (ABS), the ABS sensor is integrated into the unitized wheel bearing assembly.
   A. True
   B. False

_____ **16.** Gear lube is somewhat thicker than bearing grease.
   A. True
   B. False

_____ **17.** Automotive wheel bearing grease is a thickened lubricant, designated as a plastic solid.
   A. True
   B. False

_____ **18.** Sealed bearings and double-row bearing assemblies come from the manufacturer with the proper clearance machined into them.
   A. True
   B. False

_____ **19.** On four-wheel drive vehicles, the adjustable wheel bearing locking mechanism commonly includes a keyed washer, adjusting nut, keyed lock washer, and lock nut.
   A. True
   B. False

_____ **20.** In many instances, _____ roller bearing assemblies use the surface of the axle as the inner bearing race.
   A. Cylindrical
   B. Tapered
   C. Double row
   D. Friction

_____ 21. _____ a bearing means that the spaces between the rollers and races are completely filled with grease.
- **A.** Sealing
- **B.** Racing
- **C.** Packing
- **D.** Replacing

_____ 22. Virtually all wheel bearings using a ball bearing assembly are of the _____ ball bearing variety, and they are commonly used in automotive light vehicle applications.
- **A.** Tapered
- **B.** Double row
- **C.** Cylindrical
- **D.** Friction

_____ 23. In some applications, the _____ is press-fit into the axle housing, which is stationary, and seals against the axle shaft, which is rotating.
- **A.** Garter spring
- **B.** Axle seal
- **C.** Tapered seal
- **D.** Grease seal

_____ 24. A _____ is a soft metal pin that can be bent into shape and is used to retain the bearing adjusting nut.
- **A.** Cotter pin
- **B.** Grease pin
- **C.** Bearing pin
- **D.** Seal pin

_____ 25. In many rear-wheel drive vehicles, the wheel bearing assemblies are open to the axle housing, which is partially filled with _____ that also lubricates the differential assembly.
- **A.** Bearing grease
- **B.** Gear lube
- **C.** Engine oil
- **D.** Lithium soap

_____ 26. A _____ is usually threaded and can be removed to allow the level of a fluid to be checked and filled.
- **A.** Drain plug
- **B.** Housing plug
- **C.** Snap plug
- **D.** Fill plug

_____ 27. _____ refers to the thickness of the gear lube; the higher the number, the thicker the gear lube.
- **A.** Viscosity
- **B.** End play
- **C.** Interference
- **D.** Preload

_____ 28. The thickness of grease is graded by the _____.
- **A.** American Petroleum Institute
- **B.** National Petroleum Society
- **C.** National Lubricating Grease Institute
- **D.** Society of Automotive Engineers

_____ 29. _____ refers to the absence of clearance in the bearing and the specified amount of pressure forcing the bearing components together.
- **A.** Running clearance
- **B.** Upload
- **C.** Preload
- **D.** End play

_____ **30.** The arrow in the image is pointing to what component of the wheel bearing?

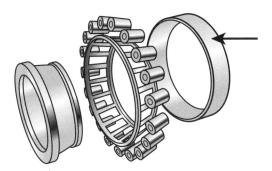

    A. Inner race
    B. Cage
    C. Rollers
    D. Outer race

_____ **31.** The arrow in the image is pointing to what component of the wheel bearing?

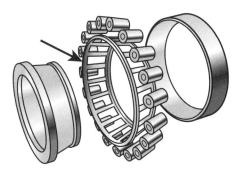

    A. Cage
    B. Rollers
    C. Outer race
    D. Inner race

_____ **32.** The arrow in the image is pointing to what component of the wheel bearing?

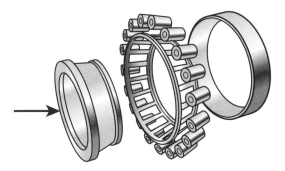

    A. Cage
    B. Inner race
    C. Outer race
    D. Rollers

_____ **33.** What type of wheel bearing is shown in the image?

    A. Unitized wheel bearing hub assembly
    B. Cylindrical roller bearing assembly
    C. Taper roller bearing assembly
    D. Double-row ball bearing assembly

_____ **34.** What type of wheel bearing is shown in the image?

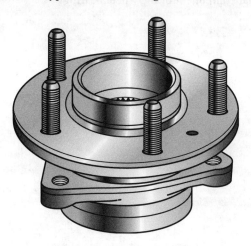

    A. Unitized wheel bearing hub assembly
    B. Taper roller bearing assembly
    C. Double-row tapered roller bearing assembly
    D. Double-row ball bearing assembly

_____ **35.** The arrow in the image is pointing to what part of the locknut-style wheel bearing locking mechanism?

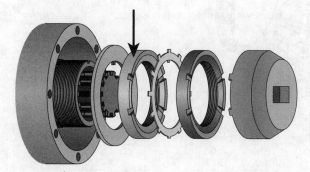

    A. Adjusting nut washer
    B. Lock washer
    C. Locknut
    D. Bearing adjusting nut

_____ **36.** The arrow in the image is pointing to what part of the locknut-style wheel bearing locking mechanism?

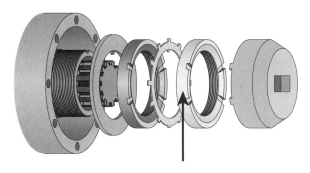

A. Locknut socket
B. Lock washer
C. Locknut
D. Bearing adjusting nut

_____ **37.** The arrow in the image is pointing to what part of the locknut-style wheel bearing locking mechanism?

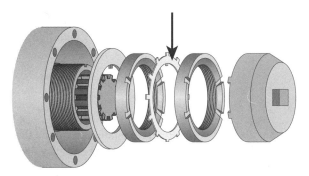

A. Locknut
B. Adjusting nut washer
C. Lock washer
D. Bearing adjusting nut

_____ **38.** Tech A says that cylindrical roller bearings can carry more weight than similarly sized ball bearings. Tech B says that tapered roller bearings, used in opposing pairs, control side thrust. Who is correct?
A. Tech A
B. Tech B
C. Both Tech A and Tech B
D. Neither Tech A nor Tech B

_____ **39.** Tech A says that a tapered roller bearing assembly has less rolling resistance than a similarly sized ball bearing assembly. Tech B says that the bearing assembly in a unitized wheel bearing assembly can normally be disassembled, cleaned, and repacked. Who is correct?
A. Tech A
B. Tech B
C. Both Tech A and Tech B
D. Neither Tech A nor Tech B

_____ **40.** Tech A says that, over time, a grease seal can wear a groove in the sealing surface of the axle or shaft. Tech B says that grease seals need to be replaced every time the bearing is removed. Who is correct?
A. Tech A
B. Tech B
C. Both Tech A and Tech B
D. Neither Tech A nor Tech B

_____ **41.** Tech A says that in a full-floating axle, the axle does not support the weight of the vehicle. Tech B says that when installing tapered wheel bearings, the final torque should be about 20 ft/lb. Who is correct?
   A. Tech A
   B. Tech B
   C. Both Tech A and Tech B
   D. Neither Tech A nor Tech B

_____ **42.** Tech A says that serviceable wheel bearings can be repacked by removing the dust cap, filling it with grease, and reinstalling it. Tech B says that the cotter pin must be replaced with a new one every time it is removed. Who is correct?
   A. Tech A
   B. Tech B
   C. Both Tech A and Tech B
   D. Neither Tech A nor Tech B

_____ **43.** Tech A says that the grease level in the final drive is okay as long as you can touch the level with your finger. Tech B says that the grease level should normally be no more than ¼" below the threads on the fill plug hole. Who is correct?
   A. Tech A
   B. Tech B
   C. Both Tech A and Tech B
   D. Neither Tech A nor Tech B

_____ **44.** Tech A says that wheel bearings need to be replaced as a set, bearing and race. Tech B says that the wheel bearings and races on both sides of the vehicle must be replaced if one side fails. Who is correct?
   A. Tech A
   B. Tech B
   C. Both Tech A and Tech B
   D. Neither Tech A nor Tech B

_____ **45.** Tech A says that unitized hubs have a wheel nut with a higher installation torque than serviceable wheel bearings. Tech B says that unitized hubs have the proper bearing end play designed into the assembly once they are torqued properly. Who is correct?
   A. Tech A
   B. Tech B
   C. Both Tech A and Tech B
   D. Neither Tech A nor Tech B

_____ **46.** Tech A says when installing a bearing race you should use a 3-lb hammer and a brass drift. Tech B says that when installing a bearing race you should use a 3-lb hammer and a 3/8″ drive extension. Who is correct?
   A. Tech A
   B. Tech B
   C. Both Tech A and Tech B
   D. Neither Tech A nor Tech B

_____ **47.** Tech A says that when a race is fully seated, a sharp metallic sound will be produced when installation is complete. Tech B says that to be sure a race is fully seated, the wheel bearing adjusting nut should be tightened to at least 100 ft-lb of torque, which will finish seating it. Who is correct?
   A. Tech A
   B. Tech B
   C. Both Tech A and Tech B
   D. Neither Tech A nor Tech B

Electronic Brake Control

At the start of each chapter, you will find the Learning Objectives from the textbook. These are your objectives as you make your way through the exercises in this workbook and the chapter in your textbook. The following activities have been designed to help you refresh your knowledge of the material in this chapter.

Learning Objectives

After reading this chapter, you will be able to:

- LO 39-01 Describe the evolution of electronic brake control systems.
- LO 39-02 Describe antilock brake system (ABS) operation.
- LO 39-03 Describe ABS master cylinder and hydraulic control unit (HCU) operation.
- LO 39-04 Describe ABS wheel speed sensor and brake switch operation.
- LO 39-05 Describe ABS electronic brake control module (EBCM) operation.
- LO 39-06 Describe traction control systems (TCS) operation.
- LO 39-07 Describe electronic stability control (ESC) system operation.

Matching

Match the following terms with the correct description or example.

A. Air gap	**K.** Nonintegral ABSs
B. Accumulator	**L.** Oversteer
C. Body control module (BCM)	**M.** Roll-rate sensor
D. Channel	**N.** Steering angle sensor
E. Common bore	**O.** Steering wheel position sensor
F. Fault codes	**P.** Tone wheel
G. High-pressure accumulator	**Q.** Understeer
H. Integral ABS	**R.** Variable reluctance sensor
I. Isolation valve	**S.** Vehicle speed sensor
J. Magneto-resistive sensor	**T.** Wheel speed sensor

_____ **1.** A type of wheel speed sensor that uses the principle of magnetic induction to create its signal.

_____ **2.** A brake system in which the master cylinder, power booster, and HCU are all combined in a common unit.

_____ **3.** A sensor that measures the amount of turning a driver desires. This information is used by the ESC system to know the driver's directional intent.

_____ **4.** An alphanumeric code system used to identify potential problems in a vehicle system.

_____ **5.** A storage container that holds pressurized brake fluid.

_____ **6.** The valve in the HCU that either allows or blocks brake fluid that comes from the master cylinder from entering the HCU hydraulic circuit.

_____ **7.** A brake system in which the master cylinder, power booster, and HCU are all separate units.

_____ **8.** An onboard computer that controls many vehicle functions, including the vehicle interior and exterior lighting, horn, door locks, power seats, and windows.

_____ **9.** The component that creates an electrical signal based on the speed of the vehicle, which is sent to the EBCM.

_____ **10.** A sensor that measures the amount of roll around the vehicle's horizontal axis that a vehicle is experiencing.

_____ **11.** A device that creates an analog or digital signal according to the speed of the wheel.

_____ **12.** A storage container designed to contain high-pressure liquids such as brake fluid.

_____ **13.** A sensor that signals to the EBCM both the position and the speed of the steering wheel.

_____ **14.** The space or clearance between two components, such as the space between the tone wheel and the pick-up coil in a wheel speed sensor.

_____ **15.** The number of wheel speed sensor circuits and hydraulic circuits the EBCM monitors and controls.

_____ **16.** A condition in which the front wheels are turned further than the direction the vehicle is moving and the front tires are slipping sideways toward the outside of the turn.

_____ **17.** A condition in which the rear wheels are slipping sideways toward the outside of the turn.

_____ **18.** The part of the wheel speed sensor that has ribs and valleys used to create an electrical signal inside of the pick-up assembly.

_____ **19.** When a single cylinder is used for two pistons. A tandem master cylinder would be an example.

_____ **20.** A type of wheel speed sensor that uses an effect similar to a Hall effect sensor to create its signal.

Multiple Choice

Read each item carefully, and then select the best response.

_____ **1.** ABSs use a computer that sends electrical signals to the _____ that momentarily hold or release hydraulic pressure to that wheel until it speeds up and starts rolling again.
  A. Wheel speed sensors
  B. Master cylinder
  C. Solenoid valves
  D. Spool valves

_____ **2.** Which component of the ABS contains electric solenoid valves controlled by the EBCM to modify hydraulic pressure in each hydraulic circuit?
  A. Master cylinder
  B. HCU
  C. Power booster
  D. Accumulator

_____ **3.** Either separate or combined into a common assembly, the isolation valve and dump valve are controlled by a(n) _____.
  A. Electric solenoid
  B. Spool valve
  C. Check valve
  D. Brake pedal sensor

_____ **4.** A _____ system uses separate speed sensors and hydraulic control circuits for each of the four wheels.
  A. Single-channel
  B. Two-channel
  C. Three-channel
  D. Four-channel

_____ **5.** If the high-pressure pump fails for any reason, the _____ holds enough brake fluid at high pressure to apply the brakes 10–20 times before the boost is used up.
  A. HCU
  B. Accumulator
  C. Master cylinder
  D. Reservoir

_____ **6.** A wheel sensor assembly consists of a toothed tone wheel (or tone ring) that rotates with the wheels and a stationary _____ attached to the hub or axle housing.
  A. Pick-up assembly
  B. Resistor
  C. Tooth
  D. Relay

_____ **7.** The height of the sine wave is called its _____.
  A. Peak
  B. Amplitude
  C. Phase
  D. Duration

_____ **8.** What style of speed sensor is sometimes called a passive system since it is self-contained and needs no outside power to function?
  A. Variable reluctance
  B. Magneto-resistive
  C. Hall effect
  D. Digital square

_____ **9.** Some ABSs provide _____ through the ABS warning lamp when a specific terminal is grounded or two specific terminals are shorted together.
  A. Blink codes
  B. Morse codes
  C. Switch codes
  D. Sensor values

_____ **10.** An ESC system may integrate all of the following sensors into the basic ABS and TCSs, EXCEPT _____.
  A. Yaw sensor
  B. Traction sensor
  C. Steering angle sensor
  D. Roll-rate sensor

_____ **11.** ESC systems take the ABSs and TCS to the next level by adding sensor information regarding the driver's directional intent.
  A. True
  B. False

_____ **12.** Understeer occurs when the vehicle is turning more sharply than the front wheels are being steered.
  A. True
  B. False

_____ **13.** Applying the brakes too hard or on a slippery surface can cause the wheels to lock.
  A. True
  B. False

_____ **14.** Maximum braking traction occurs when the wheels are rotating 20–25% slower than the vehicle speed.
  A. True
  B. False

_____ **15.** Nonintegral ABS use a fairly standard tandem master cylinder and a typical vacuum or hydraulic power booster.
  A. True
  B. False

_____ **16.** Some HCUs use a single, three-position solenoid valve per circuit, while others use dual, two-position valves per hydraulic circuit.
  A. True
  B. False

_____ **17.** The pick-up assembly and tone wheel do not touch each other; a small gap, called an air gap, must be maintained at the specified clearance.
  A. True
  B. False

_____ **18.** The magneto-resistive sensor does not function effectively below vehicle speeds of around 5 mph.
  A. True
  B. False

_____ **19.** The brake switch is a normally open switch, meaning that if the switch is not affected by any outside force, electrical current will flow through it.
  A. True
  B. False

_____ **20.** Many ESC system–equipped vehicles also monitor signals from the throttle position sensor, vehicle speed sensor, and brake pedal position sensor to help prevent a loss of control of the vehicle.
  A. True
  B. False

_____ **21.** When the ignition switch is turned on, the ABS controller illuminates the yellow ABS warning lamp and performs an automatic self-check of the system.
  A. True
  B. False

_____ **22.** The EBCM supplies the magneto-resistive or Hall effect sensor systems with a reference voltage of between 5 and 12 volts, depending on the manufacturer.
  A. True
  B. False

_____ **23.** In the quest for increased safety, manufacturers developed a series of EBC systems; the first-generation was the _____.
  A. ABS
  B. TCS
  C. ESC system
  D. Body control system

_____ **24.** The _____ system applies brake pressure to the slipping tire, which causes more of the engine's torque to be transmitted to the wheel or wheels with the most traction.
  A. Antilock brake
  B. Traction control
  C. Pressure control
  D. ESC

_____ **25.** The _____ system is designed to prevent wheels from locking or skidding, no matter how hard the brakes are applied or how slippery the road surface is, and to maintain steering control of the vehicle.
  A. ESC
  B. Antilock braking
  C. Wheel speed control
  D. Traction control

_____ **26.** Drivers need to be taught to expect ABS brake pedal _____ when in a panic stop.
  A. Release
  B. Delay
  C. Pulsation
  D. Vibration

_____ **27.** The ABS control module (or EBCM) sends commands in the form of _____ to the HCU.
  A. Audio signals
  B. Electrical signals
  C. Videos
  D. Pictures

_____ **28.** A _____ system uses one sensor circuit with the speed sensor typically located in the differential and one hydraulic control circuit to control both rear wheels.
  A. Single-channel
  B. Two-channel
  C. Three-channel
  D. Four-channel

_____ **29.** _____ accumulators hold brake fluid in a spring-loaded chamber when it is released by the dump valves during an EBC event.
- **A.** Low-pressure
- **B.** Spring-loaded
- **C.** Variable reluctance
- **D.** High pressure

_____ **30.** On a traction control system, _____ direct hydraulic pressure from the accumulator to the ABS solenoid valves so individual wheel brake units can be applied independently.
- **A.** Dump valves
- **B.** Isolation valves
- **C.** Boost valves
- **D.** Suction accumulator valves

_____ **31.** The arrow in the image is pointing to what component of the HCU solenoid valve arrangement?

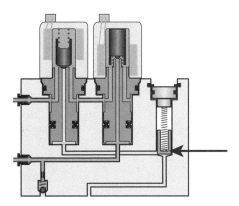

- **A.** Isolation solenoid
- **B.** High-pressure pump suction
- **C.** Dump solenoid
- **D.** Accumulator valve

_____ **32.** The arrow in the image is pointing to what component of the HCU solenoid valve arrangement?

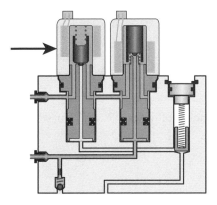

- **A.** Dump solenoid
- **B.** Isolation solenoid
- **C.** High-pressure pump discharge
- **D.** Accumulator valve

_____ **33.** The arrow in the image is pointing to what component of the HCU solenoid valve arrangement?

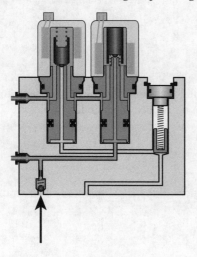

A. Isolation solenoid
B. High-pressure pump discharge
C. Accumulator valve
D. High-pressure pump suction

_____ **34.** The arrow in the image is pointing to what component of the portless ABS master cylinder?

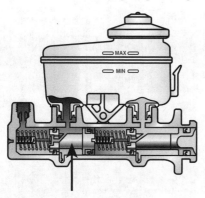

A. Primary piston
B. Primary piston seal
C. Secondary piston
D. Stopper pin

_____ **35.** The arrow in the image is pointing to what component of the portless ABS master cylinder?

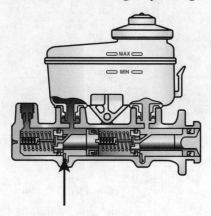

A. Reservoir
B. Primary piston
C. Primary piston seal
D. Stopper pin

_____ **36.** The arrow in the image is pointing to what component of the portless ABS master cylinder?

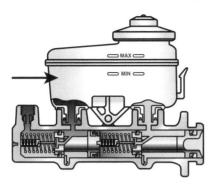

A. Primary piston
B. Primary piston seal
C. Reservoir
D. Secondary piston

_____ **37.** What type of ABS channel is shown in the image?

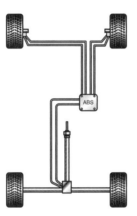

A. Single-channel system
B. Two-channel system
C. Three-channel system
D. Four-channel system

_____ **38.** What type of ABS channel is shown in the image?

A. Single-channel system
B. Two-channel system
C. Three-channel system
D. Four-channel system

_____ **39.** What type of ABS channel is shown in the image?

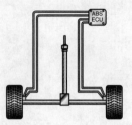

A. Single-channel system
B. Two-channel system
C. Three-channel system
D. Four-channel system

_____ **40.** The arrow in the image is pointing to what component of the high-pressure accumulator?

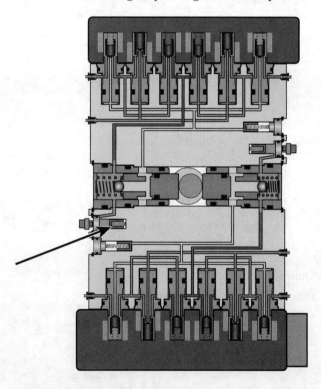

A. Suction accumulator valve
B. Isolation valve
C. High-pressure accumulator valve
D. Dump valve

_____ **41.** The arrow in the image is pointing to what component of the high-pressure accumulator?

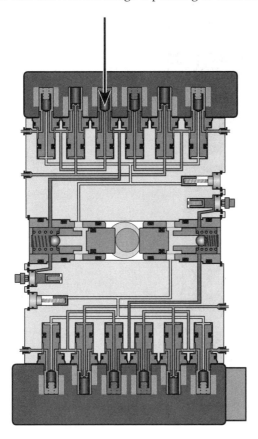

A. Inlet check valve
B. High-pressure accumulator valve
C. Dump valve
D. Boost valve

_____ **42.** The arrow in the image is pointing to what component of the high-pressure accumulator?

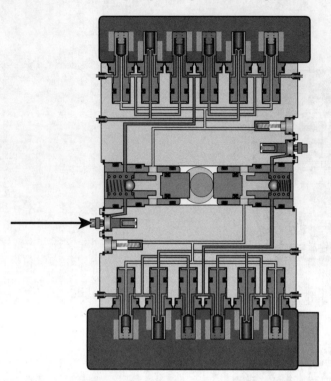

    A. Pump drive eccentric
    B. High-pressure switch
    C. High-pressure accumulator valve
    D. Control unit

_____ **43.** Tech A says that an ABS helps shorten the stopping distance during a panic stop. Tech B says that an ABS works by increasing the hydraulic pressure in the brake system so the brakes can be applied harder. Who is correct?
    A. Tech A
    B. Tech B
    C. Both Tech A and Tech B
    D. Neither Tech A nor Tech B

_____ **44.** Tech A says that traction control can reduce the power output of the engine to increase traction. Tech B says that ESC increases the risk of rollover. Who is correct?
    A. Tech A
    B. Tech B
    C. Both Tech A and Tech B
    D. Neither Tech A nor Tech B

_____ **45.** Tech A says that an EBS has sensors that monitor wheel speed. Tech B says that understeer is generally easier to recover from than oversteer. Who is correct?
    A. Tech A
    B. Tech B
    C. Both Tech A and Tech B
    D. Neither Tech A nor Tech B

_____ **46.** Tech A says that ABS controls braking every time the brakes are used. Tech B says that during an ABS event, it is normal for the brake pedal to pulsate. Who is correct?
   A. Tech A
   B. Tech B
   C. Both Tech A and Tech B
   D. Neither Tech A nor Tech B

_____ **47.** Tech A says that during antilock braking, brake fluid may be returned to the master cylinder. Tech B says that solenoid valves in the HCU will isolate the master cylinder from the brake circuit when it is in the "hold" mode. Who is correct?
   A. Tech A
   B. Tech B
   C. Both Tech A and Tech B
   D. Neither Tech A nor Tech B

_____ **48.** Tech A says that mismatched tires may cause the ABS to register a fault code. Tech B says that a TCS may automatically apply brake pressure to a wheel brake unit even if the vehicle is not being braked. Who is correct?
   A. Tech A
   B. Tech B
   C. Both Tech A and Tech B
   D. Neither Tech A nor Tech B

_____ **49.** Tech A says that an ABS key-on system test checks for faults in the vehicle's base brake system. Tech B says that on most vehicles, ABS diagnostic codes (DTCs) are stored in memory for later retrieval. Who is correct?
   A. Tech A
   B. Tech B
   C. Both Tech A and Tech B
   D. Neither Tech A nor Tech B

_____ **50.** Tech A says that a yaw sensor tells the computer the vehicle's actual direction. Tech B says that raising a vehicle's curb height has no effect on the ESC system. Who is correct?
   A. Tech A
   B. Tech B
   C. Both Tech A and Tech B
   D. Neither Tech A nor Tech B

_____ **51.** Tech A says that some hydraulic units can boost the brake pressure created by the driver. Tech B says that the dump valve can release pressure in the brake system when ABS is applied during a panic stop. Who is correct?
   A. Tech A
   B. Tech B
   C. Both Tech A and Tech B
   D. Neither Tech A nor Tech B

_____ **52.** Tech A says that some ABS wheel speed sensors create a square wave digital pattern. Tech B says that some wheel speed sensors create an alternating current (AC) sine wave pattern. Who is correct?
   A. Tech A
   B. Tech B
   C. Both Tech A and Tech B
   D. Neither Tech A nor Tech B

_____ **53.** Tech A says that rotors that are too thin cannot handle as much heat and will experience brake fade sooner. Tech B says that brake pedal pulsation is the result of air in the hydraulic system. Who is correct?
   A. Tech A
   B. Tech B
   C. Both Tech A and Tech B
   D. Neither Tech A nor Tech B

Principles of Electrical Systems

At the start of each chapter, you will find the Learning Objectives from the textbook. These are your objectives as you make your way through the exercises in this workbook and the chapter in your textbook. The following activities have been designed to help you refresh your knowledge of the material in this chapter.

Learning Objectives

After reading this chapter, you will be able to:

- LO 40-01 Describe the importance of learning electrical theory.
- LO 40-02 Explain conductor, insulator, and semiconductor materials.
- LO 40-03 Describe the process of electron movement in a simple circuit.
- LO 40-04 Explain volts, amps, ohms, power, and ground.
- LO 40-05 Describe the sources of electricity.
- LO 40-06 Describe the effects of electricity.
- LO 40-07 Use Ohm's law to calculate values.
- LO 40-08 Use Watt's law to calculate values.
- LO 40-09 Describe series circuits and use its laws to calculate values.
- LO 40-10 Describe parallel and series-parallel circuits and calculate values.
- LO 40-11 Describe direct current (DC) and alternating current (AC) and Kirchhoff's current law.
- LO 40-12 Explain how to use electrical concepts to solve problems.

Matching

Match the following terms with the correct description or example.

A. AC
B. Amp
C. Charge carrier
D. Diode
E. Electrical resistance
F. Energy
G. Ground
H. Insulator

I. Invertor
J. Ohm
K. Polarity
L. Semiconductor
M. Short
N. Sine wave
O. Volt

_____ 1. A material that has properties that prevent the easy flow of electricity. These materials are made up of atoms with five to eight electrons in the valance ring.

_____ 2. A material's property that slows down the flow of electrical current.

_____ 3. The state of charge, positive or negative.

_____ 4. A two-lead electronic component that allows current flow in one direction only.

_____ 5. A mathematical function that describes a repetitive waveform such as an AC signal.

_____ 6. A type of current flow that flows back and forth.

_____ 7. The ability to do work.

_____ 8. Also called a short circuit, the flow of current along an unintended route.

_____ 9. A mobile particle that has a positive or negative charge.

_____ 10. A device that changes DC into AC.

_____ 11. An abbreviation for amperes, the unit for current measurement.

_____ **12.** The unit for measuring electrical resistance.

_____ **13.** The unit used to measure potential difference or electrical pressure.

_____ **14.** A material used to make microchips, transistors, and diodes.

_____ **15.** The return path for electrical current in a vehicle chassis, other metal of the vehicle, or dedicated wire.

Multiple Choice

Read each item carefully, and then select the best response.

_____ **1.** Electromotive force is also referred to as _____.
  A. Voltage
  B. Resistance
  C. Amperage
  D. Electrons

_____ **2.** Which type of current flow is produced by a battery?
  A. AC
  B. Three phase
  C. DC
  D. Sine wave

_____ **3.** The term _____ describes a low-voltage circuit that does not have a complete circuit and therefore cannot conduct current.
  A. Closed
  B. Open
  C. Short
  D. Grounded

_____ **4.** Which of the following Ohm's law formulas is correct?
  A. A = V/R
  B. V = A × R
  C. R = V ÷ A
  D. All of these are correct.

_____ **5.** The rate of transforming energy is also known as _____.
  A. Power
  B. Voltage
  C. Work
  D. Magnetism

_____ **6.** In a _____ circuit, all components are connected directly to the voltage supply.
  A. Series
  B. Parallel
  C. Series/parallel
  D. Simple

_____ **7.** Which law states that current entering any junction is equal to the sum of the current flowing out of the junction?
  A. Ohm's law
  B. Kirchhoff's current law
  C. The law of conservation of energy
  D. Newton's first law of energy

_____ **8.** A _____ is made up of an electromagnet, a set of switch contacts, terminals, and the case.
  A. Thermocouple
  B. Solenoid
  C. Diode
  D. Relay

_____ **9.** A deficiency of electrons gives an atom an overall positive charge.
  A. True
  B. False

_____ **10.** Most wiring diagrams are written from the conventional theory perspective, while electronic circuits are typically designed and operate on the electron theory perspective.
   A. True
   B. False

_____ **11.** The number of charge carriers, either an excess of electrons or a deficiency of electrons (causing holes in a material), can be altered by doping or adding very large quantities of impurities to a semiconductor material.
   A. True
   B. False

_____ **12.** Hertz is the measurement of frequency and indicates the number of cycles per second.
   A. True
   B. False

_____ **13.** Voltage drop testing is the best way of finding high resistance in the feed side or ground side of the circuit.
   A. True
   B. False

_____ **14.** Since current flowing through a resistance causes heat, you can sometimes locate the high resistance in the circuit just by smelling the wires and connections.
   A. True
   B. False

_____ **15.** Ohm's law is a relationship between volts, amps, and ohms, and because they must always balance out, if we know any two of the values, then we can calculate the third.
   A. True
   B. False

_____ **16.** A light bulb uses a certain amount of electrical power, but the power used is not an indication of brightness.
   A. True
   B. False

_____ **17.** Some capacitors, and most semiconductor components, are polarity sensitive.
   A. True
   B. False

_____ **18.** _____ are only loosely held by the nucleus and are free to move from one atom to another when an electrical potential (pressure) is applied.
   A. Protons
   B. Ions
   C. Free electrons
   D. Insulators

_____ **19.** Electrical _____, measured in ohms, affects the current flow in a circuit.
   A. Force
   B. Resistance
   C. Voltage
   D. Conduction

_____ **20.** _____ tells us that if we increase current flow through a resistance, the voltage used by that resistance will increase.
   A. Electron theory
   B. Conventional theory
   C. Hole theory
   D. Ohm's law

_____ **21.** Voltage can be measured by hooking a voltmeter, or a(n) _____ set to read voltage, across two parts of a circuit where you want to measure the difference in volts.
   A. Ampmeter
   B. Multimeter
   C. Ohmmeter
   D. Ammeter

_____ **22.** Devices are available to change DC into AC and are called _____.
  A. Multimeters
  B. Conductors
  C. Insulators
  D. Inverters

_____ **23.** One _____ is produced when 1 volt causes 1 amp of current to flow.
  A. Ohm
  B. Power
  C. Electron
  D. Watt

_____ **24.** The _____ across each resistor can be found by subtracting the voltage after a resistor from the voltage before it or by measuring the difference.
  A. Electrochemical energy
  B. Current flow
  C. Voltage drop
  D. AC

_____ **25.** The arrow in the image is pointing to what part of the atom?

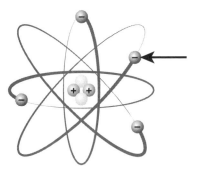

  A. Proton
  B. Neutron
  C. Electron
  D. Ion

_____ **26.** The arrow in the image is pointing to what part of the atom?

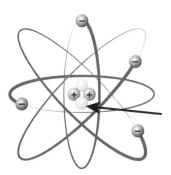

  A. Proton
  B. Neutron
  C. Electron
  D. Conductor

_____ **27.** The arrow in the image is pointing to what part of the atom?

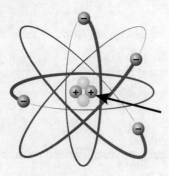

A. Proton
B. Insulator
C. Neutron
D. Electron

_____ **28.** What type of ion is shown in the image?

A. Positive ion
B. Negative ion
C. Neutral ion
D. Free ion

_____ **29.** What type of ion is shown in the image?

A. Negative ion
B. Free ion
C. Neutral ion
D. Positive ion

_____ **30.** What type of material is shown in the image?

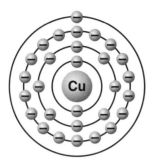

    A. Conductor
    B. Insulator
    C. Semiconductor
    D. Semi-insulator

_____ **31.** What type of material is shown in the image?

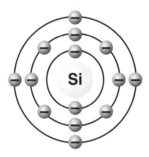

    A. Conductor
    B. Insulator
    C. Semiconductor
    D. Semi-insulator

_____ **32.** What type of material is shown in the image?

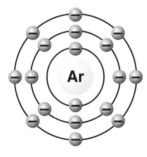

    A. Conductor
    B. Insulator
    C. Semiconductor
    D. Semi-insulator

_____ **33.** Tech A says that the movement of electrons in a circuit is called current flow. Tech B says that the movement of electrons in a circuit is measured in amps. Who is correct?
    A. Tech A
    B. Tech B
    C. Both Tech A and Tech B
    D. Neither Tech A nor Tech B

_____ **34.** Tech A says that electromotive force is also known as voltage. Tech B says that when electrons flow in one direction only, this is DC. Who is correct?
  A. Tech A
  B. Tech B
  C. Both Tech A and Tech B
  D. Neither Tech A nor Tech B

_____ **35.** Two technicians are discussing electron flow. Tech A says that in "conventional theory," current is believed to flow from positive to negative. Tech B says that in "electron theory," current is believed to flow from negative to positive. Who is correct?
  A. Tech A
  B. Tech B
  C. Both Tech A and Tech B
  D. Neither Tech A nor Tech B

_____ **36.** Tech A says that hertz is the number of cycles per second. Tech B says that hertz is the amount of current flow produced by an alternator. Who is correct?
  A. Tech A
  B. Tech B
  C. Both Tech A and Tech B
  D. Neither Tech A nor Tech B

_____ **37.** Two technicians are discussing a series circuit with four resistors of various resistances. Tech A says that current flow will be different in each resistor. Tech B says that current flow will be the same in each resistor. Who is correct?
  A. Tech A
  B. Tech B
  C. Both Tech A and Tech B
  D. Neither Tech A nor Tech B

_____ **38.** Tech A says that a voltage drop is typically used to find excessive resistance in a circuit. Tech B says that high resistance creates heat at the point of resistance in the circuit. Who is correct?
  A. Tech A
  B. Tech B
  C. Both Tech A and Tech B
  D. Neither Tech A nor Tech B

_____ **39.** Tech A says that in a series circuit with two resistors of 120 ohms each, the total circuit resistance is 240 ohms. Tech B says that in a parallel circuit with two resistors of 120 ohms each, the total circuit resistance is 240 ohms. Who is correct?
  A. Tech A
  B. Tech B
  C. Both Tech A and Tech B
  D. Neither Tech A nor Tech B

_____ **40.** Tech A says that a relay is a one-way electrical check valve used in alternators to change AC into DC. Tech B says that a relay uses electromagnetism to open or close a switch. Who is correct?
  A. Tech A
  B. Tech B
  C. Both Tech A and Tech B
  D. Neither Tech A nor Tech B

Electrical Components and Repair

At the start of each chapter, you will find the Learning Objectives from the textbook. These are your objectives as you make your way through the exercises in this workbook and the chapter in your textbook. The following activities have been designed to help you refresh your knowledge of the material in this chapter.

Learning Objectives

After reading this chapter, you will be able to:

- LO 41-01 Describe electrical switches.
- LO 41-02 Describe circuit protection devices.
- LO 41-03 Describe the operation of relays and solenoids.
- LO 41-04 Describe the basic operation of motors and transformers.
- LO 41-05 Describe the common types of resistors.
- LO 41-06 Describe wire.
- LO 41-07 Describe wire harnesses.
- LO 41-08 Use wiring diagrams to trace circuits.
- LO 41-09 Replace wire terminals.
- LO 41-10 Perform solder repairs.

Matching

Match the following terms with the correct description or example.

A. American Wire Gauge (AWG)   **G.** Potentiometer
B. Circuit breaker   **H.** Solenoid
C. Commutator   **I.** Switch
D. Connector   **J.** Terminal
E. Flasher can   **K.** Transformer action
F. Fuse   **L.** Twisted pair

_____ **1.** An electromagnet with a moving iron core that is used to cause mechanical motion.

_____ **2.** The transfer of electrical energy from one coil to another through induction in a transformer.

_____ **3.** An electrical device with contacts that turns current flow on and off.

_____ **4.** A device that trips and opens a circuit, preventing excessive current flow in a circuit. It is resettable to allow for reuse.

_____ **5.** A three-terminal resistive device with a connected movable wiper arm that moves up and down, creating a varying voltage signal.

_____ **6.** Standardized wire gauge used in North America, where the higher the number, the smaller the wire is.

_____ **7.** Two conductors that are twisted together to reduce electrical interference.

_____ **8.** Metal connectors that are attached to wire ends that are used to create electrical connections, which can be disconnected and reconnected.

_____ **9.** A plastic housing at the end of a wiring harness that holds the wire terminals in place.

_____ **10.** A mechanical device that switches the turn signal and hazard flasher bulbs on and off.

_____ **11.** A circuit protection device with a conductive metal strip that melts when excessive current flows through it.

_____ **12.** A device made on the armatures of electric generators and motors to control the direction of current flow in the armature windings.

Match the following steps with the correct sequence for stripping wire insulation.

A. Step 1   **C.** Step 3

B. Step 2

_____ **1.** Choose the correct stripping tool.

_____ **2.** Remove the insulation. To keep the strands together, give them a light twist.

_____ **3.** Select the hole that matches the diameter of the wire to be stripped. Place the wire in the hole, and close the jaws firmly around it to cut the insulation.

Match the following steps with the correct sequence for installing a solderless terminal.

A. Step 1   **D.** Step 4

B. Step 2   **E.** Step 5

C. Step 3   **F.** Step 6

_____ **1.** Lightly twist the wire strands in their normal direction, and place the terminal onto the wire.

_____ **2.** Use a proper crimping tool for the terminal you are crimping. Do not use pliers, as they have a tendency to cut through the connection. Select the proper anvil.

_____ **3.** If crimping an insulated terminal, lightly crimp the insulation tabs so that they hold the insulation firmly.

_____ **4.** Make sure that you have the correct size of terminal for the wire to be terminated and that the terminal has the correct volt/amp rating. Remove an appropriate amount of the protective insulation from the wire.

_____ **5.** If crimping a factory terminal, use the proper tool and follow the instructions.

_____ **6.** Crimp the core section first. Use firm pressure so that a good electrical contact will be made, but do not use excessive force, as this can bend the pin or terminal.

Match the following steps with the correct sequence for soldering wires and connectors.

A. Step 1   **D.** Step 4

B. Step 2   **E.** Step 5

C. Step 3

_____ **1.** Safely position the soldering iron while it is heating up. While the soldering iron is heating, use wire strippers to remove an appropriate amount of insulation from the wires.

_____ **2.** Place heat shrink tubing down one of the wires. Twist the wires together to make a good mechanical connection between them.

_____ **3.** Once the electrical connection has been made and it has cooled, slide the heat shrink tube over the joint. Use a heat gun to shrink the tubing around the joint.

_____ **4.** Tin the soldering iron tip, and gently heat up the wires while placing the solder opposite of the soldering iron. Allow the solder to be drawn into the joint.

_____ **5.** To solder a wire to a terminal connector, it is best to crimp it in place as before and use the solder to "glue" the joint together. Place the heated iron onto the terminal. Apply the solder to the end of the crimped wire tabs. Some solder will be pulled between the terminal and the wire. Cover the terminal with heat shrink tubing.

Multiple Choice

Read each item carefully, and then select the best response.

_____ **1.** An ignition coil can be described as a _____.

   A. Solenoid

   B. Relay

   C. Step-up transformer

   D. Step-down transformer

_____ **2.** Mechanical variable resistors with three connections, two fixed and one movable, are called _____.
   A. Thermistors
   B. Theostats
   C. Potentiometers
   D. Transistors

_____ **3.** All of the following are examples of the type of shielding used in shielded wiring harnesses, EXCEPT:
   A. Twisted pair
   B. Drain lines
   C. Seamless plastic
   D. Mylar tape

_____ **4.** Which of the following is an example of a crimp-type terminal?
   A. Push-on spade
   B. Butt connector
   C. Eye ring
   D. All of these are correct

_____ **5.** What is the most effective and safest tool to use to strip insulation from electrical wire?
   A. A set of wire strippers
   B. Knife
   C. A set of side cutters
   D. A set of combination pliers

_____ **6.** What is the type of solder that is safe for electrical wires and incorporates flux in the core of the solder?
   A. Rosin cored solder
   B. Acid cored solder
   C. Silver solder
   D. Tinning solder

_____ **7.** Cutting the insulation with a sharp tool may lead to ringing the wire, which results in which of the following situations?
   A. The current-carrying capacity of the wire being effectively reduced
   B. The wire being entirely stripped of its current-carrying capacity
   C. The wire being harmed if it happens too often
   D. The wire functioning normally but possibly becoming dangerous over time

_____ **8.** Solderless terminals are _____ to install and _____ at conducting electricity across joints that are designed to be disconnected.
   A. Quick, not very effective
   B. Quick, effective
   C. Difficult, not very effective
   D. Difficult, effective

_____ **9.** Flasher cans are electronic devices, while flasher controls are mechanical devices.
   A. True
   B. False

_____ **10.** A solenoid is an electromechanical device that converts electrical energy into mechanical linear movement.
   A. True
   B. False

_____ **11.** In a motor, the armature and brushes act as switches to control the current flow through the windings of the commutator.
   A. True
   B. False

_____ **12.** A thermistor is a mechanical variable resistor with two connections.
   A. True
   B. False

_____ **13.** Printed circuits are essentially a map of all of the electrical components and their connections.
   A. True
   B. False

_____ **14.** Wiring diagrams use abstract graphical symbols to represent electrical circuits and their connection or relationship to other components in the system.
  A. True
  B. False

_____ **15.** A good pair of wire strippers removes insulation without damaging the wire strands.
  A. True
  B. False

_____ **16.** Many solderless connectors are color coded for the size of wire they are designed to work with, such as yellow 12-10 AWG, blue 16-14 AWG, and red 22-18 AWG.
  A. True
  B. False

_____ **17.** The metal core of a solenoid, used by the electromagnet to strengthen the magnetic field, is referred to as a(n) _____.
  A. Butt connector
  B. Commutator
  C. Armature
  D. Fuse

_____ **18.** There are two scales used to measure the sizes of wires: the metric wire gauge and the _____.
  A. Automotive wire gauge
  B. National wire gauge
  C. American wire gauge
  D. International wire gauge

_____ **19.** Wiring _____, also known as wiring looms or cable harnesses, are used throughout the vehicle to group two or more wires together within a sheath of either insulating tape or tubing.
  A. Connectors
  B. Fuses
  C. Commutators
  D. Harnesses

_____ **20.** What wiring diagram symbol is shown in the image?

  A. Diode
  B. Battery
  C. Fuse
  D. Shielded wire

_____ **21.** What wiring diagram symbol is shown in the image?

  A. Ground connection
  B. Transformer
  C. Motor
  D. Relay

_____ **22.** What wiring diagram symbol is shown in the image?

  A. Circuit breaker
  B. Junction point
  C. Lamp
  D. Tachometer

_____ **23.** What wiring diagram symbol is shown in the image?

    A. Coaxial cable
    B. Connector
    C. Resistor
    D. Fuse

_____ **24.** Tech A says that if the specified fuse keeps blowing, it is generally OK to replace it with a larger fuse. Tech B says that a fusible link is one type of circuit protection device. Who is correct?
    A. Tech A
    B. Tech B
    C. Both Tech A and Tech B
    D. Neither Tech A nor Tech B

_____ **25.** Tech A says that 18 AWG wire is larger than 12 AWG wire. Tech B says that the larger the diameter of the conductor, the more electrical resistance it has. Who is correct?
    A. Tech A
    B. Tech B
    C. Both Tech A and Tech B
    D. Neither Tech A nor Tech B

_____ **26.** Tech A says that when soldering during wire repair, the hotter the better, as it causes the solder to wick up the wires. Tech B says that the stripped ends of wires to be soldered should be laid parallel to each other to minimize resistance of the joint. Who is correct?
    A. Tech A
    B. Tech B
    C. Both Tech A and Tech B
    D. Neither Tech A nor Tech B

_____ **27.** Tech A says a rosin core solder should be used during electrical circuit repair. Tech B says the repair should not be overheated. Who is correct?
    A. Tech A
    B. Tech B
    C. Both Tech A and Tech B
    D. Neither Tech A nor Tech B

_____ **28.** Tech A says that solder should be applied to the soldering iron tip while soldering during solder repair. Tech B says that the solder should be applied to the wire joint while soldering. Who is correct?
    A. Tech A
    B. Tech B
    C. Both Tech A and Tech B
    D. Neither Tech A nor Tech B

_____ **29.** Tech A says that electrical tape can be used to cover the joint during solder wire repair. Tech B says that heat shrink tubing can be used to cover the joint during solder wire repair. Who is correct?
    A. Tech A
    B. Tech B
    C. Both Tech A and Tech B
    D. Neither Tech A nor Tech B

_____ **30.** Tech A says that solder typically used in automotive electrical applications is composed of 20% tin and 80% lead. Tech B says soldered wire repairs should have a good mechanical connection as well as a good electrical connection. Who is correct?
    A. Tech A
    B. Tech B
    C. Both Tech A and Tech B
    D. Neither Tech A nor Tech B

Meter Usage and Circuit Diagnosis

At the start of each chapter, you will find the Learning Objectives from the textbook. These are your objectives as you make your way through the exercises in this workbook and the chapter in your textbook. The following activities have been designed to help you refresh your knowledge of the material in this chapter.

Learning Objectives

After reading this chapter, you will be able to:

- LO 42-01 Describe basic meter information.
- LO 42-02 Describe basic meter layout and ranges.
- LO 42-03 Describe special meter settings probing techniques.
- LO 42-04 Describe how to measure volts, amps, and ohms.
- LO 42-05 Perform available voltage and voltage drop measurements.
- LO 42-06 Perform resistance measurements.
- LO 42-07 Perform current measurements.
- LO 42-08 Perform series circuit measurements.
- LO 42-09 Perform parallel circuit measurements.
- LO 42-10 Perform series-parallel circuit measurements.
- LO 42-11 Perform measurements on variable resistors.
- LO 42-12 Describe electrical circuit testing.
- LO 42-13 Perform voltage and voltage drop measurements.
- LO 42-14 Locate opens, shorts, grounds, and high resistance.
- LO 42-15 Test circuits with a test light and fused jumper wire.
- LO 42-16 Test circuit protection devices, switches, and relays.

Matching

Match the following terms with the correct description or example.

A. Digital volt-ohmmeter (DVOM)　　　　**D.** Open circuit

B. High resistance　　　　**E.** Probing technique

C. Min/max setting　　　　**F.** Short to power

_____ **1.** A setting on a DVOM to display the maximum and minimum readings.

_____ **2.** A term that describes a circuit or components with more resistance than designed.

_____ **3.** A condition in which current flows from one circuit into another.

_____ **4.** A circuit that has a break that prevents current from flowing.

_____ **5.** The way in which test probes are connected to a circuit.

_____ **6.** A test instrument with a digital display for measuring voltage, resistance, and current.

Match the following steps with the correct sequence for checking a circuit with a test light.

A. Step 1　　　　**B.** Step 2　　　　**C.** Step 3

_____ **1.** Place the probe on the terminal to be tested. If voltage is present, the light will come on.

_____ **2.** Connect the end of the light with the clip on it to the negative battery terminal. Touch the probe end of the test light to the positive battery terminal. The light should come on.

_____ **3.** Connect the clip to any known good ground.

Multiple Choice

Read each item carefully, and then select the best response.

_____ **1.** A digital volt-ohmmeter is used to measure all of the following, EXCEPT _____.
   A. Voltage
   B. Amperage
   C. Resistance
   D. Wattage

_____ **2.** A digital volt-ohmmeter for three-phase fixed equipment and single-phase commercial lighting would have a _____ rating.
   A. CAT I
   B. CAT II
   C. CAT III
   D. CAT IV

_____ **3.** The _____ setting is often used to measure vehicle battery voltage while the engine is cranking or the battery is charging.
   A. Hold
   B. Sample
   C. Min/max
   D. Save

_____ **4.** The red lead for the DVOM is labeled with the _____ symbol.
   A. +
   B. V/Ω
   C. –
   D. ±

_____ **5.** All of the following are examples of common types of DVOM probes, EXCEPT _____.
   A. Ground probes
   B. Alligator clips
   C. Fine-pin probes
   D. Insulation piercing probes

_____ **6.** The sum of all the _____ in a series circuit equals the supply voltage.
   A. Resistors
   B. Batteries
   C. Voltage drops
   D. Loads

_____ **7.** The _____ is the same in all parts of a properly working series circuit.
   A. Voltage
   B. Current
   C. Resistance
   D. Wattage

_____ **8.** A(n) _____ must be connected in series within the circuit.
   A. DVOM
   B. Resistor
   C. Conductor
   D. Ammeter

_____ **9.** _____ refers to a condition where power from one circuit leaks into another circuit.
   A. Short to power
   B. Grounds
   C. High resistance
   D. Shorts

_____ **10.** Hybrid vehicles typically require meters and test leads rated as CAT III or CAT IV.
   A. True
   B. False

_____ 11. 5208 MV is the same as 5208 millivolts. It could also be called 52.08 volts.
   A. True
   B. False

_____ 12. The red lead is the positive lead, and the black lead is the negative lead.
   A. True
   B. False

_____ 13. After back-probing, always reinsulate the hole that the probe makes with room-temperature vulcanizing silicone to prevent any corrosion.
   A. True
   B. False

_____ 14. To accurately measure the resistance of a component, you should remove or isolate the component from the circuit.
   A. True
   B. False

_____ 15. A DVOM reading of OL means overload and indicates that the voltage being read is higher than the maximum allowed for the range.
   A. True
   B. False

_____ 16. Voltage drop can be measured across components, connectors, or cables, but current has to be flowing to get an accurate measurement.
   A. True
   B. False

_____ 17. Parallel circuits are commonly used in the vehicle's electrical system, especially for lights.
   A. True
   B. False

_____ 18. A series-parallel circuit occurs when wanted resistance shows up in series with a parallel circuit.
   A. True
   B. False

_____ 19. If the resistance and voltage of a circuit are known, then the theoretical current can be calculated using Ohm's law.
   A. True
   B. False

_____ 20. A basic digital _____ can measure alternating current (AC) and direct current (DC) voltage, AC and DC amperage, and resistance.
   A. Volt-ohmmeter
   B. Analog meter
   C. Ammeter
   D. Thermometer

_____ 21. _____ are used to connect the digital multimeter (DMM) to or into the circuit being tested and come in pairs: one red, the other black.
   A. Test leads
   B. Function switches
   C. Adapters
   D. Function buttons

_____ 22. When the _____ function is activated, the value on the display will "freeze" until the function or DMM is turned off.
   A. Min/max
   B. Throttle position sensor
   C. Hold
   D. Probing

_____ 23. _____ occurs when the probe is pushed in from the back of a connector to make a connection.
   A. Fine-pin probing
   B. Back-probing
   C. Insulation probing
   D. Alligator probing

_____ **24.** The _____ fastens around the conductor and measures the strength of the magnetic field produced from current flowing through the conductor.
   A. Probing lead
   B. Meter lead
   C. Alligator clip
   D. Current clamp

_____ **25.** Unwanted voltage _____ in vehicle circuits can cause real problems and faults.
   A. Increases
   B. Delays
   C. Drops
   D. Interruption

_____ **26.** Current flow is _____ proportional to resistance.
   A. Inversely
   B. Directly

_____ **27.** The sum of the current flow in each branch of a series-parallel circuit is equal to the total _____ circuit current flow.
   A. Series
   B. Parallel
   C. Series-parallel
   D. Variable

_____ **28.** _____ are used to extend connections to allow circuit readings or tests to be undertaken with a DVOM, an oscilloscope, current clamps on fuses, relays, and connector plugs on components.
   A. Common leads
   B. Probing leads
   C. Jumper leads
   D. Red leads

_____ **29.** _____ can occur anywhere in the circuit and can be difficult to locate, especially if it is intermittent.
   A. Short circuits
   B. Insulation
   C. AC
   D. Power circuits

_____ **30.** Tech A says that total resistance goes up as more parallel paths are added. Tech B says that total amperage goes up as more parallel paths are added. Who is correct?
   A. Tech A
   B. Tech B
   C. Both Tech A and Tech B
   D. Neither Tech A nor Tech B

_____ **31.** Tech A says that when reading DC voltage on a meter, a "+" before the number means that there is a higher voltage at the red lead than at the black lead. Tech B says that a "-" before the number means that there is a lower voltage at the red lead than at the black lead. Who is correct?
   A. Tech A
   B. Tech B
   C. Both Tech A and Tech B
   D. Neither Tech A nor Tech B

_____ **32.** Tech A says that to read amperage, the meter needs to be hooked up in series in a circuit. Tech B says that to read amperage at a load, place one lead of the ammeter on the input side of the load and the other lead on the output. Who is correct?
   A. Tech A
   B. Tech B
   C. Both Tech A and Tech B
   D. Neither Tech A nor Tech B

_____ **33.** Tech A says that when checking amperage on a battery, hook the red lead to positive and the black lead to ground. Tech B says that a resistance reading on a light bulb requires the DVOM to be hooked to each side of the bulb and the switch turned on. Who is correct?

   A. Tech A
   B. Tech B
   C. Both Tech A and Tech B
   D. Neither Tech A nor Tech B

_____ **34.** Tech A says that when checking a voltage drop across a closed switch, a measurement of 1.2 volts means the switch is functioning properly. Tech B says that when checking voltage drop across a closed switch, a measurement of 0.0 volts means the switch is faulty. Who is correct?

   A. Tech A
   B. Tech B
   C. Both Tech A and Tech B
   D. Neither Tech A nor Tech B

_____ **35.** A customer complains of slow engine cranking. Tech A says that the starter is faulty and should be replaced. Tech B says that performing a voltage drop test on the high current side of the starter circuit is a valid test in this situation. Who is correct?

   A. Tech A
   B. Tech B
   C. Both Tech A and Tech B
   D. Neither Tech A nor Tech B

_____ **36.** Tech A says that when resistance increases, current flow decreases. Tech B says that when voltage decreases, current increases. Who is correct?

   A. Tech A
   B. Tech B
   C. Both Tech A and Tech B
   D. Neither Tech A nor Tech B

_____ **37.** Tech A says that an open circuit will typically cause higher current flow, which will blow the fuse. Tech B says that a short to ground can typically be found by checking voltage at different points in the circuit. Who is correct?

   A. Tech A
   B. Tech B
   C. Both Tech A and Tech B
   D. Neither Tech A nor Tech B

_____ **38.** Tech A says that when a fuse has popped, it just needs to be replaced, since it performed its job. Tech B says that when a fuse has blown, the circuit needs to be diagnosed because the fuse was probably not the problem. Who is correct?

   A. Tech A
   B. Tech B
   C. Both Tech A and Tech B
   D. Neither Tech A nor Tech B

_____ **39.** Tech A says that the black voltmeter lead stays on the negative battery terminal whenever you perform a "direct voltage drop test." Tech B says that an "indirect voltage drop test" is used to accurately measure current flow in a circuit. Who is correct?

   A. Tech A
   B. Tech B
   C. Both Tech A and Tech B
   D. Neither Tech A nor Tech B

Battery Systems

At the start of each chapter, you will find the Learning Objectives from the textbook. These are your objectives as you make your way through the exercises in this workbook and the chapter in your textbook. The following activities have been designed to help you refresh your knowledge of the material in this chapter.

Learning Objectives

After reading this chapter, you will be able to:

- LO 43-01 Describe basic battery construction and operation.
- LO 43-02 Describe basic types of batteries.
- LO 43-03 Describe battery configurations, terminals, and cables.
- LO 43-04 Describe battery ratings and the charge–discharge cycle.
- LO 43-05 Describe conditions that shorten/lengthen the life of a battery.
- LO 43-06 Describe the purpose and types of battery maintenance.
- LO 43-07 Inspect, clean, fill, and replace the battery and cables.
- LO 43-08 Perform battery charging and jump-starting.
- LO 43-09 Perform battery state of charge and specific gravity tests.
- LO 43-10 Perform battery capacity tests.
- LO 43-11 Maintain and restore electronic memories.
- LO 43-12 Measure parasitic draw.

Matching

Match the following terms with the correct description or example.

A. Absorbed glass mat
B. Battery terminal configuration
C. Cold cranking amps (CCA)
D. Current clamp
E. Cranking amps (CA)
F. Electrolyte
G. Keep alive memory (KAM)
H. Thermal runaway

_____ **1.** The liquid in lead-acid battery cells. It is a mixture of about 67% water and 33% sulfuric acid.

_____ **2.** A standard for rating the ability of a vehicle battery to supply high current under cold operating conditions.

_____ **3.** The placement of positive and negative battery terminals.

_____ **4.** A certain minimum amount of parasitic current draw that is used by the vehicle's circuits to maintain memory functions and monitor systems.

_____ **5.** A device that clamps around a conductor to measure current flow. It is often used in conjunction with a digital volt-ohmmeter (DVOM).

_____ **6.** A cycle during battery charging in which heating lowers resistance, which in turn increases current flow, which in turn further increases heat created. During this cycle, dangerous gases may build up, creating the potential for an explosion or damage through excessive current flow.

_____ **7.** A type of lead-acid battery.

_____ **8.** A standard similar to CCA but that measures the battery's function at a higher temperature, 32° F (0° C).

Multiple Choice

Read each item carefully, and then select the best response.

_____ **1.** A standard 12-volt car battery consists of _____ cells connected in series.
   A. 1
   B. 2
   C. 6
   D. 12

_____ **2.** All of the following are methods used to rate automotive battery capacity, EXCEPT _____.
   A. CCA
   B. Voltage output
   C. CA
   D. Reserve capacity

_____ **3.** All of the following are types of rechargeable cell batteries, EXCEPT _____.
   A. Lead-acid
   B. Nickel–cadmium
   C. Lithium ion
   D. Disposable

_____ **4.** All of the following are advantages of lithium-ion batteries, EXCEPT _____.
   A. High energy density
   B. Cost
   C. Low internal resistance
   D. Low self-discharge

_____ **5.** Which tool is commonly used to measure parasitic draw from the battery?
   A. Refractometer
   B. Scan tool
   C. Ammeter
   D. Hydrometer

_____ **6.** Performing a battery state-of-charge test with any of the following will give a good indication of whether the battery needs to be charged or not, EXCEPT a(n) _____.
   A. Refractometer
   B. Hydrometer
   C. DVOM
   D. Ammeter

_____ **7.** The higher the _____, the higher the percentage of acid in the electrolyte, which corresponds to a high battery state of charge.
   A. Specific gravity
   B. Water level
   C. Temperature
   D. Amperage

_____ **8.** Batteries store electricity in chemical form.
   A. True
   B. False

_____ **9.** The sulfuric acid contained in batteries is highly corrosive and can also be very harmful to metal, painted surfaces, and skin.
   A. True
   B. False

_____ **10.** Sizing has to do with the electrical specifications of the battery, while ratings have to do with the battery's physical attributes.
   A. True
   B. False

_____ **11.** Reserve capacity is the time in minutes that a new fully charged battery at 80° F (27° C) will supply a constant load of 25 amps without its voltage dropping below 10.5 volts for a 12-volt battery.
   A. True
   B. False

_____ **12.** The typical cell voltage of a lithium-ion battery is 1.2 volts.
   A. True
   B. False

_____ **13.** Connecting a battery into a vehicle backward can instantly destroy onboard electronics.
   A. True
   B. False

_____ **14.** When jump-starting a vehicle, the slave battery is connected in series to the host (discharged) battery in order to provide additional capacity to crank and start the vehicle.
   A. True
   B. False

_____ **15.** Many municipalities require battery recycling and levy a "core charge" on every new automotive battery sold.
   A. True
   B. False

_____ **16.** State-of-charge testing tells us how much capacity the battery has left.
   A. True
   B. False

_____ **17.** In some cases, it may be possible to use a 9-volt memory minder or memory saver to maintain the vehicle's memory while the vehicle battery is disconnected.
   A. True
   B. False

_____ **18.** Each cell of a fully charged "12-volt" battery has a nominal _____ volts, for a total of _____ volts.
   A. 1.2, 5.6
   B. 2.1, 20.6
   C. 2.1, 12.6
   D. 5.1, 24.6

_____ **19.** The more plate _____ there is, the higher the electrical capacity of the battery.
   A. Storage area
   B. Surface area
   C. Proximity
   D. Bending

_____ **20.** Because the electrolyte in _____ batteries is a gel, which does not spill, this type of battery is especially handy for rough handling or tipping.
   A. Absorbed glass mat
   B. Nickel–cadmium
   C. Nickel-metal hydride
   D. Lithium-ion

_____ **21.** The charging process increases the amount of _____ in the electrolyte, making the electrolyte stronger.
   A. Lithium
   B. Water
   C. Lead
   D. Acid

_____ **22.** The _____ the battery temperature, the higher the rate of charging.
   A. Hotter
   B. More neutral
   C. Colder

_____ **23.** Lithium-ion batteries may suffer from _____ and cell rupture if overheated or overcharged.
   A. Gassing
   B. Thermal runaway
   C. Parasitic draw
   D. Electrical runaway

_____ **24.** Battery cables and terminals are designed to carry the _____ currents that are required during cranking of the automotive engine.
  A. Low discharge
  B. Positive
  C. High discharge
  D. Negative

_____ **25.** Battery terminals are usually _____ or _____ onto the battery cables to ensure strong, low-resistance connections.
  A. Crimped, soldered
  B. Pressed, glued
  C. Hammered, bolted
  D. Crimped, glued

_____ **26.** Always remove the _____ or ground terminal first when disconnecting battery cables.
  A. Top
  B. Left
  C. Positive
  D. Negative

_____ **27.** Some manufacturers say that their batteries should not be load tested, and instead should be _____ tested.
  A. Clamp
  B. Gas
  C. Conductance
  D. Reserve

_____ **28.** The arrow in the image is pointing to what component of the simple battery?

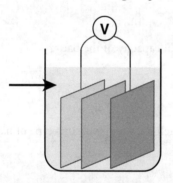

  A. Positive plate
  B. Negative plate
  C. Insulator
  D. Electrolyte

_____ **29.** The arrow in the image is pointing to what component of the simple battery?

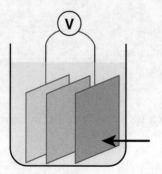

  A. Insulator
  B. Positive plate
  C. Negative plate
  D. Electrolyte

_____ **30.** The arrow in the image is pointing to what component of the simple battery?

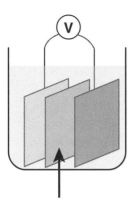

    A. Electrolyte
    B. Positive plate
    C. Insulator
    D. Negative plate

_____ **31.** The arrows in the image are pointing to what component of the wet cell battery?

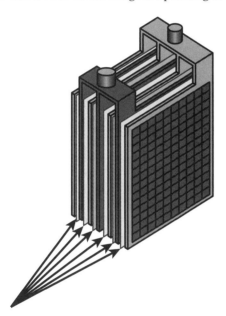

    A. Negative plates
    B. Positive plates
    C. Insulator plates
    D. Separator plates

_____ **32.** The arrows in the image are pointing to what component of the wet cell battery?

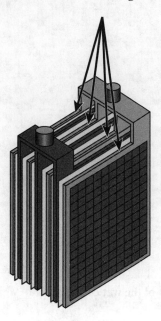

    A. Positive plates
    B. Insulator plates
    C. Negative plates
    D. Separator plates

_____ **33.** Tech A says that a battery stores electrical energy in chemical form. Tech B says that a battery creates direct current. Who is correct?
    A. Tech A
    B. Tech B
    C. Both Tech A and Tech B
    D. Neither Tech A nor Tech B

_____ **34.** Tech A says that a 12-volt battery has six cells. Tech B says that the more plates a battery cell has, the more voltage it creates. Who is correct?
    A. Tech A
    B. Tech B
    C. Both Tech A and Tech B
    D. Neither Tech A nor Tech B

_____ **35.** Tech A says that a parasitic draw is measured in volts. Tech B says that pulling fuses one at a time can help locate a parasitic draw. Who is correct?
    A. Tech A
    B. Tech B
    C. Both Tech A and Tech B
    D. Neither Tech A nor Tech B

_____ **36.** Tech A says that when disconnecting the battery, the negative terminal should be disconnected first. Tech B says that baking soda and water will remove oxidation from battery terminals. Who is correct?
    A. Tech A
    B. Tech B
    C. Both Tech A and Tech B
    D. Neither Tech A nor Tech B

_____ **37.** Tech A says that checking the specific gravity will indicate the battery's CCA. Tech B says that a battery load test should be performed when the battery is heavily discharged. Who is correct?
    A. Tech A
    B. Tech B
    C. Both Tech A and Tech B
    D. Neither Tech A nor Tech B

_____ **38.** Tech A says that before attempting to jump-start a vehicle, you need to make sure the battery is not frozen. Tech B says that it is good practice to place a load across the discharged battery, such as turning on the headlamps, to absorb any sudden rise in voltage. Who is correct?
  A. Tech A
  B. Tech B
  C. Both Tech A and Tech B
  D. Neither Tech A nor Tech B

_____ **39.** Tech A says that there is an additional method to measure parasitic draw using an ohmmeter. Tech B says that the method known as the Chesney parasitic load test is named after its creator, Sean Chesney. Who is correct?
  A. Tech A
  B. Tech B
  C. Both Tech A and Tech B
  D. Neither Tech A nor Tech B

_____ **40.** Tech A says that a thermal runaway can occur during battery charging, in which heating increases resistance and decreases the potential for an explosion or damage through excessive current flow. Tech B says that a thermal runaway can occur during battery charging, in which heating lowers resistance and increases the potential for an explosion or damage through excessive current flow. Who is correct?
  A. Tech A
  B. Tech B
  C. Both Tech A and Tech B
  D. Neither Tech A nor Tech B

_____ **41.** Tech A says that some manufacturers recommend removing the negative battery terminal cable while charging a battery to reduce risk to the vehicle's electronics. Tech B says that this is not necessary since all electronics are protected. Who is correct?
  A. Tech A
  B. Tech B
  C. Both Tech A and Tech B
  D. Neither Tech A nor Tech B

_____ **42.** Tech A says that CCA ratings are a standard for determining the ability of a vehicle battery to supply high current under cold operating conditions. Tech B says that CA is similar to CCA, but measures the battery's function at a higher temperature, 32° F (0° C). Who is correct?
  A. Tech A
  B. Tech B
  C. Both Tech A and Tech B
  D. Neither Tech A nor Tech B

Starting and Charging Systems

At the start of each chapter, you will find the Learning Objectives from the textbook. These are your objectives as you make your way through the exercises in this workbook and the chapter in your textbook. The following activities have been designed to help you refresh your knowledge of the material in this chapter.

Learning Objectives

After reading this chapter, you will be able to:

- LO 44-01 Describe starting system fundamentals.
- LO 44-02 Describe starter motor construction.
- LO 44-03 Describe starter motor engagement.
- LO 44-04 Describe armature and starter drive operation.
- LO 44-05 Describe solenoid operation.
- LO 44-06 Describe starter control circuit operation.
- LO 44-07 Describe solenoid operation.
- LO 44-08 Test starter high-current circuit voltage drop.
- LO 44-09 Test starter control circuit voltage drop.
- LO 44-10 Test starter relays and solenoids.
- LO 44-11 Remove and install a starter.
- LO 44-12 Describe idle stop–start stop system operation.
- LO 44-13 Describe charging system operation.
- LO 44-14 Describe the rotor, slip ring, and brushes.
- LO 44-15 Describe the stator, end frames, fan, and pulley.
- LO 44-16 Describe rectification.
- LO 44-17 Describe voltage regulation.
- LO 44-18 Perform a charging system output test.
- LO 44-19 Perform charging system circuit voltage and voltage drop tests.
- LO 44-20 Replace alternator.

Matching

Match the following terms with the correct description or example.

A. Counter-electromotive force (CEMF)

B. Current clamp

C. Hold-in winding

D. Pull-in winding

_____ **1.** A low-current winding found in starter solenoids that holds the plunger in the activated position.

_____ **2.** A device that clamps around a conductor to measure current flow. It is often used in conjunction with a digital volt-ohmmeter (DVOM).

_____ **3.** Voltage created in the field windings as the motor rotates, which opposes battery voltage and limits motor speed.

_____ **4.** A high-current winding found in starter solenoids that pulls the solenoid plunger into the activated position.

Multiple Choice

Read each item carefully, and then select the best response.

_____ **1.** The starter motor is mounted on the transmission or cylinder block in a position to engage a _____ around the outside edge of the engine flywheel, flex plate, or torque converter.
   A. Magnet
   B. Commutator
   C. Ring gear
   D. Sleeve

_____ **2.** In the direct-drive system, the starter drive is mounted directly on one end of the _____.
   A. Flywheel
   B. Armature shaft
   C. Transmission
   D. Reduction gear

_____ **3.** The commutator end frame carries copper-impregnated carbon _____, which conduct current through the armature when it is being rotated in operation.
   A. Spur gears
   B. Brushes
   C. Coils
   D. Strips

_____ **4.** A(n) _____ consists of two semicircular segments that are connected to the two ends of the loop and are insulated from each other.
   A. Helix
   B. Armature
   C. Commutator
   D. Electromagnet

_____ **5.** The _____ is typically a cylindrical device mounted on the starter motor that switches the high-current flow required by the starter motor on and off and engages the starter drive with the ring gear.
   A. Solenoid
   B. Armature
   C. Commutator
   D. Pinion

_____ **6.** For many years, manufacturers have placed switches _____ with the starter solenoid windings, which prevents the starter from being activated.
   A. Parallel
   B. In series
   C. Series/parallel
   D. Redundantly

_____ **7.** The _____ prevents the starter motor from being driven by the engine once the engine starts, which would spin the armature faster than it could handle.
   A. Flywheel
   B. Starter control circuit
   C. Pole shoe
   D. Overrunning clutch

_____ **8.** The _____ converts mechanical energy into electrical energy by electromagnetic induction.
   A. Direct current (DC) generator
   B. Inverter
   C. Alternator
   D. Rectifier

_____ **9.** The voltage potential induced by an AC generator is called _____.
   A. Electromotive force
   B. Electromagnetic induction
   C. Counter-electromotive force
   D. Rotating force

_____ **10.** To change AC to DC, automotive alternators use a rectifier assembly consisting of _____ in a specific configuration.
   A. Transistors
   B. Diodes
   C. Resistors
   D. Semiconductors

_____ **11.** All vehicles equipped with an automatic transmission use a neutral safety switch or a similar device.
   A. True
   B. False

_____ **12.** The starter motor converts mechanical energy to electrical energy for the purpose of cranking the engine over.
   A. True
   B. False

_____ **13.** A conductor loop that can freely rotate within the magnetic field is the most efficient motor design.
   A. True
   B. False

_____ **14.** The hold-in winding draws a higher current and creates a stronger magnetic field than the pull-in winding.
   A. True
   B. False

_____ **15.** Some Ford vehicles use a separate starter relay in the engine compartment, instead of a solenoid, to control the high current for the starter motor.
   A. True
   B. False

_____ **16.** Some immobilizer systems now use keyless starting. The vehicle has a start button on the dash and does not require the key to be inserted into an ignition switch.
   A. True
   B. False

_____ **17.** A diode bridge gets its name from two diodes in series bridged with a wire.
   A. True
   B. False

_____ **18.** Rectification is a process of converting DC into the AC that is required by the battery and nearly all of the automobile systems.
   A. True
   B. False

_____ **19.** When the ignition is on, current flows from the positive battery terminal through the ignition switch and charge indicator lamp to the L terminal of the alternator.
   A. True
   B. False

_____ **20.** Alternators need to be replaced whenever they are electrically or mechanically faulty.
   A. True
   B. False

_____ **21.** The _____ system provides a method of rotating the vehicle's internal combustion engine to begin the combustion cycle.
   A. Combustion
   B. Low-amperage
   C. Starting
   D. High-amperage

_____ 22. _____ starters use an extra gear between the armature and the starter drive mechanism.
- **A.** Gear reduction
- **B.** Direct drive
- **C.** Rotating
- **D.** Electromagnetic

_____ 23. The _____ activates the solenoid winding to draw the plunger forward.
- **A.** Security system
- **B.** Drive gear
- **C.** Pinion gear
- **D.** Starter control circuit

_____ 24. A _____ system is a computer-managed security system that disables the vehicle starter and engine systems by using an electronic system to uniquely identify each vehicle key by a security code system.
- **A.** PCM-controlled
- **B.** Solenoid
- **C.** Vehicle immobilization
- **D.** Relay

_____ 25. Most hybrid vehicles use a _____ electric motor for engine start-up, auxiliary power, and regenerative braking functions.
- **A.** High-voltage
- **B.** 12-volt
- **C.** Low-voltage
- **D.** Low-amperage

_____ 26. A _____ AC generator has only one stationary coil, which creates a single sine wave.
- **A.** Four-phase
- **B.** Single-phase
- **C.** Three-phase
- **D.** Two-phase

_____ 27. The _____ monitors battery voltage and adjusts the current flow through the rotor appropriately.
- **A.** Alternator
- **B.** Voltage regulator
- **C.** Battery
- **D.** Charge-warning light

_____ 28. The _____ must not be operated with the battery disconnected or with the terminals at the back of the alternator disconnected.
- **A.** DC generator
- **B.** Commutator
- **C.** Voltage regulator
- **D.** Alternator

_____ 29. _____ leads to electrical systems that do not function fully, and the battery may not fully charge, leading to an early death due to sulfation.
- **A.** Overcharging
- **B.** Undercharging
- **C.** Not charging
- **D.** Rectifying

_____ **30.** The arrow in the image is pointing to what part of the gear reduction starter?

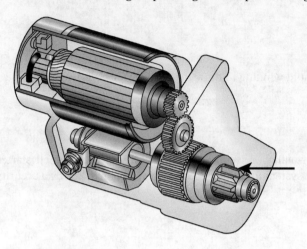

    A. Starter clutch
    B. Armature
    C. Pinion gear
    D. Drive gear

_____ **31.** The arrow in the image is pointing to what part of the gear reduction starter?

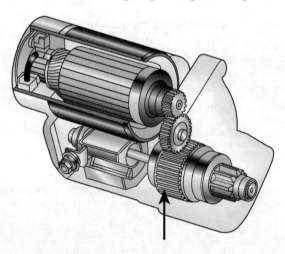

    A. Armature
    B. Drive gear
    C. Pinion gear
    D. Starter clutch

_____ **32.** The arrow in the image is pointing to what part of the gear reduction starter?

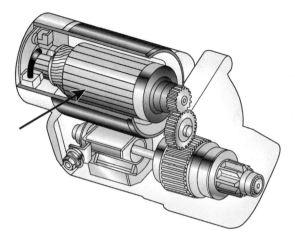

    A. Armature
    B. Starter clutch
    C. Pinion gear
    D. Drive gear

_____ **33.** The arrow in the image is pointing to what part of the series-wound starter?

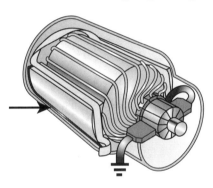

    A. Armature
    B. Pole shoe
    C. Armature shaft
    D. Commutator

_____ **34.** The arrow in the image is pointing to what part of the series-wound starter?

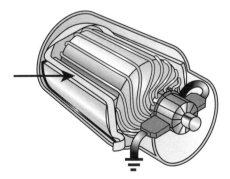

    A. Pole shoe
    B. Field windings
    C. Brushes
    D. Commutator

_____ **35.** The arrow in the image is pointing to what part of the series-wound starter?

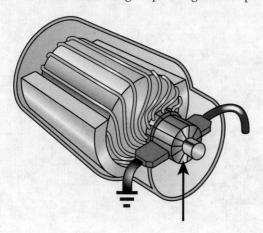

    A. Commutator
    B. Armature
    C. Pole shoe
    D. Armature shaft

_____ **36.** The arrow in the image is pointing to what part of the simple multiloop motor and electromagnetic fields?

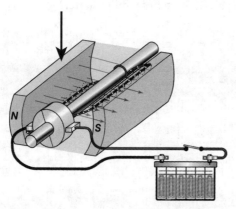

    A. Commutator
    B. Armature
    C. Permanent magnets
    D. Looped conductor

_____ **37.** The arrow in the image is pointing to what part of the simple multiloop motor and electromagnetic fields?

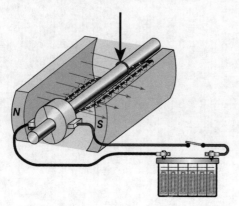

    A. Commutator
    B. Looped conductor
    C. Permanent magnets
    D. Armature

_____ **38.** The arrow in the image is pointing to what part of the simple multiloop motor and electromagnetic fields?

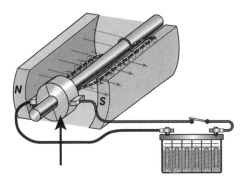

A. Magnetic field
B. Armature
C. Looped conductor
D. Commutator

_____ **39.** The arrow in the image is pointing to what part of the basic starter control circuit in the automatic transmission?

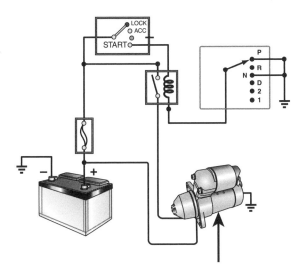

A. Ignition switch
B. Starter relay
C. Neutral safety switch
D. Starter motor

_____ **40.** The arrow in the image is pointing to what part of the basic starter control circuit in the automatic transmission?

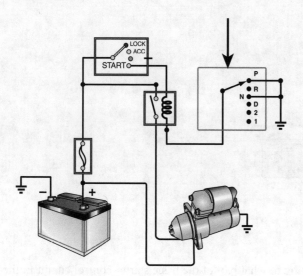

   A. Ignition switch
   B. Starter relay
   C. Neutral safety switch
   D. Fusible link

_____ **41.** The arrow in the image is pointing to what part of the basic starter control circuit in the manual transmission?

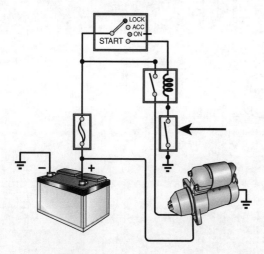

   A. Clutch switch
   B. Ignition switch
   C. Fusible link
   D. Battery

_____ **42.** The arrow in the image is pointing to what part of the basic starter control circuit in the manual transmission?

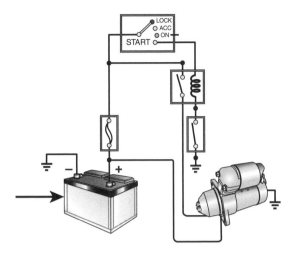

    A. Fusible link
    B. Battery
    C. Starter motor
    D. Starter relay

_____ **43.** The arrow in the image is pointing to what part of the starter drive one-way clutch?

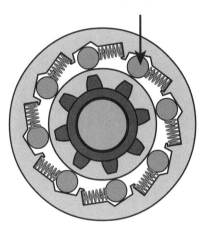

    A. Pinion gear
    B. Clutch housing
    C. Clutch roller
    D. Spring

_____ **44.** The arrow in the image is pointing to what part of the starter drive one-way clutch?

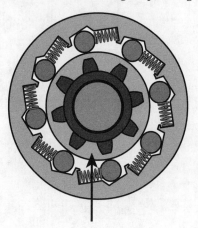

A. Pinion gear
B. Plunger
C. Spring
D. Inner race

_____ **45.** The arrow in the image is pointing to what part of the starter drive one-way clutch?

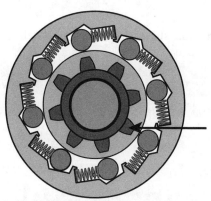

A. Pinion gear
B. Clutch roller
C. Plunger
D. Inner race

_____ **46.** The arrow in the image is pointing to what part of the alternator?

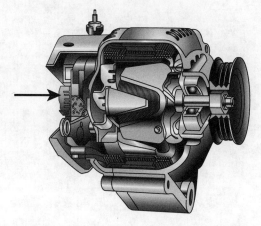

A. Slip ring
B. Rotor
C. Fan
D. Stator

_____ **47.** The arrow in the image is pointing to what part of the alternator?

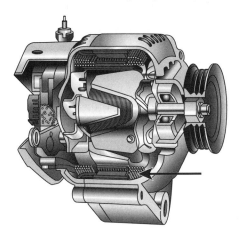

A. Fan
B. Slip ring
C. Stator
D. Rotor

_____ **48.** The arrow in the image is pointing to what part of the alternator?

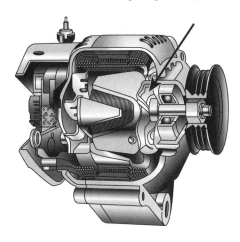

A. Slip ring
B. Rotor
C. Stator
D. Fan

_____ **49.** Tech A says that some starters use gear reduction to improve efficiency. Tech B says that a starter converts electrical energy to mechanical energy. Who is correct?
A. Tech A
B. Tech B
C. Both Tech A and Tech B
D. Neither Tech A nor Tech B

_____ **50.** Tech A says that the pull-in winding is short-circuited when the solenoid is fully engaged. Tech B says that the starter drive has a built-in one-way clutch. Who is correct?
A. Tech A
B. Tech B
C. Both Tech A and Tech B
D. Neither Tech A nor Tech B

_____ 51. Tech A says that a voltage drop of 0.8 volts on the starter ground circuit is within specifications. Tech B says that high starter draw current could be caused by a spun main bearing in the engine. Who is correct?
  A. Tech A
  B. Tech B
  C. Both Tech A and Tech B
  D. Neither Tech A nor Tech B

_____ 52. Tech A says that the first item to check if an engine does not crank is the voltage to the S-terminal on the starter. Tech B says that the first item to check if an engine does not crank is battery voltage. Who is correct?
  A. Tech A
  B. Tech B
  C. Both Tech A and Tech B
  D. Neither Tech A nor Tech B

_____ 53. Tech A says that the voltage regulator controls the strength of the rotor's magnetic field. Tech B says that the voltage regulator is installed between the output terminal of the alternator and the positive terminal of the battery. Who is correct?
  A. Tech A
  B. Tech B
  C. Both Tech A and Tech B
  D. Neither Tech A nor Tech B

_____ 54. Tech A says that overcharging can lead to short life of bulbs and other electrical devices. Tech B says that maintaining a proper charge is critical to long life and proper operation of the electrical system. Who is correct?
  A. Tech A
  B. Tech B
  C. Both Tech A and Tech B
  D. Neither Tech A nor Tech B

_____ 55. Tech A says that an excessive voltage drop in the alternator output circuit will likely cause a high charging system voltage. Tech B says that the alternator output circuit voltage drop must be checked with the charging system under heavy load. Who is correct?
  A. Tech A
  B. Tech B
  C. Both Tech A and Tech B
  D. Neither Tech A nor Tech B

_____ 56. Tech A says that most hybrid vehicles use the high-voltage electric motor to start the engine. Tech B says that most hybrid engines need to crank over more slowly than regular ones. Who is correct?
  A. Tech A
  B. Tech B
  C. Both Tech A and Tech B
  D. Neither Tech A nor Tech B

_____ 57. Tech A says that failure to crank over properly is always caused by mechanical problems. Tech B says that failure to crank over properly is always caused by electrical problems. Who is correct?
  A. Tech A
  B. Tech B
  C. Both Tech A and Tech B
  D. Neither Tech A nor Tech B

_____ 58. Tech A says that the starter drive uses a pinion gear with an overrunning clutch. Tech B says that on some engines the starter motor is mounted on top of the engine under the intake manifold. Who is correct?
  A. Tech A
  B. Tech B
  C. Both Tech A and Tech B
  D. Neither Tech A nor Tech B

Lighting Systems

At the start of each chapter, you will find the Learning Objectives from the textbook. These are your objectives as you make your way through the exercises in this workbook and the chapter in your textbook. The following activities have been designed to help you refresh your knowledge of the material in this chapter.

Learning Objectives

After reading this chapter, you will be able to:

- LO 45-01 Describe the purpose of the lighting system.
- LO 45-02 Describe the types of lights.
- LO 45-03 Describe light bulb configurations.
- LO 45-04 Describe park, tail, marker, and license lights.
- LO 45-05 Describe driving, fog, and cornering lights.
- LO 45-06 Describe brake and backup lights.
- LO 45-07 Describe turn signal and hazard lights.
- LO 45-08 Describe headlights and headlight systems.
- LO 45-09 Describe lighting system testing and precautions.
- LO 45-10 Perform peripheral lighting service.
- LO 45-11 Perform headlight service.

Matching

Match the following terms with the correct description or example.

A. Halogen lamp   **D.** Light-emitting diode (LED)

B. High-intensity discharge (HID)   **E.** Vacuum tube fluorescent (VTF)

C. Incandescent lamp

_____ **1.** A type of lighting that produces light with an electric arc rather than a glowing filament.

_____ **2.** A type of lighting used for instrumentation displays on vehicle instrument panel clusters. This type of lighting emits a very bright light with high contrast and can shine in various colors. Also called vacuum fluorescent display.

_____ **3.** A type of bulb that produces a bright white light.

_____ **4.** A type of lighting used in various automotive applications, such as warning indicators and alphanumeric displays.

_____ **5.** The traditional bulb that uses a heated filament to produce light.

Match the following steps with the correct sequence for checking and changing an exterior light bulb.

A. Step 1   **C.** Step 3

B. Step 2

_____ **1.** Remove the cover to expose the bulb. If the bulb is pin mounted, gently grip the bulb and push it inward. Turn the bulb slightly counterclockwise, and remove it from the bulb holder. Some bulbs pull straight out.

_____ **2.** Insert the new bulb into the bulb holder, depress it fully, turn it slightly clockwise, and release it. Test it by switching it on and off. Then replace the cover, and test it again.

_____ **3.** Inspect the bulb holder to make sure there is no corrosion. If there is, clean it with a bulb socket wire brush or emery cloth.

Match the following steps with the correct sequence for checking and changing a headlight bulb.

A. Step 1

B. Step 2

C. Step 3

D. Step 4

_____ **1.** Replace the unit and the retaining ring or bulb assembly, and then plug in the connector. Switch on the lights again to confirm that they are both operating correctly.

_____ **2.** Remove the old bulb, and replace it with the new one. Handle the new bulb only by its base or, if supplied, by the card cover.

_____ **3.** Switch the headlights on at low beam, then to high beam. Check that the high-beam indicator is operating. If one of the lights does not operate or is dim, that headlight will have to be diagnosed and potentially replaced.

_____ **4.** Test the vehicle headlights. Obtain the replacement lamp for the vehicle. Unplug the electrical connector at the back of the lamp unit.

Match the following steps with the correct sequence for aiming headlights.

A. Step 1

B. Step 2

C. Step 3

_____ **1.** On the types of aligners that require the headlights to be on during alignment, turn the headlights on to a low-beam setting. The center of the illuminating beams should be in the lower right quadrants of the chart or wall markings or as specified by the manufacturer.

_____ **2.** Make sure the tires are inflated properly, the wheels point straight ahead, and there is no extra weight in the vehicle. Position the vehicle correctly in relation to the headlamp aligner unit, following the equipment manufacturer's instructions. Calibrate the aligner for any floor slope and for the vehicle being tested.

_____ **3.** The high beam should be centered, falling on the intersections of the horizontal and vertical marks or as specified by the manufacturer. If necessary, turn the adjustment screws on the headlight so the lights point to the correct places or bubbles on the levels are centered, depending on the type of aligner equipment you are using.

Multiple Choice

Read each item carefully, and then select the best response.

_____ **1.** Incandescent bulbs are inefficient, converting only about _____ of the electricity to visible light.

    A. 10%

    B. 15%

    C. 25%

    D. 30%

_____ **2.** What kind of incandescent lamps are filled with bromine or iodine gas?

    A. LED

    B. VTF

    C. Halogen

    D. HID

_____ **3.** LEDs are often required to give off a specified amount of light; to achieve this, they are usually connected in groups called _____.

    A. Clusters

    B. Series strings

    C. Parallel sets

    D. Packs

_____ **4.** Which type of bulb works well as a combination taillight and brake light because it has one filament that emits a small amount of light and a second filament that emits more light?

    A. LED cluster

    B. Festoon-style bulb

    C. Tandem bulb

    D. Dual-filament bulb

_____ **5.** Which type of lamp base gets its name from the two retaining pins on the side of the base?
   A. Bayonet-style
   B. Festoon-style
   C. Wedge-style
   D. Pin-type

_____ **6.** Which type of lights usually light up as part of a self-test when the ignition initially comes on to show they are in working order?
   A. Marker lights
   B. Warning lamps
   C. Taillights
   D. Courtesy lights

_____ **7.** For safety reasons, _____ continue to operate when the light switch is moved to the headlight position.
   A. Park lights
   B. License plate illumination lamps
   C. Cornering lights
   D. Fog lights

_____ **8.** Many vehicles, by law, now have an additional, higher third brake light mounted on top of the trunk lid or on the rear window called a _____ light.
   A. Pedestal light
   B. Courtesy light
   C. Center high mount stop light (CHMSL)
   D. Festoon light

_____ **9.** White lights mounted at the rear of a vehicle that provide the driver with vision behind the vehicle at night are called _____.
   A. Backup lights
   B. Taillights
   C. License lights
   D. Marker lights

_____ **10.** Which type of lights simultaneously cause a pulsing in all exterior indicator lights and both indicator lights on the instrument panel?
   A. Hazard warning lights
   B. Turn signal lights
   C. Backup lights
   D. Driving lights

_____ **11.** Existing lights that turn on when the vehicle is running and turn off when the engine stops in order to improve the vehicle's visibility to other drivers in all weather conditions are called _____.
   A. Hazard warning lights
   B. Dual-filament lights
   C. Courtesy lights
   D. Daytime running lights

_____ **12.** A(n) _____ headlight has a highly polished aluminized glass reflector that is fused to the optically designed lens.
   A. Dual-filament
   B. Semi-sealed beam
   C. Sealed-beam
   D. HID

_____ **13.** Which type of headlight is filled with xenon gas and does not use a filament-style lamp?
   A. LED
   B. HID
   C. Sealed beam
   D. Halogen

_____ **14.** Which type of lights are used with other vehicle lighting in poor weather such as thick fog, driving rain, or blowing snow?
  A. Driving lights
  B. Cornering lights
  C. Fog lights
  D. Smart lights

_____ **15.** To improve visibility during night driving, some vehicle manufacturers provide _____.
  A. Night vision
  B. Driving lights
  C. Fog lights
  D. Cornering lights

_____ **16.** Wattage is found by multiplying the voltage used by the lamp by the current flowing through it.
  A. True
  B. False

_____ **17.** High-intensity halogen lamp light comes from metallic salts that are vaporized within an arc chamber.
  A. True
  B. False

_____ **18.** Some countries mandate that HID headlamps may only be installed on vehicles with lens-cleaning systems and automatic self-leveling systems.
  A. True
  B. False

_____ **19.** Wattage is an indication of light output.
  A. True
  B. False

_____ **20.** Some lights use festoon-style lights, which have a base on each end of a cylindrical light bulb.
  A. True
  B. False

_____ **21.** Government regulations control the height of taillights and their brightness.
  A. True
  B. False

_____ **22.** Taillights are the only white lights on the rear of the vehicle.
  A. True
  B. False

_____ **23.** Imported vehicles tend to have separate amber-colored turn signal lamps on both the front and the rear of the vehicle.
  A. True
  B. False

_____ **24.** The beam selector switch is a double-pole, single-throw switch, meaning it has two movable poles but makes contact in one position.
  A. True
  B. False

_____ **25.** Some vehicles use LED lights as headlights.
  A. True
  B. False

_____ **26.** Passive night vision systems use a heat-sensing camera to pick up thermal radiation emitted by objects.
  A. True
  B. False

_____ **27.** Cornering lights are white lights usually installed into the bumper or fender and are designed to provide side lighting when the vehicle is turning corners.
  A. True
  B. False

_____ **28.** Each lighting circuit has particular operating characteristics based on its purpose and the system design.
- **A.** True
- **B.** False

_____ **29.** HID lamps produce a very bright red light.
- **A.** True
- **B.** False

_____ **30.** Not all wiring diagrams use the same symbols or the same numbering system.
- **A.** True
- **B.** False

_____ **31.** One of the advantages of a(n) _____ is that it turns on instantly.
- **A.** VTF
- **B.** HID
- **C.** HED
- **D.** LED

_____ **32.** In a bulb marked 12V/21W, the _____ will consume 21 watts of power when 12 volts is applied across it.
- **A.** Ballast
- **B.** Filament
- **C.** LED
- **D.** Halogen gas

_____ **33.** Many newer bulbs use a _____ base made either from the glass bulb itself or with a built-in plastic base.
- **A.** Wedge
- **B.** Dual-filament
- **C.** Bayonet
- **D.** Festoon

_____ **34.** _____ lights are usually controlled by the vehicle body computer with inputs from the ignition and door switches in the handle, latch, or door pillar.
- **A.** Stop
- **B.** Tail
- **C.** Head
- **D.** Courtesy

_____ **35.** _____ lights are used to mark the sides of some vehicles and are often located down the sides of the vehicle or trailer.
- **A.** Marker
- **B.** Park
- **C.** Tail
- **D.** License

_____ **36.** Today, computer-controlled brake lights are activated by the _____ when the computer senses an input from the brake pedal switch.
- **A.** Taillights
- **B.** Light switch
- **C.** Body control module (BCM)
- **D.** Circuit breaker

_____ **37.** The canceling mechanism of the _____ lights operates to return the switch to its central or "off" position after a turn has been completed and the steering wheel is returned to the straight-ahead position.
- **A.** Fog
- **B.** Turn signal
- **C.** Brake
- **D.** Backup

_____ **38.** Some domestic vehicles use the rear brake lamps as turn signals by flashing the _____ on one side to indicate the turn.
  A. Hazard warning lights
  B. Brake lamp
  C. Headlights
  D. Courtesy lights

_____ **39.** Most vehicle _____ require two beams to provide for a high-beam and low-beam operation.
  A. Headlights
  B. License lights
  C. Marker lights
  D. Brake lights

_____ **40.** A(n) _____ headlight system produces a high-intensity forward beam and uses a lens system to project the light forward.
  A. Passive
  B. Reflector-type
  C. Semi-sealed
  D. Projection-type

_____ **41.** Active _____ systems use an infrared light generator that projects infrared light in front of and to the side of the roadway ahead.
  A. LED
  B. Night vision
  C. HID
  D. Daytime running light

_____ **42.** _____ are installed on the front of the vehicle and provide higher-intensity illumination over longer distances than standard headlight systems.
  A. Parking lights
  B. Taillights
  C. Cornering lights
  D. Driving lights

_____ **43.** _____ turn on only when the headlights and turn signal switch on, and turn off automatically when the turn signal cancels.
  A. Cornering lights
  B. Backup lights
  C. Brake lights
  D. Hazard warning lights

_____ **44.** What battery and fuse symbol is shown in the image?

  A. Batteries
  B. Fuses
  C. Fusible link
  D. Circuit breaker

_____ **45.** What battery and fuse symbol is shown in the image?

  A. Circuit breaker
  B. Fuses
  C. Batteries
  D. Fusible link

_____ **46.** What ground and connector symbol is shown in the image?

    A. Chassis ground
    B. Device ground
    C. Internal engine control unit (ECU) ground
    D. Connected wires

_____ **47.** What ground and connector symbol is shown in the image?

    A. Internal ECU ground
    B. Crossing wires with no connection
    C. Connected wires
    D. Wiring connector

_____ **48.** What switch symbol is shown in the image?

    A. Single pole switch
    B. Double-pole switch
    C. Momentary contact switch
    D. Reed switch

_____ **49.** What switch symbol is shown in the image?

    A. Double-pole switch
    B. Breaker switch
    C. Momentary contact switch
    D. Rotary switch

_____ **50.** What resistor, coil, and relay symbol is shown in the image?

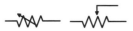

    A. Coil
    B. Transformer
    C. Variable resistor
    D. Thermistor

_____ **51.** What resistor, coil, and relay symbol is shown in the image?

    A. Resistor
    B. Transformer
    C. Normally open relay
    D. Dual pole relay

_____ **52.** What semiconductor symbol is shown in the image?

A. Zener diode
B. LED
C. Photo diode
D. Photo transistor

_____ **53.** What semiconductor symbol is shown in the image?

A. Photo diode
B. Negative–positive–negative (NPN) transistor
C. Positive–negative–positive (PNP) transistor
D. Photo transistor

_____ **54.** What capacitor and device symbol is shown in the image?

A. Piezoelectric device
B. Capacitor
C. Electrolytic capacitor
D. Spark gap

_____ **55.** What capacitor and device symbol is shown in the image?

A. Single filament globe
B. Capacitor
C. Dual-filament globe
D. Spark gap

_____ **56.** What motor, generator, and solenoid symbol is shown in the image?

A. Motor
B. Alternator
C. Pulse generator
D. Solenoid

_____ **57.** What motor, generator, and solenoid symbol is shown in the image?

A. Motor
B. Motor with permanent magnet
C. Motor with wound magnet
D. Alternator

_____ **58.** What gauge and warning device symbol is shown in the image?

A. Voltmeter
B. Ammeter
C. Buzzer
D. Speaker

_____ **59.** What gauge and warning device symbol is shown in the image?

A. Buzzer
B. Chime
C. Horn
D. Speaker

_____ **60.** Tech A says that a voltage drop on the ground side of a bulb will not affect its brightness because the electricity has already been used up. Tech B says that a voltage drop is one way to determine if there is unwanted resistance in a circuit. Who is correct?
A. Tech A
B. Tech B
C. Both Tech A and Tech B
D. Neither Tech A nor Tech B

_____ **61.** Tech A says that incandescent bulbs resist vibration well. Tech B says that HID headlamps require up to approximately 25,000 volts to start. Who is correct?
A. Tech A
B. Tech B
C. Both Tech A and Tech B
D. Neither Tech A nor Tech B

_____ **62.** Tech A says that many automotive light bulbs have more than one filament inside. Tech B says that some lights use a filament made of quartz. Who is correct?
A. Tech A
B. Tech B
C. Both Tech A and Tech B
D. Neither Tech A nor Tech B

_____ **63.** Tech A says that LED brake lights illuminate faster than incandescent bulbs. Tech B says that LED brake lights have more visibility and last longer. Who is correct?
A. Tech A
B. Tech B
C. Both Tech A and Tech B
D. Neither Tech A nor Tech B

_____ **64.** Tech A says that if the brake light switch is open, neither brake light will illuminate. Tech B says that the backup lights are connected in parallel with the taillights. Who is correct?
A. Tech A
B. Tech B
C. Both Tech A and Tech B
D. Neither Tech A nor Tech B

_____ **65.** Tech A says that some brake lights get power from the brake switch through the turn signal switch. Tech B says many turn signals use amber lights. Who is correct?
- **A.** Tech A
- **B.** Tech B
- **C.** Both Tech A and Tech B
- **D.** Neither Tech A nor Tech B

_____ **66.** Tech A says that daytime running light systems illuminate the taillights to enhance visibility. Tech B says that the CHMSL is illuminated when the taillights are activated. Who is correct?
- **A.** Tech A
- **B.** Tech B
- **C.** Both Tech A and Tech B
- **D.** Neither Tech A nor Tech B

_____ **67.** Tech A says that some turn signals are flashed by a flasher can. Tech B says that some turn signals are flashed by the BCM. Who is correct?
- **A.** Tech A
- **B.** Tech B
- **C.** Both Tech A and Tech B
- **D.** Neither Tech A nor Tech B

_____ **68.** Tech A says that setting tire pressure is part of a headlight adjustment. Tech B says that some headlight aligners require the headlight to be on during the alignment process. Who is correct?
- **A.** Tech A
- **B.** Tech B
- **C.** Both Tech A and Tech B
- **D.** Neither Tech A nor Tech B

_____ **69.** Tech A says that if a bulb is dim when operated, the circuit likely has a short to ground. Tech B says that if a bulb is dim, you should perform a voltage drop test on the power and ground side of the bulb. Who is correct?
- **A.** Tech A
- **B.** Tech B
- **C.** Both Tech A and Tech B
- **D.** Neither Tech A nor Tech B

Body Electrical System

At the start of each chapter, you will find the Learning Objectives from the textbook. These are your objectives as you make your way through the exercises in this workbook and the chapter in your textbook. The following activities have been designed to help you refresh your knowledge of the material in this chapter.

Learning Objectives

After reading this chapter, you will be able to:

- LO 46-01 Describe the reasons for vehicle networks.
- LO 46-02 Describe the types of vehicle networks.
- LO 46-03 Use scan tool to check for module communication and software updates.
- LO 46-04 Test electric motor circuits.
- LO 46-05 Test power door locks and remove door panels.
- LO 46-06 Describe the operation of keyless entry/remote start systems.
- LO 46-07 Test horn systems.
- LO 46-08 Test wiper and washer systems.
- LO 46-09 Describe supplemental restraint systems and how to disable or enable them.

Matching

Match the following terms with the correct description or example.

A. Blower motor

B. Bus

C. Brushless direct current (DC) motor

D. Controller area network (CAN)

E. Data link connector (DLC)

F. Permanent magnet electric motor

G. Stepper motor

H. Supplemental restraint system (SRS)

_____ **1.** A type of brushless motor with a key difference: It is designed to rotate in fixed steps through a set number of degrees.

_____ **2.** The port to which a scan tool can be connected.

_____ **3.** An electric motor that does not have any brushes and is sometimes called an "electronically commutated motor." In this type of motor, an electronic control module replaces the brushes and commutator.

_____ **4.** The data transport system for electronic control modules.

_____ **5.** The network that connects the vehicle's electronic control modules.

_____ **6.** A passenger safety system, such as airbags and seat belt pretensioners.

_____ **7.** An electric motor, usually the permanent magnet type that moves air over the air-conditioning evaporator and heater core.

_____ **8.** An electric motor in which the magnetic field in the casing is produced by permanent magnets, while the armature has an electromagnetic field generated by passing electrical current through loops or windings, thereby producing the motor action.

Multiple Choice

Read each item carefully, and then select the best response.

_____ **1.** A group of eight bits of binary data is called a _____.
   A. Microbyte
   B. Byte
   C. Kilobyte
   D. Megabyte

_____ **2.** What type of motor can be used to rotate a short distance and then stop or go in the reverse direction for a set number of degrees by controlling each coil individually through the microcontroller?
   A. Brush-type motor
   B. Potentiometer
   C. Stepper motor
   D. Induction motor

_____ **3.** The horn switch on vehicles with a driver's side airbag is usually mounted in the steering wheel and requires a _____ to maintain an electrical connection to the circuit as the steering wheel rotates.
   A. Clock spring
   B. Slot switch
   C. Slip ring and brush assembly
   D. Transmitter

_____ **4.** All of the following control the intermittent operation of a windshield wiper timer circuit, EXCEPT the _____.
   A. Discrete electronic timer
   B. Timer circuit
   C. Wiper relay
   D. Vehicle body computer

_____ **5.** Which of the following is an example of a secondary vehicle safety system?
   A. Seat belt
   B. Collapsible steering column
   C. SRS airbag
   D. Seat belt pretensioners

_____ **6.** To prevent incorrect and unnecessary airbag deployment, systems include a(n) _____ mounted within the SRS control unit.
   A. Mercury switch
   B. Accelerometer
   C. Safing sensor
   D. Proximity sensor

_____ **7.** All of the following can be controlled by remote start systems, EXCEPT _____.
   A. Heating systems
   B. Air-conditioning systems
   C. Window defrost systems
   D. Wiper/washer systems

_____ **8.** Controller area network bus (CAN-bus)-compliant diagnostic systems have been required on all vehicles sold in the United States since 2008.
   A. True
   B. False

_____ **9.** Some networks transmit and receive signals over a single wire, but most have dual wires and are commonly called "CAN-bus high (H)" and "CAN-bus low (L)," with the same message sent on both lines.
   A. True
   B. False

_____ 10. A brushless DC motor has brushes and is sometimes called an "electronically commutated motor."
- **A.** True
- **B.** False

_____ 11. A stepper motor is designed to rotate in fixed steps through a set number of degrees.
- **A.** True
- **B.** False

_____ 12. Slower motor speed can be controlled by using a number of resistors connected in series and a switch to select between combinations of resistors or a more complex electronic speed control module.
- **A.** True
- **B.** False

_____ 13. Cooling fans in modern vehicles are usually switched by relays, which are controlled by an electric control unit based on information from the coolant temperature sensor.
- **A.** True
- **B.** False

_____ 14. The clock spring is a device within the steering wheel with a flexible ribbon cable that can rotate endlessly in a single direction.
- **A.** True
- **B.** False

_____ 15. Power door locks use an electrical actuator to move the mechanical door lock mechanism.
- **A.** True
- **B.** False

_____ 16. The wiper motor will usually have a high and low speed and a time delay or intermittent operation.
- **A.** True
- **B.** False

_____ 17. There are two types of airbag triggering mechanisms: electrical and automatic.
- **A.** True
- **B.** False

_____ 18. In electrical and electronic systems, a _____ is a means of connecting many electrical or electronic components for either data or power sharing.
- **A.** Node
- **B.** Body control module
- **C.** Bus
- **D.** CAN

_____ 19. Regardless of how the network is physically connected, the CAN-bus-H and CAN-bus-L systems will have terminating resistors connected to the data lines to form a _____ through the resistors.
- **A.** Line
- **B.** Loop
- **C.** Series
- **D.** Wave

_____ 20. The _____ in a vehicle is usually the permanent magnet type and moves air over the air-conditioning evaporator and heater core.
- **A.** Stepper motor
- **B.** Brush motor
- **C.** Brushless motor
- **D.** Blower motor

_____ 21. _____ in modern vehicles are usually switched by relays, which are controlled by an electronic control unit based on information from the coolant temperature sensor.
- **A.** Blower motors
- **B.** Cooling fans
- **C.** Electric locks
- **D.** Horns

_____ **22.** The sound of a horn is produced by the _____ of a metal diaphragm, which is operated by an electromagnet switched by a set of contacts.
   A. Vibration
   B. Electrification
   C. Illumination
   D. Operation

_____ **23.** The driver's side door is usually a _____ door lock, which means that when it is locked or unlocked with the key (or the remote fob) the other locks follow suit.
   A. Power
   B. Master
   C. Slave
   D. Secondary

_____ **24.** Removing _____ requires the removal of the arm rest, door lever, window crank, and any switch panels mounted on the door.
   A. Door locks
   B. Horns
   C. Wipers
   D. Door panels

_____ **25.** Seat belt _____ are used to tighten the seat belt in a severe frontal accident.
   A. Sensors
   B. Pretensioners
   C. Control modules
   D. Motors

_____ **26.** What network configuration is shown in the image?

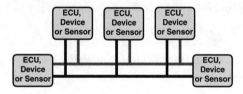

   A. Star parallel
   B. Bus parallel
   C. Looped series
   D. Linked series

_____ **27.** What network configuration is shown in the image?

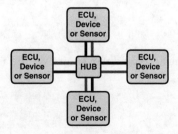

   A. Star parallel
   B. Sun parallel
   C. Bus parallel
   D. Looped series

_____ **28.** What network configuration is shown in the image?

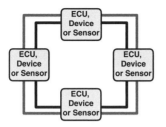

A. Star parallel
B. Bus parallel
C. Looped series
D. Bus series

_____ **29.** The arrow in the image is pointing to what part of the airbag assembly?

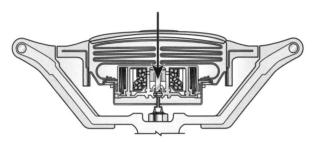

A. Nylon bag
B. Ignitor
C. Squib
D. Gas generator

_____ **30.** The arrow in the image is pointing to what part of the airbag assembly?

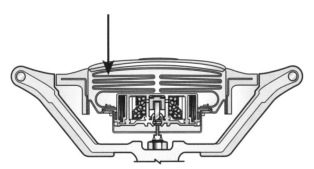

A. Gas generator
B. Ignitor
C. Nylon bag
D. Squib

_____ **31.** The arrow in the image is pointing to what part of the seat belt pretensioner?

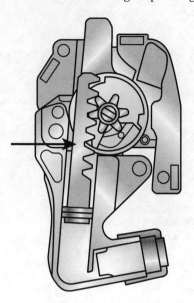

    A. Gas generator
    B. Piston
    C. Pinion gear
    D. Squib

_____ **32.** The arrow in the image is pointing to what part of the seat belt pretensioner?

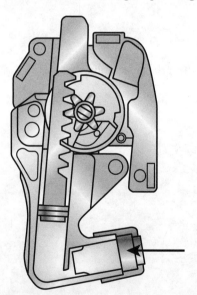

    A. Gas generator
    B. Pretension motor
    C. Piston
    D. Pinion gear

_____ **33.** Tech A says that local interconnect network is a low-speed data network. Tech B says that CAN is a high-speed network. Who is correct?
  A. Tech A
  B. Tech B
  C. Both Tech A and Tech B
  D. Neither Tech A nor Tech B

_____ **34.** Tech A says that CAN-bus system has the ability to accommodate up to 100 nodes. Tech B says that CAN-bus systems use two 120-ohm resistors with a total circuit resistance of 240 ohms. Who is correct?
  A. Tech A
  B. Tech B
  C. Both Tech A and Tech B
  D. Neither Tech A nor Tech B

_____ **35.** Tech A says that a twisted pair of wires in a wire loom makes it easier to route wires to their destination. Tech B says that the twisted pair of wires transmit signals of opposite polarity on the two lines. Who is correct?
  A. Tech A
  B. Tech B
  C. Both Tech A and Tech B
  D. Neither Tech A nor Tech B

_____ **36.** Two technicians are discussing windshield washers that are not working. Tech A says that the windshield washer motor could be faulty. Tech B says that the washer nozzles could be plugged up. Who is correct?
  A. Tech A
  B. Tech B
  C. Both Tech A and Tech B
  D. Neither Tech A nor Tech B

_____ **37.** Tech A says that stepper motors are used as blower motors to achieve various speeds controlled by the driver. Tech B says that stepper motors are used where precise movement must occur, such as in electronic throttle controls. Who is correct?
  A. Tech A
  B. Tech B
  C. Both Tech A and Tech B
  D. Neither Tech A nor Tech B

_____ **38.** Tech A says that some airbags are dual stage and can have two charges in them. Tech B says that airbags must be manually deflated after being deployed. Who is correct?
  A. Tech A
  B. Tech B
  C. Both Tech A and Tech B
  D. Neither Tech A nor Tech B

_____ **39.** Tech A says that the clock spring is a device used in the steering wheel to transmit signals across the rotating electrical connection. Tech B says that the clock spring must be wound correctly during installation, as it allows turning both ways only a certain number of rotations. Who is correct?
  A. Tech A
  B. Tech B
  C. Both Tech A and Tech B
  D. Neither Tech A nor Tech B

_____ **40.** Tech A says that most SRS systems use a safing sensor to reduce the possibility of accidental deployment of the airbags. Tech B says that airbags are designed to act like a nice soft pillow in an accident. Who is correct?

A. Tech A

B. Tech B

C. Both Tech A and Tech B

D. Neither Tech A nor Tech B

_____ **41.** Tech A says that if airbags have been deployed, the affected seat belts and pretensioners need to be replaced. Tech B says that there are relatively large holes in the rear of an airbag that allow it to quickly deflate after deployment. Who is correct?

A. Tech A

B. Tech B

C. Both Tech A and Tech B

D. Neither Tech A nor Tech B

_____ **42.** Tech A says that in some vehicles, airbag deployment can be based on the speed of the impact and the weight of the driver. Tech B says that the SRS seat belt pretensioners tighten the seat belt when you connect the seat belt. Who is correct?

A. Tech A

B. Tech B

C. Both Tech A and Tech B

D. Neither Tech A nor Tech B

Principles of Heating and Air-Conditioning Systems

At the start of each chapter, you will find the Learning Objectives from the textbook. These are your objectives as you make your way through the exercises in this workbook and the chapter in your textbook. The following activities have been designed to help you refresh your knowledge of the material in this chapter.

Learning Objectives

After reading this chapter, you will be able to:

- LO 47-01 Describe the history of air conditioning and the required licensure.
- LO 47-02 Explain the physics that allows air conditioning.
- LO 47-03 Explain the qualities of a refrigerant and the refrigerant cycle.
- LO 47-04 Explain the main air-conditioning components and their operation.
- LO 47-05 Explain the types of refrigerant and refrigerant oils.
- LO 47-06 Describe the heating and ventilation system operation.
- LO 47-07 Describe the operation of the defroster and blower motor operation.
- LO 47-08 Performance test and check a heating, ventilation, and air-conditioning system.

Matching

Match the following terms with the correct description or example.

A. Accumulator
B. Atmospheric pressure
C. British thermal unit (Btu)
D. Chlorofluorocarbon (CFC)
E. Closed-loop system
F. Condensation
G. Convection
H. Dichlorodifluoromethane (R-12)
I. Electric servo
J. Evaporator

K. Fin
L. Fixed-orifice tube system
M. Heat transfer
N. Latent heat of condensation
O. Latent heat of evaporation
P. Refrigerant
Q. Restriction
R. Tetrafluoroethane (R-134a)
S. Tubes
T. Vaporization

_____ 1. The name given to a chemical compound designed to meet the needs of the refrigeration system.

_____ 2. The process of transferring heat by the circulatory movement that occurs in a gas or fluid as areas of differing temperatures exchange places owing to variations in density and the action of gravity.

_____ 3. A system with a fixed-orifice tube that uses an accumulator between the evaporator and the compressor.

_____ 4. An air-conditioning door actuator controlled by electricity.

_____ 5. Metal pipes running side to side or up and down that the coolant or refrigerant travels through.

_____ 6. A totally self-contained system with no materials entering or exiting.

_____ 7. The amount of heat required to change the state from a liquid to a gas without raising the actual gauge temperature.

_____ 8. The changing of a gas into a liquid through cooling.

_____ 9. The changing of a liquid to a gas through boiling.

_____ 10. An inert, colorless gas that can be used as a refrigerant. It is stored in white containers.

_____ 11. An inert colorless gas that can be used as a refrigerant. It is stored in light blue containers.

_____ **12.** The pressure of the air surrounding everything caused by gravity and the weight of air.

_____ **13.** A small, flat piece of metal placed between the tubes to help with the transfer of heat from the coolant to the air, refrigerant to air, or air to refrigerant.

_____ **14.** The flow of heat from a hotter part to a cooler part; it can occur in solids, liquids, or gases.

_____ **15.** A manufactured compound designed to be used as a refrigerant. It is now illegal owing to the high chlorine content.

_____ **16.** The air-conditioning component used on fixed-orifice systems to protect the compressor by storing liquid refrigerant so it does not reach the compressor.

_____ **17.** The air-conditioning component normally located in the passenger compartment designed to allow low-pressure refrigerant liquid to change states to a gas.

_____ **18.** A blockage that partially stops or slows the flow of a material such as a refrigerant.

_____ **19.** The amount of heat removal necessary to change the state from a gas to a liquid without changing the actual gauge temperature.

_____ **20.** A measure of heat energy.

Match the following steps with the correct sequence for performance testing an air-conditioning system.

A. Step 1   **C.** Step 3

B. Step 2

_____ **1.** Raise the engine rpm to 1200–2000. Check the vent temperature using a thermometer. Compare the temperature recorded with the diagnostic chart in the service information.

_____ **2.** Close all windows. Turn the air conditioner to its maximum cold setting.

_____ **3.** Turn on the vehicle. Place a fan in front of the vehicle to simulate the airflow that occurs when driving.

Multiple Choice

Read each item carefully, and then select the best response.

_____ **1.** The federal law that regulates air emissions from stationary and mobile sources and contains all of the motor vehicle air-conditioning requirements and laws is called the _____.

   A. National Air Quality Standards Act

   B. Clean Air Act

   C. Air-Conditioning Act

   D. Environmental Protection Act

_____ **2.** The amount of water vapor in the air, expressed as a percentage, is called the _____.

   A. Saturation level

   B. Condensation point

   C. Relative humidity

   D. Evaporation point

_____ **3.** The process of transferring heat through matter by the movement of heat energy through solids from one particle to another is called _____.

   A. Convection

   B. Conduction

   C. Radiation

   D. Evaporation

_____ **4.** Heat energy is measured in _____.

   A. Degrees Fahrenheit

   B. Degrees Celsius

   C. British thermal units

   D. Relative humidity

_____ **5.** The heat required to change the water at its maximum temperature from a liquid to a gas is called _____.

    A. Latent heat of evaporation

    B. Latent heat of condensation

    C. Latent heat of freezing

    D. Latent heat of vaporization

_____ **6.** The _____ changes a low-pressure refrigerant gas to a high-pressure gas and provides the needed refrigerant movement in the system.

    A. Condenser

    B. Compressor

    C. Evaporator

    D. Restrictor

_____ **7.** Hot high-pressure gas enters the _____ from the compressor; the gas flows through a series of coils, and ambient air removes the heat from the hot high-pressure gas, which begins to change into a liquid.

    A. Condenser

    B. Evaporator

    C. Restriction

    D. Orifice tube

_____ **8.** Dichlorodifluoromethane, commonly known as _____, is a member of the CFC family of gases and was the first common refrigerant to be used in an automotive air conditioner.

    A. R-10

    B. R-12

    C. R-124b

    D. R-134a

_____ **9.** All of the following are common oils used in refrigeration systems, EXCEPT _____.

    A. Polyoxyethylene (POE) oil

    B. Mineral oil

    C. Polyalkylene glycol (PAG) oil

    D. Copaiba oil

_____ **10.** Fixed-orifice tube systems use a(n) _____ to ensure that a pure gas is delivered to the compressor.

    A. Receiver filter drier

    B. Accumulator

    C. Schrader valve

    D. Bernoulli filter

_____ **11.** The _____ is located inside the cabin in the plastic box with the air-conditioning evaporator.

    A. Defroster

    B. Condenser

    C. Heater core

    D. Accumulator

_____ **12.** The most common type of blower motor control is a _____ that uses resistors in series to regulate the speed of the fan.

    A. Resistor pack

    B. Potentiometer

    C. Module

    D. Relay

_____ **13.** The method of controlling a fan using "on" time compared with "off" time on the ground side of the circuit is known as _____ modulation.

    A. Pulse width

    B. Potential

    C. Resistance

    D. Current

_____ 14. The term used to describe the standard air-conditioning testing process is _____ testing.
   A. Condensation
   B. Latent heat
   C. Performance
   D. Temperature

_____ 15. Automotive air-conditioning service and repair technicians need to have a special license.
   A. True
   B. False

_____ 16. When pressure is lowered on a gas, its temperature rises; when pressure is raised, its temperature drops.
   A. True
   B. False

_____ 17. The amount of heat given up by steam to turn back into a liquid is 970 Btu and is called latent heat of condensation.
   A. True
   B. False

_____ 18. With the use of refrigerants, maintaining vaporization and condensing points at normal ambient temperatures is simply done by raising or reducing the pressure of the refrigerant.
   A. True
   B. False

_____ 19. PAG oil mixed with mineral oil creates a hazardous gas that can corrode the air-conditioning system.
   A. True
   B. False

_____ 20. The fixed-orifice tube is adjustable based on the temperature of the outlet pipe from the evaporator, whereas the thermostatic expansion valve (TXV) provides a nonadjustable passage for the refrigerant to pass through.
   A. True
   B. False

_____ 21. Many of the components used to warm the passenger cabin of a vehicle also function in removing heat from the engine.
   A. True
   B. False

_____ 22. Attached to the spinning shaft of the blower motor is what is called the "bird cage," a plastic cage with fins to pull or push air.
   A. True
   B. False

_____ 23. The heater core is a small radiator consisting of tubes, fins, and tanks.
   A. True
   B. False

_____ 24. The compressor is the most common source of abnormal noises arising from the air conditioner.
   A. True
   B. False

_____ 25. Automotive air conditioning is federally regulated by the _____.
   A. National Ambient Air Quality Agency
   B. Environmental Protection Agency
   C. Mainstream Engineering
   D. Mobile Air-conditioning Society

_____ 26. Heat always transfers from _____ objects to _____ objects.
A. Warm, hot
B. Cold, warm
C. Hot, cold
D. Cold, hot

_____ 27. The transfer of heat through the emission of energy in the form of invisible waves is called _____.
A. Conduction
B. Convection
C. Condensation
D. Radiation

_____ 28. One pound of water at _____ and at a temperature of 32° F (0° C) would require 1 Btu of energy added to increase the temperature by 1° F.
A. Refrigerant pressure
B. Atmospheric pressure
C. Thermal pressure
D. Sensible pressure

_____ 29. At 32° F, the latent heat of _____ begins.
A. Freezing
B. Condensation
C. Evaporation
D. Convection

_____ 30. If the velocity of a liquid rises, the pressure of the liquid must drop according to the _____ principle.
A. Ackermann
B. Helmholtz
C. Ohm's
D. Bernoulli

_____ 31. Tetrafluoroethane, commonly known as _____, was the replacement for R-12 refrigerant.
A. R-134a
B. HFC 1234
C. HFO-1234yf
D. H-134a

_____ 32. Attached to the spinning shaft of the blower motor is a plastic cage with fins to pull or push air called a _____ cage.
A. Shark
B. Mouse
C. Squirrel
D. Bird

_____ 33. Some manufacturers use an activated charcoal cabin _____, which helps trap odors and airborne pollutants such as carbon monoxide and oxides of nitrogen.
A. Air conditioner
B. Air heater
C. Air box
D. Air filter

_____ **34.** The arrow in the image is pointing to what component of the air-conditioning unit?

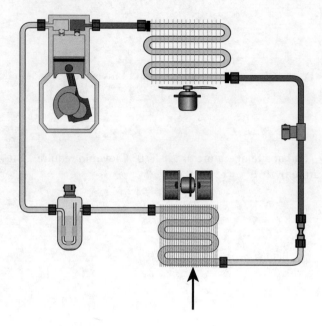

    A. Accumulator
    B. Compressor
    C. Condenser
    D. Evaporator

_____ **35.** The arrow in the image is pointing to what component of the air-conditioning unit?

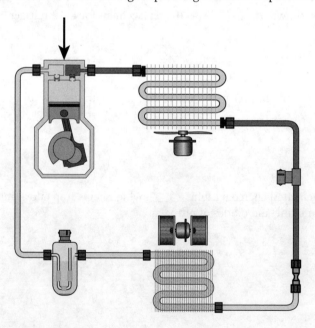

    A. Compressor
    B. Condenser
    C. Condenser fan
    D. Blower motor and fan

_____ **36.** The arrow in the image is pointing to what component of the air-conditioning unit?

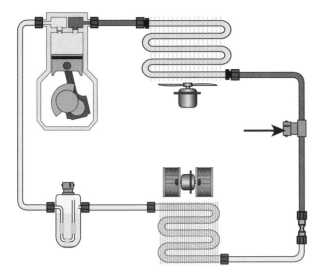

   A. Low-pressure switch
   B. Condenser fan
   C. High-pressure switch
   D. Orifice tube

_____ **37.** The arrow in the image is pointing to what part of the heating system?

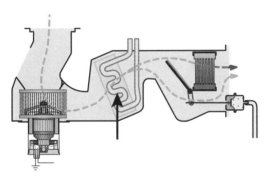

   A. Evaporator
   B. Blend door
   C. Heater core
   D. Vacuum servo

_____ **38.** The arrow in the image is pointing to what part of the heating system?

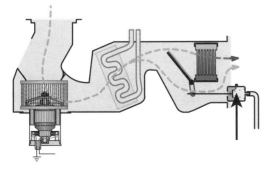

   A. Blend door
   B. Evaporator
   C. Vacuum servo
   D. Distribution ducts

_____ **39.** The arrow in the image is pointing to what part of the heating system?

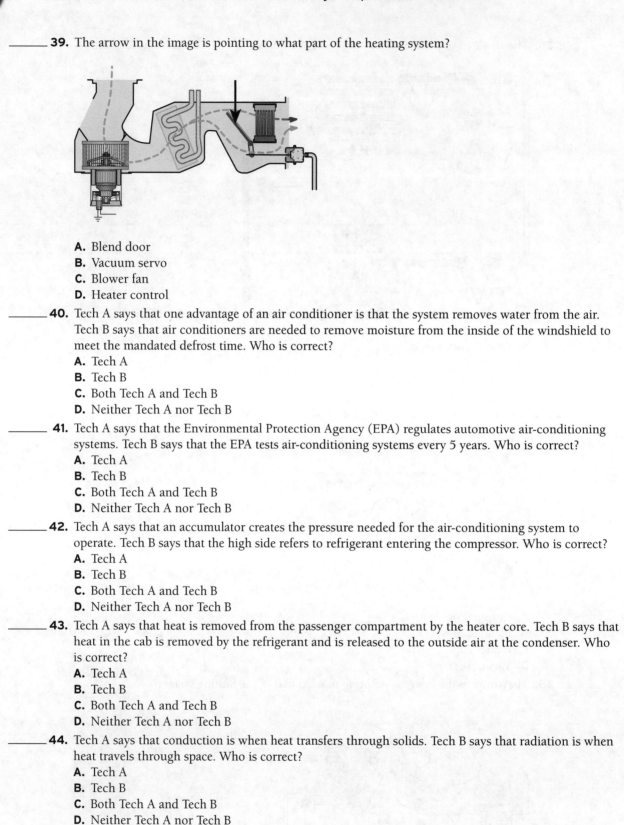

   A. Blend door
   B. Vacuum servo
   C. Blower fan
   D. Heater control

_____ **40.** Tech A says that one advantage of an air conditioner is that the system removes water from the air. Tech B says that air conditioners are needed to remove moisture from the inside of the windshield to meet the mandated defrost time. Who is correct?
   A. Tech A
   B. Tech B
   C. Both Tech A and Tech B
   D. Neither Tech A nor Tech B

_____ **41.** Tech A says that the Environmental Protection Agency (EPA) regulates automotive air-conditioning systems. Tech B says that the EPA tests air-conditioning systems every 5 years. Who is correct?
   A. Tech A
   B. Tech B
   C. Both Tech A and Tech B
   D. Neither Tech A nor Tech B

_____ **42.** Tech A says that an accumulator creates the pressure needed for the air-conditioning system to operate. Tech B says that the high side refers to refrigerant entering the compressor. Who is correct?
   A. Tech A
   B. Tech B
   C. Both Tech A and Tech B
   D. Neither Tech A nor Tech B

_____ **43.** Tech A says that heat is removed from the passenger compartment by the heater core. Tech B says that heat in the cab is removed by the refrigerant and is released to the outside air at the condenser. Who is correct?
   A. Tech A
   B. Tech B
   C. Both Tech A and Tech B
   D. Neither Tech A nor Tech B

_____ **44.** Tech A says that conduction is when heat transfers through solids. Tech B says that radiation is when heat travels through space. Who is correct?
   A. Tech A
   B. Tech B
   C. Both Tech A and Tech B
   D. Neither Tech A nor Tech B

_____ **45.** Tech A says that the air-conditioning compressor creates high-pressure liquid. Tech B says that the metering device creates a physical change in refrigerant from a liquid to a gas. Who is correct?
   A. Tech A
   B. Tech B
   C. Both Tech A and Tech B
   D. Neither Tech A nor Tech B

_____ **46.** Tech A says that when an air-conditioning system freezes up, the refrigerant freezes and stops the cooling process. Tech B says that when an air-conditioning unit freezes up, the evaporator coils are covered with frozen water molecules that stop airflow across the coils, thus preventing cooling. Who is correct?
   A. Tech A
   B. Tech B
   C. Both Tech A and Tech B
   D. Neither Tech A nor Tech B

_____ **47.** Tech A says that the evaporator is on the high side of the system. Tech B says that the condenser transfers heat to the atmosphere. Who is correct?
   A. Tech A
   B. Tech B
   C. Both Tech A and Tech B
   D. Neither Tech A nor Tech B

_____ **48.** Tech A says that R-134a replaced R-12 as an approved refrigerant. Tech B says that HFO-1234yf refrigerant is flammable. Who is correct?
   A. Tech A
   B. Tech B
   C. Both Tech A and Tech B
   D. Neither Tech A nor Tech B

_____ **49.** Tech A says that the engine thermostat directly controls the available coolant temperature for the heater system. Tech B says that the heater core can have engine coolant or refrigerant flowing through it, depending on whether heat or coolness is commanded. Who is correct?
   A. Tech A
   B. Tech B
   C. Both Tech A and Tech B
   D. Neither Tech A nor Tech B

_____ **50.** Tech A states that an air-conditioning performance test usually requires that an auxiliary condenser fan be used during the test. Tech B states that a performance test will show if the air-conditioning system is contaminated with sealer. Who is correct?
   A. Tech A
   B. Tech B
   C. Both Tech A and Tech B
   D. Neither Tech A nor Tech B

Ignition Systems

At the start of each chapter, you will find the Learning Objectives from the textbook. These are your objectives as you make your way through the exercises in this workbook and the chapter in your textbook. The following activities have been designed to help you refresh your knowledge of the material in this chapter.

Learning Objectives

After reading this chapter, you will be able to:

- LO 48-01 Describe ignition system preliminaries.
- LO 48-02 Describe the operation of the primary and secondary ignition systems.
- LO 48-03 Explain required voltage and available voltage.
- LO 48-04 Explain spark timing.
- LO 48-05 Explain common ignition components.
- LO 48-06 Describe spark plugs.
- LO 48-07 Describe contact breaker point ignition systems.
- LO 48-08 Describe mechanical spark timing systems.
- LO 48-09 Describe electronic ignition systems.
- LO 48-10 Describe distributorless ignition systems.
- LO 48-11 Perform ignition system maintenance.

Matching

Match the following terms with the correct description or example.

A. Advance mechanism
B. Breaker plate
C. Center electrode
D. Cranking
E. Direct ignition system
F. Distributor base plate
G. Electronic ignition system
H. Event cylinder
I. Heat range
J. Ignition coil

K. Ignition switch
L. Inductive current
M. Primary winding
N. Rotor
O. Secondary circuit
P. Spark plug
Q. Spark timing
R. Throttle body
S. Waste cylinder

_____ **1.** The low-voltage coiled copper wiring found in an ignition coil.

_____ **2.** The cylinder in a waste spark ignition system that receives a spark near the top of its exhaust stroke.

_____ **3.** A high-voltage rotating switch that transfers voltage from the distributor cap's center terminal to the outer terminals.

_____ **4.** The point at which a spark occurs at the spark plug relative to the position of the piston.

_____ **5.** A device used to amplify an input voltage into the much higher voltage needed to jump the electrodes of a spark plug.

_____ **6.** The housing on an intake manifold that is used to control the amount of filtered air that enters the cylinders.

_____ **7.** The rating of a spark plug's operating temperature.

_____ **8.** A device used to trigger an earlier spark based on engine conditions.

_____ **9.** An ignition system that uses a nonmechanical method of triggering the ignition coil's primary circuit.

_____ **10.** The part of an ignition system that operates on higher voltage and delivers the necessary high voltage to the spark plugs.

_____ **11.** The movable plate the breaker points are mounted on that pivots as the vacuum advance pulls on it.

_____ **12.** May refer to a waste spark ignition or a coil-on-plug ignition system, in which the coils are directly attached to the spark plugs.

_____ **13.** The current that has been created across a conductor by moving it through a magnetic field.

_____ **14.** The electrode located in the center of a spark plug. It is the hottest part of the spark plug.

_____ **15.** A round metal plate near the top of the distributor that is attached to a distributor housing; also called a breaker plate.

_____ **16.** Rotating the engine by turning the ignition key to the start position.

_____ **17.** A switch operated by a key or start/stop button and used to turn on or off a vehicle's electrical and ignition system.

_____ **18.** The cylinder that uses the spark to ignite the air–fuel mixture on a waste spark ignition system.

_____ **19.** A device that provides a gap for the high-voltage spark to occur in each cylinder.

Multiple Choice

Read each item carefully, and then select the best response.

_____ **1.** For an engine to run smoothly and efficiently, the high-voltage spark must cross the spark plug electrode as the piston approaches top dead center of the _____.
  A. Intake stroke
  B. Compression stroke
  C. Power stroke
  D. Exhaust stroke

_____ **2.** What type of ignition system eliminates the distributor by using dedicated ignition coils—one coil for each pair of cylinders?
  A. Contact breaker point ignition system
  B. Electronic ignition system
  C. Waste spark ignition system
  D. Direct ignition system

_____ **3.** The _____ replaced the contact breaker points with an electronic switching device but still used a distributor to dispense the spark to the various cylinders.
  A. Electronic ignition system (distributor type)
  B. Waste spark ignition system
  C. Direct ignition system
  D. Coil-on-plug ignition system

_____ **4.** The ignition system uses a(n) _____ to convert relatively low voltage and high-current flow into very high voltage and very low-current flow.
  A. Spark plug
  B. Ignition module
  C. Induction coil
  D. Reluctor

_____ **5.** Faraday's Law states that relative movement between a conductor and a magnetic field allows _____ ways by which voltage can be induced in a conductor.
  A. Three
  B. Four
  C. Five
  D. Six

_____ **6.** The correct spark timing varies according to all of the following, EXCEPT _____.
  A. Transmission gear selected
  B. Detected knock
  C. Engine temperature
  D. Brake pedal operation

_____ 7. When the ignition switch is turned to the _____ position, most warning lamps on the instrument panel should illuminate.
  A. Accessory
  B. On/run
  C. Start/crank
  D. Lock and off

_____ 8. When battery voltage pushes current through the _____ of an ignition coil, a magnetic field is created.
  A. Primary winding
  B. Secondary winding
  C. Iron core
  D. High-tension terminal

_____ 9. Spark plugs are identified by all of the following, EXCEPT _____.
  A. Thread size or diameter
  B. Reach or length of the thread
  C. Heat range or operating temperature
  D. Size of gap

_____ 10. Contact breaker points are opened by _____ on the distributor shaft and closed by a spring.
  A. Rubbing blocks
  B. Sensors
  C. Cam lobes
  D. Gears

_____ 11. The _____ charges to the peak value of the primary winding voltage.
  A. Rotor
  B. Capacitor
  C. Contact breaker
  D. Ballast

_____ 12. It is the main function of the _____ to distribute the spark to the spark plugs in the correct sequence and at the correct time in the engine cycle.
  A. Condenser
  B. Ballast resistor
  C. Distributor
  D. High-tension terminal

_____ 13. The vacuum advance mechanism is designed to operate on all of the following vacuum types, EXCEPT _____.
  A. Ported
  B. Manifold
  C. Venturi
  D. Spark

_____ 14. A _____ is an example of an inductive pickup.
  A. Hall-effect switch
  B. Throttle position sensor
  C. Mass airflow sensor
  D. Manifold absolute pressure sensor

_____ 15. The _____ has one tooth for each cylinder.
  A. Reluctor
  B. Conductor
  C. Piston
  D. Stator

_____ 16. The _____ is shaped like a very shallow cup with slits, or windows, cut into it at evenly spaced intervals and is made of a ferrous metal.
  A. Optical-type sensor
  B. Interrupter ring
  C. Reluctor
  D. Stator

_____ **17.** What type of sensor, located inside the distributor, can be used to sense the position of the crankshaft and send an appropriate voltage signal to the ignition module or PCM?
  A. Hall-effect
  B. Induction-type
  C. Optical-type
  D. Variable-resistance

_____ **18.** A distributorless ignition system uses a _____ to calculate engine speed and determine engine position.
  A. Crankshaft position sensor
  B. Cam lobe
  C. Breaker plate
  D. Spark plug

_____ **19.** Newer vehicles use high-performance spark plugs that are made with all of the following materials, EXCEPT _____.
  A. Platinum
  B. Silver
  C. Iridium
  D. Gold palladium

_____ **20.** The ignition coil's function is to amplify the battery's low voltage and high current into very high voltage and low current.
  A. True
  B. False

_____ **21.** The low-voltage side of the induction coil is called the secondary circuit, and the high-voltage side is called the primary circuit.
  A. True
  B. False

_____ **22.** Available voltage is the maximum amount of voltage available to try to push current to jump the spark plug gap if the gap were infinite.
  A. True
  B. False

_____ **23.** In early vehicles that used a distributor, base spark timing was set at idle speeds by positioning the distributor body in relation to its rotating cam.
  A. True
  B. False

_____ **24.** Ignition coils are basically step-up transformers.
  A. True
  B. False

_____ **25.** The secondary ignition coil winding has 15,000 to 30,000 turns of very thin insulated aluminum wire wound around its core.
  A. True
  B. False

_____ **26.** High-tension leads conduct the high-output voltage generated in the secondary ignition circuit between the high-tension terminal(s) of the ignition coil, the distributor cap, and the spark plugs.
  A. True
  B. False

_____ **27.** The same spark plug can sometimes be used in different engines with different gap settings.
  A. True
  B. False

_____ **28.** The condenser is a critical component of the primary circuit in point-type ignition systems. Without it, the points arc as they are opened.
  A. True
  B. False

_____ **29.** The high voltage induced in the secondary winding rises to a value great enough to bridge the gap across the spark plug.
   A. True
   B. False

_____ **30.** The distributor cap covers the end of the distributor to protect the components and provide a connection point between the rotor and the spark plug leads.
   A. True
   B. False

_____ **31.** The centrifugal advance mechanism controls ignition advance in relation to engine load.
   A. True
   B. False

_____ **32.** In electronic ignition systems, the contact breaker points are eliminated, and the primary circuit is switched or triggered electronically with a power transistor located in the ignition module.
   A. True
   B. False

_____ **33.** The stator has one tooth for each cylinder; as it spins, the teeth interact with the reluctor to trigger the ignition module.
   A. True
   B. False

_____ **34.** In a distributor, the Hall-effect generator and its integrated circuit are located on one leg of a U-shaped assembly, mounted on the distributor base plate.
   A. True
   B. False

_____ **35.** The optical sensor is very precise owing to the small slits that the light passes through.
   A. True
   B. False

_____ **36.** In direct ignition systems, there is one ignition coil for each pair of companion cylinders.
   A. True
   B. False

_____ **37.** In a waste spark system, the cylinder on the compression stroke is said to be the event cylinder, and the cylinder on the exhaust stroke is the waste cylinder.
   A. True
   B. False

_____ **38.** A spark test should be performed when an engine will not start, has one or more cylinders misfiring, or has other drivability problems, such as a lack of power or stalling.
   A. True
   B. False

_____ **39.** Many technicians have been fooled by using one of the engine's spark plugs as a spark tester.
   A. True
   B. False

_____ **40.** The purpose of the ignition system is to create the _____ and deliver it at the right time to each cylinder.
   A. High-tension stroke
   B. Low-voltage spark
   C. Compression stroke
   D. High-voltage spark

_____ **41.** The original ignition system was called the _____ ignition system, a mechanical system with a switch that was opened and closed as the engine was running.
   A. Contact breaker point
   B. Waste spark
   C. Direct
   D. Electronic

_____ **42.** The amount of voltage required to initially get current to jump the spark plug gap is called _____ voltage.
   A. Self-induced
   B. Available
   C. Required
   D. Maximum

_____ **43.** Spark timing is almost always indexed to the _____ cylinder.
   A. Number one
   B. Number two
   C. Number three
   D. Number four

_____ **44.** The _____ plate, also called a breaker plate, is a movable metal plate located in the distributor, beneath the distributor cap, on which the contact breaker points are mounted.
   A. Interruptor
   B. Condenser
   C. Distributor base
   D. Cam

_____ **45.** The _____ supplies the electrical energy to the ignition circuit during start-up.
   A. Ignition switch
   B. Ignition coil
   C. Capacitor
   D. Battery

_____ **46.** _____ connect the secondary ignition components together, such as the coil to the distributor cap and the distributor cap to the spark plugs.
   A. Infrared transmitters
   B. Breaker points
   C. Power transistors
   D. High-tension leads

_____ **47.** If the insulation around the cable is insufficient and the leads are not spaced far enough apart and run parallel to each other, then a(n) _____ can be generated in the other wires.
   A. Compression stroke
   B. Induced voltage
   C. Dwell angle
   D. Ignition advance

_____ **48.** Spark plug _____ is the distance from the seat of the spark plug to the end of the spark plug threads.
   A. Size
   B. Heat range
   C. Reach
   D. Breaker point

_____ **49.** The amount of time, in degrees of _____, that the contacts are closed is called the dwell angle.
   A. Distributor rotation
   B. Coil rotation
   C. Condenser rotation
   D. Capacitor rotation

_____ **50.** The _____ advance mechanism controls ignition timing in relation to engine speed.
   A. Vacuum
   B. Centrifugal
   C. Distributor
   D. High-voltage

_____ **51.** Ignition _____ are used in virtually every type of ignition system, except the contact breaker point system.
   A. Switches
   B. Coils
   C. Circuits
   D. Modules

_____ **52.** _____ systems are electronic systems that use a magnetic pulse generator, also called a variable reluctor sensor, to generate an alternating current signal.
   A. Induction-type
   B. Optical-type
   C. Distributor-type
   D. Hall-effect

_____ **53.** The _____ has one tooth for each cylinder.
   A. Stator
   B. Distributor
   C. Reluctor
   D. Stationary winding

_____ **54.** In a distributor, the Hall-effect generator and its _____ are located on one leg of a U-shaped assembly mounted on the distributor base plate.
   A. Position sensor
   B. Event cylinder
   C. Integrated circuit
   D. Center electrode

_____ **55.** In _____ ignition systems, also known as electronic ignition systems, the distributor is eliminated and replaced by multiple ignition coils.
   A. Contact breaker
   B. Distributor
   C. Direct
   D. Distributorless

_____ **56.** While the _____ voltage is very high (up to 100,000 V), the current flow is very low.
   A. Available
   B. Primary
   C. Required
   D. Secondary

_____ **57.** If the spark plug wires have a weak spot in the _____, you should be able to see and hear the spark jumping to another wire, the test lead, or a conductive engine part.
   A. Shell
   B. Insulation
   C. Resistor
   D. Terminal nut

_____ **58.** The arrow in the image is pointing to what component of the breaker point ignition system?

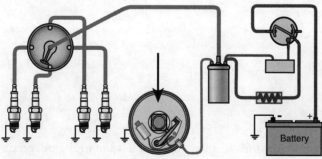

   A. Condenser
   B. Distributor cap
   C. Breaker plate
   D. Battery

_____ **59.** The arrow in the image is pointing to what component of the breaker point ignition system?

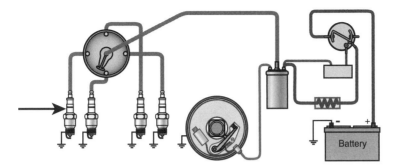

A. Spark plugs
B. High-tension leads
C. Ignition points (breaker points)
D. Ballast resistors

_____ **60.** The arrow in the image is pointing to what component of the breaker point ignition system?

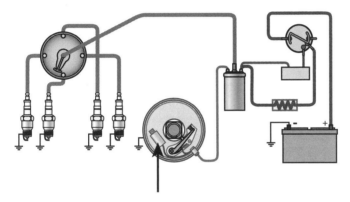

A. Condenser
B. Rotor button
C. Distributor camshaft
D. Ignition switch

_____ **61.** The arrow in the image is pointing to what component of the electronic ignition system?

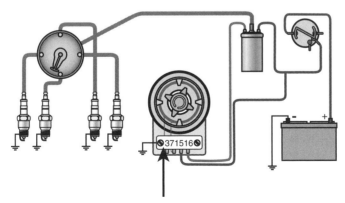

A. Electronic module
B. Reluctor
C. Pickup coil
D. Ignition coil

_____ **62.** The arrow in the image is pointing to what component of the electronic ignition system?

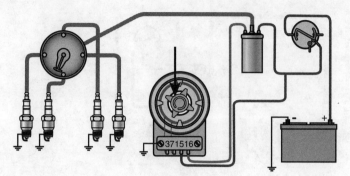

    A. Electronic module
    B. Distributor cap
    C. Reluctor
    D. Battery

_____ **63.** The arrows in the image are pointing to what components of the waste spark ignition system?

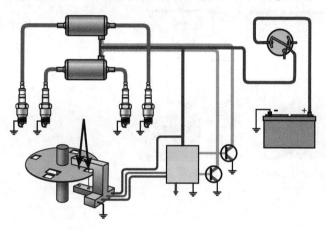

    A. Infrared receivers
    B. Infrared transmitters
    C. Spark plugs
    D. Power transistors

_____ **64.** The arrows in the image are pointing to what components of the waste spark ignition system?

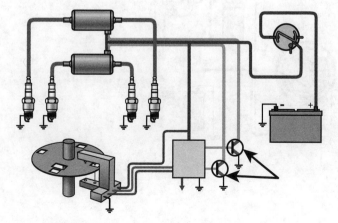

    A. Infrared receivers
    B. Spark plugs
    C. High-tension leads
    D. Power transistors

_____ **65.** The arrow in the image is pointing to what component of the coil-on-plug ignition system?

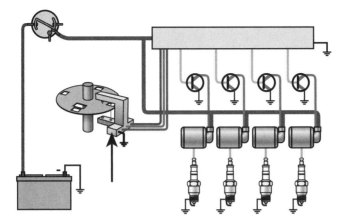

A. Ignition coil
B. Optical sensor
C. Battery
D. Interruptor plate

_____ **66.** The arrow in the image is pointing to what component of the coil-on-plug ignition system?

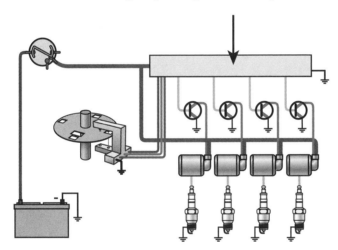

A. Spark plug
B. Engine control unit (ECU)
C. Interruptor plate
D. Ignition switch

_____ **67.** The arrow in the image is pointing to what component of the vacuum advance unit?

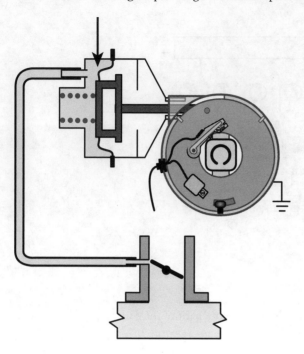

  A. Intake manifold
  B. Vacuum chamber
  C. Diaphragm
  D. Atmospheric chamber

_____ **68.** The arrow in the image is pointing to what component of the vacuum advance unit?

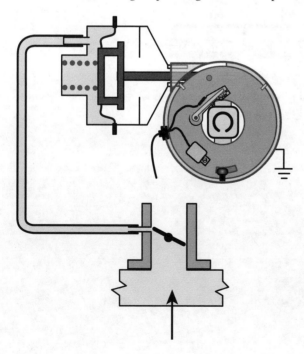

  A. Intake manifold
  B. Vacuum advance body
  C. Distributor body
  D. Vacuum port

_____ **69.** The arrow in the image is pointing to what component of the vacuum advance unit?

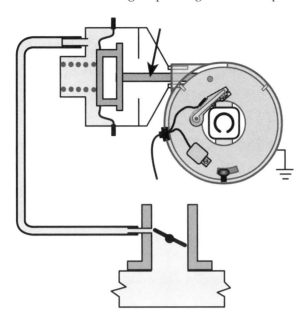

   A. Throttle
   B. Spring
   C. Advance leg
   D. Reluctor

_____ **70.** The arrow in the image is pointing to what component of the ignition coil?

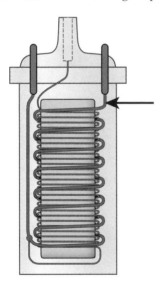

   A. Secondary winding
   B. High-tension (HT) tower
   C. Primary winding
   D. Laminated soft iron core

_____ **71.** The arrow in the image is pointing to what component of the ignition coil?

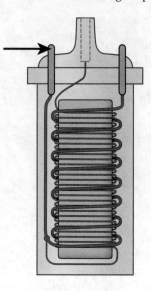

    A. Secondary winding
    B. Negative (–) terminal
    C. Positive (+) terminal
    D. Primary winding

_____ **72.** The arrow in the image is pointing to what part of the HT lead?

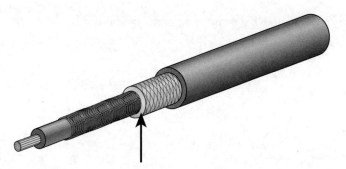

    A. Ferrite or ethylene propylene diene monomer (EPDM) core
    B. Glass fiber core
    C. High-dielectric strength insulator
    D. Silicone insulation layer

_____ **73.** The arrow in the image is pointing to what part of the HT lead?

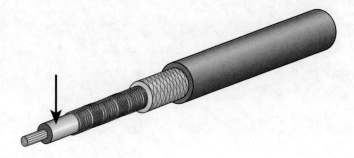

    A. Ferrite or EPDM core
    B. Braided glass
    C. Glass fiber core
    D. Variable pitch coil

_____ **74.** The arrow in the image is pointing to what part of the spark plug?

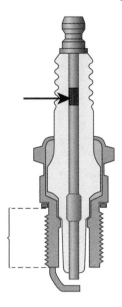

    A. Metal shell
    B. Resistor
    C. Insulator
    D. Powder filling

_____ **75.** The arrow in the image is pointing to what part of the spark plug?

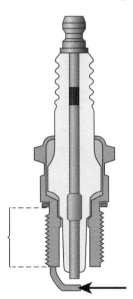

    A. Flashover rib
    B. Gasket
    C. Center electrode
    D. Ground electrode

_____ **76.** The arrow in the image is pointing to what part of contact breaker point operation?

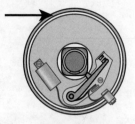

- **A.** Condenser
- **B.** Cam lobe
- **C.** Breaker plate
- **D.** Distributor body

_____ **77.** The arrow in the image is pointing to what part of contact breaker point operation?

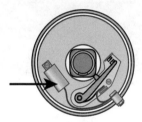

- **A.** Condenser
- **B.** Distributor camshaft
- **C.** Rubbing block
- **D.** Insulated terminal

_____ **78.** The arrow in the image is pointing to what part of the distributor?

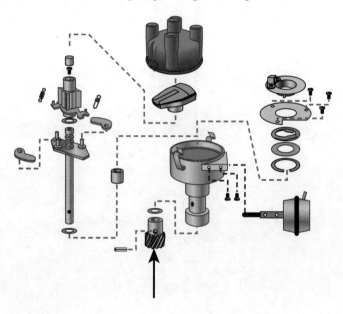

- **A.** Shim
- **B.** Reluctor
- **C.** Drive gear
- **D.** Rotor

_____ **79.** The arrows in the image are pointing to what parts of the distributor?

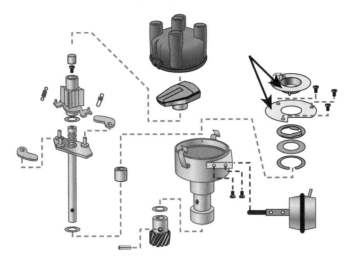

   A. Roll pins
   B. Bushings
   C. Breaker plates
   D. Snap rings

_____ **80.** The arrow in the image is pointing to what part of the distributor?

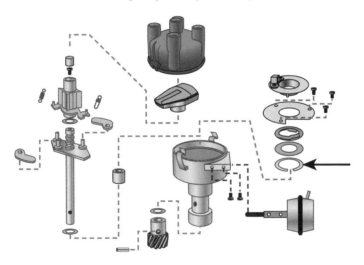

   A. Primary spring
   B. Snap ring
   C. Secondary spring
   D. Shim

_____ **81.** The arrow in the image is pointing to what vacuum source?

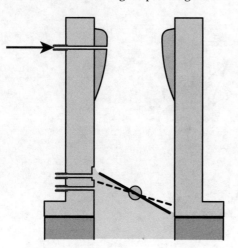

  A. Manifold vacuum
  B. Ported vacuum
  C. Throttle vacuum
  D. Venturi vacuum

_____ **82.** The arrow in the image is pointing to what vacuum source?

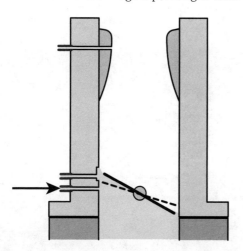

  A. Flow vacuum
  B. Manifold vacuum
  C. Ported vacuum
  D. Venturi vacuum

_____ **83.** Tech A says that contact breaker points are a mechanical switch that opens and closes once for every ignition spark that is created. Tech B says that contact breaker points send high voltage directly from the points to the spark plugs. Who is correct?
  A. Tech A
  B. Tech B
  C. Both Tech A and Tech B
  D. Neither Tech A nor Tech B

_____ **84.** Tech A says that as engines gain miles, the spark plug gap increases, which raises the ignition system's available voltage. Tech B says that misfire occurs when required voltage is higher than available voltage. Who is correct?
  A. Tech A
  B. Tech B
  C. Both Tech A and Tech B
  D. Neither Tech A nor Tech B

_____ **85.** Tech A says that as engine revolutions per minute (rpm) increases, spark timing generally increases. Tech B says that as engine load increases, spark timing generally increases. Who is correct?

A. Tech A
B. Tech B
C. Both Tech A and Tech B
D. Neither Tech A nor Tech B

_____ **86.** Tech A says that a Hall-effect switch uses light to turn a circuit on and off. Tech B says that waste spark systems do not need distributors. Who is correct?

A. Tech A
B. Tech B
C. Both Tech A and Tech B
D. Neither Tech A nor Tech B

_____ **87.** Tech A says that coil-on-plug ignition systems use one coil to fire two cylinders. Tech B says that you should twist spark plug boots before removing them. Who is correct?

A. Tech A
B. Tech B
C. Both Tech A and Tech B
D. Neither Tech A nor Tech B

_____ **88.** Tech A says that the ignition system will maintain spark at the spark plug for approximately 23 degrees of crankshaft rotation. Tech B says that the duration of spark in the spark plug lasts only 2 to 3 degrees of crankshaft rotation. Who is correct?

A. Tech A
B. Tech B
C. Both Tech A and Tech B
D. Neither Tech A nor Tech B

_____ **89.** Tech A says that the positive side of the coil primary circuit is typically switched by the module. Tech B says that the negative side of the coil primary circuit is typically switched by the module. Who is correct?

A. Tech A
B. Tech B
C. Both Tech A and Tech B
D. Neither Tech A nor Tech B

_____ **90.** Tech A says that one main advantage of distributorless ignition systems is the absence of moving parts to maintain. Tech B says that the coil-on-plug ignition system uses a rotor to distribute the spark to each cylinder. Who is correct?

A. Tech A
B. Tech B
C. Both Tech A and Tech B
D. Neither Tech A nor Tech B

_____ **91.** When removing spark plugs, Tech A says that you should blow around each spark plug to remove any debris that could fall into the cylinder. Tech B says that on engines that have more than about 100,000 miles, you should adjust the spark plug gap to about half the specified gap. Who is correct?

A. Tech A
B. Tech B
C. Both Tech A and Tech B
D. Neither Tech A nor Tech B

_____ **92.** Tech A says that dielectric grease applied to the inside of the spark plug boot is used to keep the boot from sticking to the spark plug porcelain. Tech B says that the spark plug boot may need to be "burped" so that the air does not push the boot off the spark plug. Who is correct?

A. Tech A
B. Tech B
C. Both Tech A and Tech B
D. Neither Tech A nor Tech B

Gasoline Fuel Systems

At the start of each chapter, you will find the Learning Objectives from the textbook. These are your objectives as you make your way through the exercises in this workbook and the chapter in your textbook. The following activities have been designed to help you refresh your knowledge of the material in this chapter.

Learning Objectives

After reading this chapter, you will be able to:

- LO 49-01 Describe the types of fuel systems.
- LO 49-02 Describe gasoline fuel characteristics.
- LO 49-03 Describe normal and abnormal combustion.
- LO 49-04 Describe the fuel–air requirements for internal combustion.
- LO 49-05 Describe the fuel tank components.
- LO 49-06 Describe the fuel pump components.
- LO 49-07 Describe fuel lines, fuel filters, and fuel rails.
- LO 49-08 Describe how fuel pressure is regulated.
- LO 49-09 Describe fuel injectors and their operation.
- LO 49-10 Describe throttle body and multipoint fuel injection.
- LO 49-11 Describe the operation of gas direct injection (GDI).
- LO 49-12 Describe the drawbacks of GDI systems.
- LO 49-13 Describe basic carburetor operation.
- LO 49-14 Describe the circuits on a carburetor.
- LO 49-15 Describe the barrels on a carburetor.
- LO 49-16 Remove and replace a fuel filter.
- LO 49-17 Test fuel pressure and volume.
- LO 49-18 Check for fuel contaminants and alcohol content.
- LO 49-19 Test fuel injectors.

Matching

Match the following terms with the correct description or example.

A. Air supply system	**F.** Knocking
B. Dieseling	**G.** Pressure
C. Element	**H.** Throttle
D. Fuel system	**I.** Vacuum
E. Idle	**J.** Vapor lock

_____ **1.** A condition in which the engine continues to run after the ignition key is turned off. Also referred to as run-on.

_____ **2.** The speed at which an engine runs without any throttle applied.

_____ **3.** The force per unit area applied to the surface of an object.

_____ **4.** A device used to produce acceleration by controlling the air–fuel mixture.

_____ **5.** A situation in which vapor forms in the fuel line and the bubbles of vapor block the flow of fuel and stop the engine.

_____ **6.** Equipment in a motor vehicle that delivers fuel to the engine.

_____ **7.** The replaceable portion of a filter, such as an air filter or oil filter.

_____ **8.** The pressure in an enclosed area that is lower than atmospheric pressure.

_____ **9.** A noise heard when the air–fuel mixture spontaneously ignites before the spark plug is fired at the optimum ignition moment.

_____ **10.** Equipment in a motor vehicle that delivers air to the engine.

Multiple Choice

Read each item carefully, and then select the best response.

_____ **1.** If liquid gasoline vaporizes in the fuel pump, bubbles of vapor can block the flow of fuel and stop the engine, causing _____.
   A. Engine knock
   B. Dieseling
   C. Choking
   D. Vapor lock

_____ **2.** A violent collision of flame fronts in the cylinder, caused by uncontrolled combustion, is called _____.
   A. Dieseling
   B. Detonation
   C. Misfire
   D. Throttling

_____ **3.** The stoichiometric air–fuel ratio is represented by the Greek letter _____.
   A. Alpha
   B. Delta
   C. Omega
   D. Lambda

_____ **4.** All of the following are key components of internal combustion, EXCEPT _____.
   A. Air
   B. Fuel
   C. Friction
   D. Pressure/vacuum differential

_____ **5.** Atmospheric pressure at sea level is calculated as _____.
   A. 14.7 psi
   B. 1 lambda
   C. 101 psi
   D. 30" vacuum

_____ **6.** Vapor from the fuel tank is vented through a _____, where fuel vapors are stored until they are burned in the engine.
   A. Dryer
   B. Charcoal canister
   C. Desiccant
   D. Filter sock

_____ **7.** The _____ is a variable resistor that is attached to a float mechanism in the fuel tank.
   A. Sending unit
   B. Fuel pump relay
   C. Pressure regulator
   D. Accelerometer

_____ **8.** A _____ is incorporated into the end of the fuel pickup tube and is the first line of defense against fuel contamination.
   A. Fuel filter
   B. Fuel rail
   C. Filter sock
   D. Fuel separator

_____ **9.** The _____ system uses one or two fuel injectors located centrally on the intake manifold, right above the throttle plates.
  A. Throttle body injection
  B. Fuel rail injection
  C. Diesel fuel direct injection
  D. Gasoline direct injection

_____ **10.** The _____ system uses a fuel injector for each cylinder located in the intake manifold near each intake valve that sprays fuel toward the valve.
  A. Single-point injection
  B. Multipoint fuel injection
  C. Gasoline direct injection
  D. Throttle body injection

_____ **11.** The more effectively liquid gasoline is changed into vapor, the more efficiently it burns in the engine.
  A. True
  B. False

_____ **12.** There are two different methods used to measure the octane rating of a fuel—the Research Octane Number and the Motor Octane Number.
  A. True
  B. False

_____ **13.** A lean air–fuel mixture has less air in proportion to the amount of fuel.
  A. True
  B. False

_____ **14.** Dieseling may cause an engine to run backward for a brief time when it comes to a stop.
  A. True
  B. False

_____ **15.** A vacuum gauge can be calibrated in inches of mercury in a scale reading from 0" to 30", or if using millimeters of mercury, the scale reads from 0 to 760 mm.
  A. True
  B. False

_____ **16.** Modern vehicles are required by the Environmental Protection Agency to have a vented gas cap.
  A. True
  B. False

_____ **17.** The fuel filter typically consists of a pleated paper filter housed in a sealed container, and its primary function is to prevent contaminants from reaching the injectors.
  A. True
  B. False

_____ **18.** Some vehicles use a pressure regulator on the fuel rail, some are in the fuel tank, and others control fuel pressure by controlling the speed of the fuel pump.
  A. True
  B. False

_____ **19.** Gasoline direct injection engines can run fuel mixtures as lean as 65 to 1, much leaner than the stoichiometric ratio.
  A. True
  B. False

_____ **20.** Sequential injection means that the injectors operate twice per cycle, once each crankshaft revolution, each time delivering half the fuel for the cycle.
  A. True
  B. False

_____ **21.** The _____ system draws in gasoline from the gas tank (fuel cell) and delivers it under pressure to a fuel metering device.
  A. Fuel supply
  B. Air supply
  C. Fuel metering
  D. Induction

_____ **22.** The less easily the fuel ignites, the higher the _____.
  A. Fuel supply
  B. Pressure rating
  C. Octane rating
  D. Stoichiometric ratio

_____ **23.** The term _____ describes the chemically correct air–fuel ratio necessary to achieve complete combustion of the fuel and air.
  A. Octane ratio
  B. Stoichiometric ratio
  C. Median ratio
  D. Vacuum ratio

_____ **24.** _____ is one of the essential components of the internal combustion engine.
  A. Water
  B. Carbon dioxide
  C. Air
  D. Flame

_____ **25.** At _____ speed, the pistons draw air away from the manifold at a faster rate than it can pass the throttle plate into the manifold, creating a high vacuum or low pressure.
  A. Full
  B. Off-idle
  C. Fast
  D. Idle

_____ **26.** The fuel _____ can incorporate the use of a blowback ball valve to prevent fuel from leaking from the vehicle during fill-ups and to deter gas theft.
  A. Filler neck
  B. Filter
  C. Pressure regulator
  D. Pump relay

_____ **27.** The primary job of the _____ is to send constant electrical signals to the gas gauge located in the driver information center or to the body control module, which then controls the gas gauge.
  A. Filler neck
  B. Sending unit
  C. Fuel rail
  D. Fuel injector

_____ **28.** A _____ is a special manifold designed to provide a reservoir of pressurized fuel for the fuel injectors.
  A. Fuel line
  B. Gas cap
  C. Fuel injector
  D. Fuel rail

_____ **29.** The modern _____ is simply a spring-loaded, electric-solenoid spray nozzle.
  A. Fuel tank
  B. Fuel pump
  C. Fuel filter
  D. Fuel injector

_____ **30.** There are two types of _____ fuel injection systems. One uses a pressure regulator in the fuel tank, and the other controls the speed of the fuel pump to modify pressure.
  A. Circulation
  B. Returnless
  C. Solenoid-operated
  D. Pressurized

_____ **31.** The arrow in the image is pointing to what compound in the fractional distillation tower?

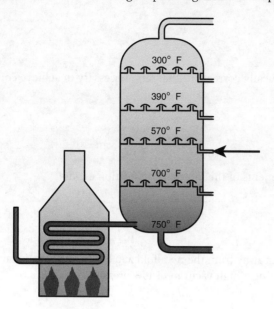

A. Kerosene
B. Diesel oil
C. Fuel oil
D. Lubricating oil

_____ **32.** The arrow in the image is pointing to what compound in the fractional distillation tower?

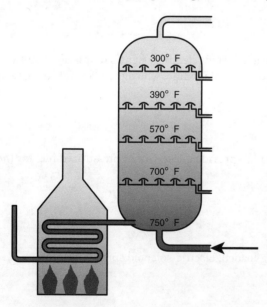

A. Crude oil
B. Gasoline (petrol)
C. Diesel oil
D. Paraffin wax

_____ **33.** The arrow in the image is pointing to what compound in the fractional distillation tower?

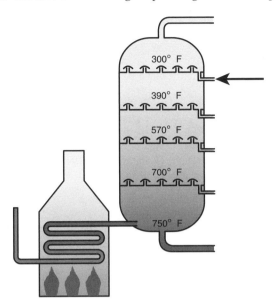

   A. Gasoline (petrol)
   B. Kerosene
   C. Diesel oil
   D. Fuel oil

_____ **34.** The arrow in the image is pointing to what part of the fuel tank?

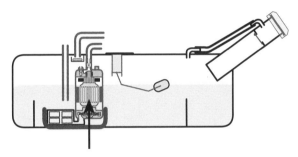

   A. Fuel pump
   B. Fuel cell
   C. Fuel return
   D. Fuel gauge sender

_____ **35.** The arrow in the image is pointing to what part of the fuel tank?

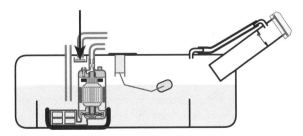

   A. Fuel cell
   B. Baffle
   C. Vapor separator valve
   D. Fuel gauge sender

_____ **36.** The arrow in the image is pointing to what part of the fuel tank?

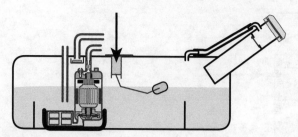

   A. Vapor separator valve
   B. Charcoal canister
   C. Pressure/vacuum cap
   D. Fuel gauge sender

_____ **37.** Tech A says that most electric fuel pumps are now mounted inside of the fuel tank. Tech B says that fuel flows through the center of the electric pump and is used to cool the pump. Who is correct?
   A. Tech A
   B. Tech B
   C. Both Tech A and Tech B
   D. Neither Tech A nor Tech B

_____ **38.** Tech A says that the stoichiometric ratio is 14.7 gallons of air to 1 gallon of fuel. Tech B says it is 14.7 lb of air to 1 lb of fuel. Who is correct?
   A. Tech A
   B. Tech B
   C. Both Tech A and Tech B
   D. Neither Tech A nor Tech B

_____ **39.** Tech A says that some throttle plates are not mechanically connected to the gas pedal, but are operated by an electric motor. Tech B says that too much alcohol in gasoline can cause the engine to not run properly. Who is correct?
   A. Tech A
   B. Tech B
   C. Both Tech A and Tech B
   D. Neither Tech A nor Tech B

_____ **40.** Tech A says that high octane fuel is harder to ignite than lower octane. Tech B says that using high octane fuel in an engine designed for lower octane will produce better fuel economy and power. Who is correct?
   A. Tech A
   B. Tech B
   C. Both Tech A and Tech B
   D. Neither Tech A nor Tech B

_____ **41.** Tech A says that dieseling in a car today indicates a possible leaking injector. Tech B says that a tripped inertia switch means that there is a possible short in the fuel pump circuit. Who is correct?
   A. Tech A
   B. Tech B
   C. Both Tech A and Tech B
   D. Neither Tech A nor Tech B

_____ **42.** Tech A says that a lambda greater than 1 means the engine is running rich. Tech B says that the amount of oxygen in the exhaust indicates how rich or lean the air–fuel mixture is. Who is correct?
   A. Tech A
   B. Tech B
   C. Both Tech A and Tech B
   D. Neither Tech A nor Tech B

_____ **43.** Tech A says that gauge pressure indicates pressure above atmospheric pressure. Tech B says that atmospheric pressure increases as elevation increases. Who is correct?

   A. Tech A
   B. Tech B
   C. Both Tech A and Tech B
   D. Neither Tech A nor Tech B

_____ **44.** Tech A says that fuel pressure is regulated by a fuel pressure regulator on some vehicles. Tech B says that fuel pressure is regulated by controlling the speed of the fuel pump on some vehicles. Who is correct?

   A. Tech A
   B. Tech B
   C. Both Tech A and Tech B
   D. Neither Tech A nor Tech B

_____ **45.** Tech A says that measuring the alcohol content in gasoline involves using water. Tech B says that as long as the fuel pressure is correct, you do not have to worry about fuel pump volume. Who is correct?

   A. Tech A
   B. Tech B
   C. Both Tech A and Tech B
   D. Neither Tech A nor Tech B

_____ **46.** When testing fuel injectors, Tech A says that the resistance can be tested and compared with specifications. Tech B says that resistance can be tested by measuring the fuel pressure drop when activating them with an injector pulse tool. Who is correct?

   A. Tech A
   B. Tech B
   C. Both Tech A and Tech B
   D. Neither Tech A nor Tech B

Engine Management System

At the start of each chapter, you will find the Learning Objectives from the textbook. These are your objectives as you make your way through the exercises in this workbook and the chapter in your textbook. The following activities have been designed to help you refresh your knowledge of the material in this chapter.

Learning Objectives

After reading this chapter, you will be able to:

- LO 50-01 Explain analog and digital signals.
- LO 50-02 Explain potentiometer-based sensor operation.
- LO 50-03 Explain thermistor-based sensor operation.
- LO 50-04 Explain position sensor and speed sensor operation.
- LO 50-05 Explain oxygen sensor operation.
- LO 50-06 Explain how airflow is measured.
- LO 50-07 Explain how air pressure is measured.
- LO 50-08 Explain engine knock and how it is detected.
- LO 50-09 Explain how switches are used in engine management.
- LO 50-10 Explain the sections in a powertrain control module (PCM).
- LO 50-11 Describe controlled devices in engine management systems.
- LO 50-12 Explain feedback and looping.
- LO 50-13 Explain short- and long-term fuel trim and fuel shutoff mode.

Multiple Choice

Read each item carefully, and then select the best response.

_____ 1. What device is used by some sensors to create a variable voltage signal that corresponds to a specific condition?
   A. Voltmeter
   B. Ohmmeter
   C. Potentiometer
   D. Ammeter

_____ 2. About how much signal voltage is generated by a throttle position sensor at closed throttle?
   A. 1 volt
   B. 10 volts
   C. 50 volts
   D. 100 volts

_____ 3. Which of the following is a resistor that changes its resistance with changes in temperature?
   A. Transistor
   B. Potentiometer
   C. Rheostat
   D. Thermistor

_____ 4. In the engine management system, which of the following components create input signals to the PCM?
   A. Sensors
   B. Relays
   C. Control modules
   D. Solenoids

_____ **5.** Which of the following tools is best at measuring the voltage output from a throttle position sensor?
  A. Scan tool
  B. Digital volt-ohmmeter (DVOM)
  C. Digital multimeter (DMM)
  D. Digital storage oscilloscope (DSO)

_____ **6.** In a vehicle, thermistors are used in various sensors related to which of the following?
  A. Pressure
  B. Temperature
  C. Voltage
  D. Speed

_____ **7.** Which of the following sensors is most involved in the control of ignition timing and injection sequencing?
  A. Fuel temperature sensor
  B. Accelerator pedal position sensor
  C. Crankshaft position sensor
  D. Oxygen sensor

_____ **8.** Which of the following sensors provides information that is the most critical for proper air–fuel mixture, as well as ignition timing?
  A. Manifold absolute pressure (MAP) sensor
  B. Barometric pressure (BARO) sensor
  C. Fuel pressure sensor
  D. Knock sensor

_____ **9.** Which of the following sensors monitors the noise that is created by a pressure spike in the combustion chamber?
  A. BARO sensor
  B. MAP sensor
  C. Detonation sensor
  D. Knock sensor

_____ **10.** Which of the following sensors works in conjunction with the BARO sensor?
  A. Mass airflow (MAF) sensor
  B. MAP sensor
  C. Vehicle speed sensor (VSS)
  D. Intake air temperature sensor

_____ **11.** The PCM of an engine performs all of the following functions, EXCEPT _____.
  A. Gathers information
  B. Processes information based on stored data
  C. Performs self-diagnosis and repairs
  D. Commands components to operate in specified ways

_____ **12.** The PCM of an engine is comprised of all of the following sections, EXCEPT _____.
  A. A memory and storage section
  B. A sensor input/output signal processing section
  C. A data processing section
  D. A feedback control section

_____ **13.** Which section of a PCM is responsible for sending the proper reference voltage to many of the sensors?
  A. Output drivers
  B. Input signal processing system
  C. Data processing section
  D. Memory and storage section

_____ **14.** Which node of the PCM translates information from or to the other networks so that it can be shared between networks?
  A. Gateway node
  B. Program node
  C. Memory node
  D. Data node

_____ **15.** Which part of a PCM retains diagnostic trouble codes (DTCs), freeze-frame data, and learned data?
  A. Memory and storage section
  B. Sensor input/output signal processing section
  C. Data processing section
  D. Output drivers section

_____ **16.** In an engine management system, work is done by the actuators, which consist of all of the following, EXCEPT _____.
  A. Relays
  B. Solenoids
  C. Modules
  D. Capacitors

_____ **17.** Which of the following components is often used as an actuator to control other actuators when the computer needs to control a high-current load?
  A. Solenoids
  B. Thermistors
  C. Motors
  D. Relays

_____ **18.** The PCM must control a high-current fuel pump. What is the best actuator to use?
  A. Bipolar transistor
  B. Metal oxide semiconductor field effect transistor (MOSFET)
  C. Set of drivers
  D. Relay

_____ **19.** A variation in pulse width at a fixed frequency is called a(n) _____.
  A. Analog variation
  B. Analog cycle
  C. Pulse-width modulation
  D. Digital cycle

_____ **20.** The speed-density method of determining the amount of air entering the engine takes into account all of the following, EXCEPT _____.
  A. Engine rpm
  B. MAP
  C. Air temperature
  D. Spark knock

_____ **21.** The feedback looping system in a three-way catalytic converter serves to adjust _____.
  A. Air–fuel ratio
  B. Engine speed
  C. Temperature in the passenger cabin
  D. Fuel consumption

_____ **22.** Which of the following modes is used by technicians to fool the engine into cranking without starting during a cranking sound diagnosis test?
  A. Fuel shutoff mode
  B. Clear flood mode
  C. Key on, engine off mode
  D. Stoichiometric mode

_____ **23.** The gasoline engines from the 1970s were about _____ efficient.
  A. 25%
  B. 45%
  C. 65%
  D. 85%

_____ **24.** _____ signals are continuously variable, typically changing in strength and time.
  A. Digital
  B. Analog
  C. Discrete time
  D. Periodic

_____ **25.** The potentiometer can be linear or circular in construction and has _____ electrical connecting points.
   A. Two
   B. Three
   C. Four
   D. Five

_____ **26.** The signal voltage generated by a throttle position sensor rises as the throttle is opened, reaching a maximum reading of almost _____ volts.
   A. 1
   B. 5
   C. 10
   D. 15

_____ **27.** When cycled very quickly, digital signals can be used to create a(n) _____, similar to the variable control provided by an analog signal.
   A. Orthogonal signal
   B. Variable resistance
   C. Varying output control signal
   D. Alternating wavelength

_____ **28.** The range of human hearing is approximately _____ cycles per second.
   A. 20–200
   B. 20–2000
   C. 200–20,000
   D. 20–20,000

_____ **29.** The rate at which sound waves reach our ears is called _____.
   A. Decibel
   B. Frequency
   C. Wavelength
   D. Pitch

_____ **30.** A(n) _____ sensor directly measures the mass of filtered air entering the engine.
   A. PCM
   B. MAF
   C. TBI
   D. MAP

_____ **31.** The purpose of the VSS is to measure the _____.
   A. Turbulence in the incoming airstream
   B. Throttle position
   C. Speed of the vehicle
   D. Coolant temperature

_____ **32.** The _____ section of a PCM contains the software for the computer, as well as a place to retain DTCs, freeze-frame data, and learned data.
   A. Memory and storage
   B. Sensor input/output signal processing
   C. Data processing
   D. Output drivers

_____ **33.** In many cases, the software in the PCM's memory can be updated by uploading software updates; this is called a _____ process.
   A. Reinstallation
   B. Reflash
   C. Debugging
   D. Rebooting

_____ **34.** The _____ section is hardware related and makes up the physical portion of the PCM.
   A. Output drivers
   B. Memory and storage
   C. Data processing
   D. Signal processing

_____ **35.** The data processor function of a PCM receives the data from the input signal processing section and compares it with _____ in the memory.
  A. Control module charts
  B. Adaptive learning
  C. Data maps
  D. Self-diagnostics

_____ **36.** The _____ section of a PCM typically sends either simple on/off signals or pulse width–modulated signals to the actuators, depending on the actuator being controlled.
  A. Output drivers
  B. Memory and storage
  C. Data processing
  D. Signal processing

_____ **37.** In an engine management system, _____ monitor various operating conditions.
  A. Memory devices
  B. Sensors
  C. Control modules
  D. Data processors

_____ **38.** In an engine management system, the information about various operating conditions captured by devices is sent to the _____.
  A. Output driver
  B. Body control module (BCM)
  C. PCM
  D. Body electrical system

_____ **39.** Once an input signal is sent to the PCM for processing, an output signal is sent to the corresponding _____.
  A. BCM
  B. Controlled device
  C. Module
  D. Memory and storage section

_____ **40.** The on and off time of a stepper motor relates to its duty cycle, which is generally expressed as a _____.
  A. Percentage
  B. Numeric value
  C. Binary number
  D. Hexi-decimal value

On-Board Diagnostics

At the start of each chapter, you will find the Learning Objectives from the textbook. These are your objectives as you make your way through the exercises in this workbook and the chapter in your textbook. The following activities have been designed to help you refresh your knowledge of the material in this chapter.

Learning Objectives

After reading this chapter, you will be able to:

- LO 51-01 Describe why on-board diagnostic systems are needed.
- LO 51-02 Describe on-board diagnostics generation (OBD) I and II.
- LO 51-03 Describe diagnostic trouble codes (DTCs).
- LO 51-04 Describe malfunctioning indicator lamp (MIL) operation and freeze-frame data.
- LO 51-05 Describe the purpose of drive cycles and system readiness monitors.
- LO 51-06 Describe the purpose of a scan tool.
- LO 51-07 Retrieve DTCs, monitor status, and freeze-frame data.

Matching

Match the following terms with the correct description or example.

A. Aftermarket
B. Bidirectional scanners
C. Body control module (BCM)
D. Continuous monitoring
E. Controller area network (CAN)
F. Data link connector (DLC)
G. DTC
H. Drive cycle
I. Electronic brake control module (EBCM)
J. Emissions

K. Emission analyzer
L. Freeze frame
M. History code
N. MIL
O. Module
P. Monitor
Q. Powertrain control module (PCM)
R. Scan tool
S. Transmission control module (TCM)

_____ **1.** Tailpipe and volatile organic compound pollutants emitted by the automobile.

_____ **2.** An OBD II test run to ensure that a specific component or system is working properly.

_____ **3.** A series of prescribed automobile operating conditions during which emissions testing is performed.

_____ **4.** A plug-in electronic device for extracting and interpreting fault codes (DTCs) in the automobile, plus much more.

_____ **5.** A localized (on-board) vehicle network that enables computers and components to send and receive signals across a shielded twisted pair of wires.

_____ **6.** An electronic computer that controls transmission function; it may include adaptive learning capabilities for driver preferences.

_____ **7.** A computer-driven on-board report regarding component or system faults following the running of monitors.

_____ **8.** An electronic computer or circuit board that controls specific functions.

_____ **9.** The module that controls and monitors the antilock braking system.

_____ **10.** A feature of OBD II that records events before, during, and after a fault occurs.

_____ **11.** Scanners used to monitor engine compression, vacuum, internal engine anomalies, etc. by causing various components and systems to operate for test purposes.

_____ **12.** The module that controls and monitors the engine ignition, fuel, and emission system functions.

_____ **13.** A term that describes OBD II monitors that run continuously throughout the drive cycle.

_____ **14.** That segment of the trade that supplies parts, services, and repair for vehicles outside of the original equipment manufacturer (OEM) or the dealer network.

_____ **15.** A fault code that has occurred but is not current and is saved in the PCM's memory for 40 drive cycles.

_____ **16.** A device that enables a scan tool to access data stored in the vehicle's various computers.

_____ **17.** A service bay or lab device used for detecting or measuring vehicle emissions.

_____ **18.** An electronic unit that monitors and regulates electronic devices in the vehicle.

_____ **19.** An indicator located in the instrument cluster that illuminates when the PCM detects a fault in one of the vehicle systems. Formerly called a check engine or service engine soon light.

Match the following steps with the correct sequence for retrieving and recording DTCs, OBD monitor status, and freeze-frame data.

A. Step 1
B. Step 2
C. Step 3
D. Step 4
E. Step 5
F. Step 6

_____ **1.** Retrieve and record the DTCs.

_____ **2.** Retrieve and record freeze-frame data applicable to DTCs and monitors.

_____ **3.** Select the scan tool to provide the best coverage for the type and make of vehicle. Locate the DLC and connect the scan tool.

_____ **4.** Power on the scan tool, and turn the ignition on. Establish scan tool communications with the vehicle.

_____ **5.** Power off the scan tool, turn the ignition off, and disconnect the scan tool.

_____ **6.** Retrieve and record OBD monitor status.

Multiple Choice

Read each item carefully, and then select the best response.

_____ **1.** When a vehicle's _____ is illuminated, it means the vehicle is not complying with clean air regulations.
   A. MIL
   B. Volatile organic compounds (VOC) indicator lamp
   C. Red engine lamp
   D. Emission indicator lamp

_____ **2.** All of the following pollutants are monitored and controlled by on-board systems using state-of-the-art electronics, EXCEPT _____.
   A. Hydrocarbons
   B. Carbon dioxide
   C. Oxides of nitrogen
   D. Sulfur dioxide

_____ **3.** Oxidizing catalytic converters were introduced in _____.
   A. 1968
   B. 1971
   C. 1975
   D. 1981

_____   **4.** The second generation of on-board diagnostic (OBD II) systems operates under standards set by the _____.
    A. Association des Constructeurs Européens d'Automobiles
    B. Society of Automotive Engineers (SAE)
    C. Japanese Automotive Standards Organization
    D. Environmental Protection Organization

_____   **5.** The complete listing of OBD I and OBD II standardized nomenclature for parts and systems used by engineers and technicians is known as _____.
    A. SAE J2012
    B. SAE J1930
    C. SAE J1962
    D. SAE J1968

_____   **6.** To simplify wiring, _____ have become commonplace in today's vehicles.
    A. CANs
    B. DLCs
    C. OBD I systems
    D. Closed-loop systems

_____   **7.** All DTCs have been standardized by the SAE and are listed in _____.
    A. SAE J1972
    B. SAE J1930
    C. SAE J2012
    D. SAE J1962

_____   **8.** If the first character of an OBD II diagnostic code is the letter *U*, then the fault is located in the _____.
    A. Body
    B. Communication system
    C. Chassis controller
    D. Emission system

_____   **9.** Snapshots that are automatically recorded in the vehicle's PCM when a vehicle fault occurs are referred to as _____ data.
    A. Freeze-frame
    B. Continuously monitored
    C. Keep alive
    D. OBD I

_____   **10.** The _____ includes the following events: A vehicle starts, warms up, is accelerated, cruises, slows down, accelerates once more, decelerates, stops, and cools down.
    A. Warm-up cycle
    B. Scan cycle
    C. Drive cycle
    D. Fault cycle

_____   **11.** Carbon monoxide and sulfur dioxide react together to create ground-level ozone, which is considered a health hazard.
    A. True
    B. False

_____   **12.** Evaporative emission control systems were first introduced in 1971.
    A. True
    B. False

_____   **13.** OBD I monitored mainly for parts and wiring malfunctions. OBD II is an enhanced diagnostic system that identifies faults in anything that may affect the vehicle's emission system.
    A. True
    B. False

_____   **14.** Emission-related DTCs are the same across all vehicle makes and models, as are the SAE-recommended names used to describe components and systems.
    A. True
    B. False

_____ 15. If the catalytic converter is at risk, such as from overfueling or a continuous misfire, the MIL will flash.
    A. True
    B. False

_____ 16. DTCs are saved even if the vehicle's battery is disconnected.
    A. True
    B. False

_____ 17. OBD I systems monitor all emission-related components and circuits for opens, shorts, and abnormal operation.
    A. True
    B. False

_____ 18. There are hundreds of possible DTCs, and the list grows constantly as on-board systems become more sophisticated and unique.
    A. True
    B. False

_____ 19. An evaporative emission monitor will not run if the fuel tank is nearly full or nearly empty.
    A. True
    B. False

_____ 20. The bidirectional scan tool is like a "magic bullet." The fault code points directly to the problem.
    A. True
    B. False

_____ 21. _____ may be fuel or oil vapors emitted from the fuel tank, fuel lines, engine crankcase, or elsewhere.
    A. VOCs
    B. Oxides of nitrogen
    C. Volatile inorganic compounds
    D. Tailpipe emissions

_____ 22. OBD II faults and background data can be accessed and read by OEM test equipment or by _____ test equipment made by other companies.
    A. Diagnostic
    B. Freeze-frame
    C. Bidirectional
    D. Aftermarket

_____ 23. The first generation of _____ diagnostic systems operated under manufacturer standards, starting with California vehicles and becoming nationwide in 1981.
    A. On-board
    B. Online
    C. Crankcase
    D. Closed-loop

_____ 24. The _____ is a 16-pin connector with a common (SAE J1962) size and shape.
    A. Transmission control connector
    B. Diagnostic trouble connector
    C. DLC
    D. Body control connector

_____ 25. Diagnostic trouble _____ are "set" in one or more of the vehicle's on-board computers once a fault is detected.
    A. Codes
    B. Videos
    C. Alarms
    D. Loops

_____ 26. Possible faults, including engine misfiring and an incorrect air–fuel mixture, are monitored on a continuous basis by the _____ control module.
    A. Electronic brake
    B. Body
    C. Transmission
    D. Powertrain

_____ **27.** Lower-priority faults are monitored on a(n) _____ basis, meaning they are checked only once during each engine warm-up cycle, or even less often, depending on certain circumstances such as ambient temperatures or fuel level.
   A. Continuous
   B. Weekly
   C. Noncontinuous
   D. Annual

_____ **28.** A DTC will remain logged in the PCM's memory as a _____ for a set period.
   A. Pending code
   B. Current code
   C. History code
   D. Fault code

_____ **29.** A(n) _____ is a device able to electronically communicate with and extract data from the vehicle's one or more on-board computers.
   A. Emission analyzer
   B. Scan tool
   C. Noncontinuous monitor
   D. OBD I

_____ **30.** More sophisticated than simple scan tools, _____ scanners are used to cause various components and systems to operate for test purposes.
   A. Emission
   B. Bidirectional
   C. Unidirectional
   D. Freeze-frame

_____ **31.** Tech A says that on OBD II vehicles, it is a good idea to clear the codes before diagnosis and see if they reset. Tech B says that the DTC will tell you what part needs to be changed. Who is correct?
   A. Tech A
   B. Tech B
   C. Both Tech A and Tech B
   D. Neither Tech A nor Tech B

_____ **32.** Tech A says that OBD I and OBD II use different DLCs. Tech B says that OBD II standardizes the designations for DTCs. Who is correct?
   A. Tech A
   B. Tech B
   C. Both Tech A and Tech B
   D. Neither Tech A nor Tech B

_____ **33.** Tech A says that monitors are designed to test if emission systems are working properly. Tech B says that monitors are designed to store sensor data about a fault if a DTC is set. Who is correct?
   A. Tech A
   B. Tech B
   C. Both Tech A and Tech B
   D. Neither Tech A nor Tech B

_____ **34.** Tech A says that control modules communicate back and forth today using a CAN, which is a bundle of many individual wires, specially designed for fast communication speed. Tech B says that OBD II mandates that a DTC must be set if the emissions are more than 1.5 times the Environmental Protection Agency Federal Test Procedure. Who is correct?
   A. Tech A
   B. Tech B
   C. Both Tech A and Tech B
   D. Neither Tech A nor Tech B

_____ **35.** Tech A says that OBD II will allow a technician to hook up a generic scan tool to read DTCs and clear DTCs. Tech B says that when the MIL is on, the vehicle should be repaired as soon as possible. Who is correct?
   A. Tech A
   B. Tech B
   C. Both Tech A and Tech B
   D. Neither Tech A nor Tech B

_____ **36.** Tech A says that something as simple as a loose gas tank fill cap will turn on the MIL. Tech B says that every digit of a DTC has identifying characteristics. Who is correct?
   A. Tech A
   B. Tech B
   C. Both Tech A and Tech B
   D. Neither Tech A nor Tech B

_____ **37.** Tech A says that a "P" in the DTC stands for a powertrain code. Tech B says a "U" stands for an undetermined code. Who is correct?
   A. Tech A
   B. Tech B
   C. Both Tech A and Tech B
   D. Neither Tech A nor Tech B

_____ **38.** Tech A says that on OBD II systems, the keep alive memory (KAM) stores DTCs forever. Tech B says that on OBD II systems, the KAM is erased when the battery is disconnected. Who is correct?
   A. Tech A
   B. Tech B
   C. Both Tech A and Tech B
   D. Neither Tech A nor Tech B

_____ **39.** Tech A says that the MIL can turn off if the fault does not reappear in a certain number of tests. Tech B says that when an MIL is activated, the PCM stores the DTC until a predetermined number of drive cycles have been performed or cleared by a scan tool. Who is correct?
   A. Tech A
   B. Tech B
   C. Both Tech A and Tech B
   D. Neither Tech A nor Tech B

_____ **40.** Tech A says some mechanical engine problems can cause OBD II DTCs to be set. Tech B says that OBD II codes monitor only the emission control system components. Who is correct?
   A. Tech A
   B. Tech B
   C. Both Tech A and Tech B
   D. Neither Tech A nor Tech B

Induction and Exhaust

At the start of each chapter, you will find the Learning Objectives from the textbook. These are your objectives as you make your way through the exercises in this workbook and the chapter in your textbook. The following activities have been designed to help you refresh your knowledge of the material in this chapter.

Learning Objectives

After reading this chapter, you will be able to:

- LO 52-01 Describe the intake system.
- LO 52-02 Describe the air cleaner assembly.
- LO 52-03 Describe heated air intake systems.
- LO 52-04 Describe differences in intake manifold designs.
- LO 52-05 Describe volumetric efficiency and forced induction.
- LO 52-06 Describe turbocharger and intercooler operation.
- LO 52-07 Describe the exhaust system and exhaust pipe.
- LO 52-08 Describe catalytic converters and connecting pipes.
- LO 52-09 Describe mufflers and resonators.
- LO 52-10 Inspect air filters, housings, and duct work.
- LO 52-11 Inspect the intake system for leaks.
- LO 52-12 Inspect the integrity of exhaust system components.

Matching

Match the following terms with the correct description or example.

A. Blow-off valve
B. Compressor surge
C. Forced induction
D. Helmholtz resonator
E. Mandrel forming
F. Mass airflow (MAF) sensor
G. Plenum chamber

_____ 1. The pressurization of airflow going into the cylinder through the use of a turbocharger or supercharger.

_____ 2. The special bending of pipe to ensure the pipe does not collapse. The use of this pipe bender allows for very tight bends without creating kinks or reducing the size of the pipe.

_____ 3. The powertrain control module input sensor that tells the computer the amount of air coming into the engine. The sensor allows the correct calculated amount of fuel delivery to the engine. It is usually located in the ducting of the air cleaner system.

_____ 4. The backup of air against the throttle plate as it is closed. The turbocharger is still spinning, pressurizing air, when the throttle plate is closed. Air will stack up, creating a rapid slowing of the turbocharger compressor wheel. This can damage the compressor wheel.

_____ 5. A valve that allows the release of excessive boost pressure from the turbocharger when the throttle plate is quickly closed.

_____ 6. A device that uses the principle of noise cancelation through the collision of sound waves. When necessary, it is used in addition to the muffler to cancel additional sounds. This device may also be used on the induction system to muffle noise of airflow through the induction system.

_____ 7. A large portion of the intake manifold after the throttle plate and before the intake runner tubes. The manifold provides a reservoir of air and helps prevent interference with the flow of air between individual branches.

Match the following steps with the correct sequence for inspecting the throttle body, air induction system, and intake manifold.

A. Step 1   **D.** Step 4

B. Step 2   **E.** Step 5

C. Step 3   **F.** Step 6

_____ **1.** Start the engine and let the idle stabilize. Using an acetylene- or propane-testing tool, place the hose near the suspected leak area.

_____ **2.** On vehicles with an idle control system, connect a scan tool and select data stream. Observe the oxygen, short-term fuel trims, and injector pulse width as you are moving the acetylene or propane around.

_____ **3.** Open the fuel valve to slowly release the acetylene or propane.

_____ **4.** On vehicles without an idle control system, observe the revolutions per minute (rpm) and smoothness of the engine. If the engine speed increases or smooths out, then a vacuum leak is present. Determine any necessary repairs.

_____ **5.** Using a bright light, look around the engine compartment for traces of smoke. Any smoke coming from the air cleaner inlet is normal. Determine any necessary action(s).

_____ **6.** Shut off the engine and connect a smoke machine to the intake manifold. Start the smoke machine and inject smoke into the intake manifold.

Match the following steps with the correct sequence for inspecting the exhaust system for leaks.

A. Step 1   **C.** Step 3

B. Step 2

_____ **1.** Use large adjustable pliers to test the integrity of the pipes by moderately squeezing the pipes. Determine necessary repairs.

_____ **2.** Safely lift and secure the vehicle on a hoist. Inspect for leaks or rust holes in the exhaust system, including the exhaust manifolds. Inspect brackets, hangers, clamps, and heat shields.

_____ **3.** Have a helper hold a rag against the exhaust pipe(s) while the engine is idling to increase the pressure in the system. Listen and feel for any leaks.

Multiple Choice

Read each item carefully, and then select the best response.

_____ **1.** The air cleaner element or filter is manufactured using any of the following, EXCEPT _____.

     A. Pleated paper

     B. Oil-impregnated cloth or felt

     C. An oil bath configuration

     D. Fiberglass fibers

_____ **2.** A(n) _____ is a container that is sealed and specially shaped to cancel noise created by pressure waves.

     A. Air cleaner

     B. Baffle

     C. Helmholtz resonator

     D. Intake manifold

_____ **3.** Cylinder heads that have intake and exhaust manifolds on opposite sides of the engine are known as _____ heads.

     A. Parallel

     B. Cross-flow

     C. In-line

     D. Variable-intake

_____ **4.** Manifolds that respond to changes in engine load and speed by changing their effective length in two or three stages are called _____ systems.
   A. Variable inertia
   B. Cross-flow
   C. Runner control
   D. Adjustable

_____ **5.** The method of warming the incoming air by passing it through a shroud around the exhaust manifold is called a(n) _____ system.
   A. Early fuel evaporation
   B. Manifold preheating
   C. MAF heating
   D. Heated induction

_____ **6.** The volume of air entering a cylinder during intake in relation to the internal volume of the cylinder when the piston is at bottom dead center is called _____ and is usually expressed as a percentage.
   A. Cylinder volume
   B. Induction volume
   C. Volumetric efficiency
   D. Compression efficiency

_____ **7.** _____ increases air pressure in the intake manifold above atmospheric pressure.
   A. Back pressure
   B. Forced induction
   C. Intake air heating
   D. Intake air cooling

_____ **8.** When air pressure in the intake manifold reaches a preset level in a turbocharged engine, the _____ automatically directs the exhaust gases so they bypass the turbine.
   A. Wastegate
   B. Bypass valve
   C. Blow-by valve
   D. Aspirator

_____ **9.** When manifold pressure rises above normal and works against the spinning exhaust turbine wheel, slowing it considerably, it is called _____.
   A. Back pressure
   B. Blowback
   C. Forced induction
   D. Compressor surge

_____ **10.** When the outgoing pulse from one cylinder is timed to arrive at the header junction at exactly the right time to help draw out the pulse from another cylinder, it is known as a _____ exhaust.
   A. Balanced
   B. Cross-flow
   C. Tuned
   D. Mandrel-formed

_____ **11.** The _____ is attached to the exhaust manifold and connects to the catalytic converter.
   A. Engine pipe
   B. Flexible connector
   C. Muffler
   D. Tailpipe

_____ **12.** The _____ helps with the alignment of the exhaust pipes as the engine moves under load.
   A. Engine pipe
   B. Flexible connector
   C. Exhaust manifold
   D. Intermediate pipe

_____ **13.** A _____ catalytic converter converts oxides of nitrogen back into nitrogen and oxygen, and hydrocarbons and carbon monoxide to water and carbon dioxide.
  A. Two-way
  B. Three-way
  C. Four-way
  D. One-way

_____ **14.** The _____ connects the catalytic converter to the muffler.
  A. Flexible connector
  B. Exhaust bracket
  C. Intermediate pipe
  D. Engine pipe

_____ **15.** A(n) _____ is sometimes used between the muffler and the exhaust outlet to reduce any resonance levels that the muffler could not adequately suppress.
  A. Baffle chamber
  B. Intermediate pipe
  C. Resonator
  D. Secondary muffler

_____ **16.** Vaporization starts when the fuel system atomizes the fuel by breaking it up into very small particles by spraying it into the charge of air.
  A. True
  B. False

_____ **17.** The throttle body controls airflow with a butterfly valve or valves, also called throttle valves.
  A. True
  B. False

_____ **18.** In many vehicles, the air cleaner is mounted where it can obtain cool, clean air.
  A. True
  B. False

_____ **19.** Intake manifolds were originally made from light alloy metal castings and later from cast iron.
  A. True
  B. False

_____ **20.** Cylinder heads that have intake and exhaust manifolds on opposite sides of the engine are known as cross-flow heads.
  A. True
  B. False

_____ **21.** The air intake manifold for an electronic fuel injection multipoint engine normally has short branches of variable length.
  A. True
  B. False

_____ **22.** Port fuel-injected or gasoline direct-injected intake manifolds do not normally need to be heated, since the manifold does not carry fuel.
  A. True
  B. False

_____ **23.** Electric heaters can be placed between the intake manifold and the throttle body and used to preheat the air.
  A. True
  B. False

_____ **24.** In a stock naturally aspirated engine, one without forced induction, volumetric efficiency can almost never be 100%.
  A. True
  B. False

_____ **25.** In a turbocharged engine, a bypass valve may be installed so that if the wastegate should fail, it can prevent an abnormal rise in manifold pressure.
  A. True
  B. False

_____ **26.** In many current vehicles, the exhaust manifold is often replaced with a header.
    A. True
    B. False

_____ **27.** Leaded fuel must not be used in an engine with a catalytic converter, because lead will coat the catalyst and prevent it from doing its job.
    A. True
    B. False

_____ **28.** Noise cancelation refers to putting a sound material around a perforated pipe that the exhaust gases flow through.
    A. True
    B. False

_____ **29.** Adding a variable-flow exhaust to the baffle or chamber system reduces emission noise.
    A. True
    B. False

_____ **30.** Exhaust gaskets can be found between the engine cylinder head and the exhaust manifold, the engine pipe and the catalytic converter, and possibly the catalytic converter and the muffler.
    A. True
    B. False

_____ **31.** The _____ system ensures that clean, dry air is supplied to the engine.
    A. Exhaust
    B. Heating
    C. Intake
    D. Induction

_____ **32.** The _____ system provides a path for the burned exhaust gases to safely exit the engine and travel out the rear of the vehicle.
    A. Induction
    B. Exhaust
    C. Intake
    D. Cooling

_____ **33.** The _____ filters the incoming air.
    A. Exhaust manifold
    B. Engine pipe
    C. Catalytic converter
    D. Air cleaner

_____ **34.** A(n) _____ sensor can measure air entering the engine down to tenths of a gram per second.
    A. MAF
    B. Oxygen
    C. False air
    D. Clean air

_____ **35.** A _____ resonator is a container that is sealed and specially shaped to cancel noise created by pressure waves.
    A. Turbocharge
    B. Von Herman
    C. Helmholtz
    D. Wastegate

_____ **36.** The _____ has several tubular branches and carries air and/or air/fuel mixture from the air cleaner to the cylinder head.
    A. Intake manifold
    B. Exhaust manifold
    C. Intercooler
    D. Air induction system

_____ **37.** On vehicles with emission controls, air cleaners use a(n) _____ valve to control how much hot air enters the air cleaner.
   A. Intake
   B. Butterfly
   C. Throttle
   D. Thermostatic

_____ **38.** A buildup of pressure in the system that interferes with the outward flow of exhaust gases is known as _____.
   A. Scavenging
   B. Back pressure
   C. Volumetric efficiency
   D. Atmospheric pressure

_____ **39.** A _____ compresses the air in the intake system to above atmospheric pressure, which increases the density of the air entering the engine.
   A. Supercharger
   B. Turbocharger
   C. Wastegate
   D. Blow-off valve

_____ **40.** A(n) _____ is a forced induction system that uses wasted kinetic energy from the exhaust gases to increase the intake pressure.
   A. Wastegate
   B. Supercharger
   C. Turbocharger
   D. Intercooler

_____ **41.** The purpose of a(n) _____ is to reduce the intake air temperature up to a few hundred degrees Fahrenheit before it enters the intake manifold.
   A. Supercharger
   B. Turbocharger
   C. Intercooler
   D. Header

_____ **42.** Headers provide a(n) _____ effect to help remove exhaust gases from the cylinders.
   A. Scavenging
   B. Supercharged
   C. Air dampening
   D. Suctioning

_____ **43.** There may be a flexible connection between the _____ and an intermediate pipe.
   A. Catalytic converter
   B. Tailpipe
   C. Muffler
   D. Engine pipe

_____ **44.** The exhaust components are supported by _____-mounted exhaust brackets that prevent vibrations from being felt by the driver.
   A. Steel
   B. Rubber
   C. Nickel chromium
   D. Copper

_____ **45.** The function of a vehicle's _____ is to minimize the sounds coming from the exhaust system.
   A. Muffler
   B. Catalytic converter
   C. Exhaust gasket
   D. Resonator

_____ **46.** Noise _____ is a system that prevents the sound waves from leaving the exhaust system by canceling them out inside the muffler.
   A. Reduction
   B. Pressure
   C. Absorption
   D. Cancelation

_____ **47.** The _____-type muffler uses a perforated tube or baffle that is wrapped in a sound deadening material, and this will absorb the noise of combustion as gases flow past.
   A. Noise-canceling
   B. Absorptive
   C. Expansion
   D. Harmonic

_____ **48.** The _____ takes the exhaust gases away from the vehicle and must not allow any of the exhaust gases to enter the vehicle.
   A. Engine pipe
   B. Intermediate pipe
   C. Tailpipe
   D. Muffler

_____ **49.** The arrow in the image is pointing to what part of the air induction system?

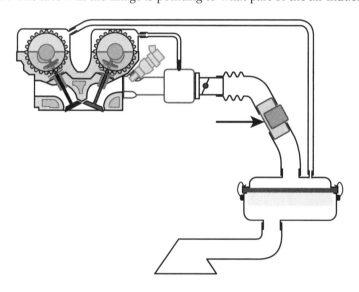

   A. Outside air intake duct
   B. Air cleaner housing
   C. Air flow sensor
   D. Engine breather

_____ **50.** The arrow in the image is pointing to what part of the air induction system?

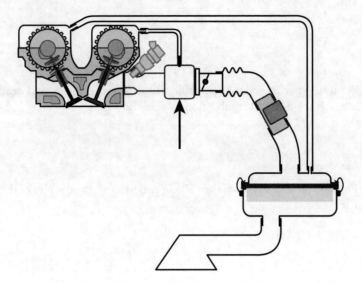

    A. Air cleaner housing
    B. Engine intake duct
    C. Throttle body
    D. Plenum chamber

_____ **51.** The arrow in the image is pointing to what part of the air induction system?

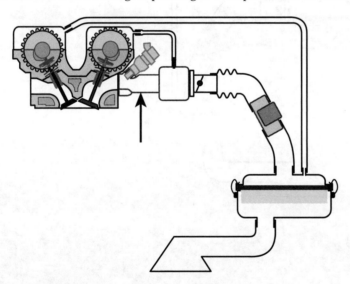

    A. Outside air intake duct
    B. Engine intake duct
    C. Intake manifold
    D. PCV valve

_____ **52.** What exhaust system component is shown in the image?

    A. Resonator
    B. Engine pipe
    C. Intercooler
    D. Muffler

_____ **53.** What exhaust system component is shown in the image?

    A. Catalytic converter
    B. Flexible connector
    C. Engine pipe
    D. Resonator

_____ **54.** What exhaust system component is shown in the image?

    A. Flexible connector
    B. Intercooler
    C. Muffler
    D. Exhaust manifold

_____ **55.** What exhaust system component is shown in the image?

    A. Resonator
    B. Catalytic converter
    C. Muffler
    D. Engine pipe

_____ **56.** What exhaust system component is shown in the image?

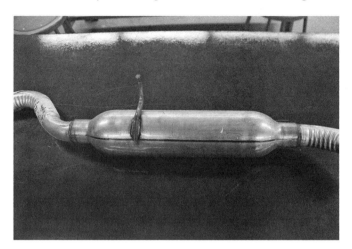

    A. Catalytic converter
    B. Engine pipe
    C. Intercooler
    D. Resonator

_____ **57.** Tech A says that coolant circulates through some intake manifolds to help warm them up. Tech B says that some intake manifolds use an electric heater grid to warm up the intake air. Who is correct?
    A. Tech A
    B. Tech B
    C. Both Tech A and Tech B
    D. Neither Tech A nor Tech B

_____ **58.** Tech A says that in some gas engines, the intake manifold runner length can be changed. Tech B says that some intake systems use resonators to quiet intake noise. Who is correct?
    A. Tech A
    B. Tech B
    C. Both Tech A and Tech B
    D. Neither Tech A nor Tech B

_____ **59.** Tech A says that a good way to check the integrity of exhaust pipes is to tap on them with the handle of a screwdriver. Tech B says that small exhaust leaks can be found with a smoke machine. Who is correct?
    A. Tech A
    B. Tech B
    C. Both Tech A and Tech B
    D. Neither Tech A nor Tech B

_____ **60.** Tech A says that scavenging creates a low pressure in the cylinder by means of the exhaust gases flowing through the exhaust pipe. Tech B says that back pressure in the exhaust system increases the scavenging effect. Who is correct?
    A. Tech A
    B. Tech B
    C. Both Tech A and Tech B
    D. Neither Tech A nor Tech B

_____ **61.** Tech A says that shorter intake manifolds produce higher torque at lower rpm. Tech B says that longer intake manifolds produce higher torque at high engine speeds. Who is correct?
    A. Tech A
    B. Tech B
    C. Both Tech A and Tech B
    D. Neither Tech A nor Tech B

_____ **62.** Tech A says that air becomes heated as a supercharger compresses it. Tech B says that a turbocharger pulls the exhaust gases out of the engine, thereby increasing the scavenging effect. Who is correct?
   A. Tech A
   B. Tech B
   C. Both Tech A and Tech B
   D. Neither Tech A nor Tech B

_____ **63.** Tech A says that a wastegate is designed to direct waste gases to the intake manifold. Tech B says that a blow-off valve vents excess boost pressure from the turbocharger when the throttle is closed quickly. Who is correct?
   A. Tech A
   B. Tech B
   C. Both Tech A and Tech B
   D. Neither Tech A nor Tech B

_____ **64.** Tech A says that the exhaust system is usually suspended from the underbody by rubber mounts. Tech B says some engines use a flexible pipe between the engine pipe and the intermediate pipe. Who is correct?
   A. Tech A
   B. Tech B
   C. Both Tech A and Tech B
   D. Neither Tech A nor Tech B

_____ **65.** Tech A says that a catalytic converter operates best at cold temperatures. Tech B says that the catalytic converter is typically located after the muffler. Who is correct?
   A. Tech A
   B. Tech B
   C. Both Tech A and Tech B
   D. Neither Tech A nor Tech B

_____ **66.** Tech A says that small vacuum leaks can be found with a vacuum gauge. Tech B says that the catalytic converter can be ruined if it becomes contaminated with lead or silicone. Who is correct?
   A. Tech A
   B. Tech B
   C. Both Tech A and Tech B
   D. Neither Tech A nor Tech B

Emission Control

At the start of each chapter, you will find the Learning Objectives from the textbook. These are your objectives as you make your way through the exercises in this workbook and the chapter in your textbook. The following activities have been designed to help you refresh your knowledge of the material in this chapter.

Learning Objectives

After reading this chapter, you will be able to:

- LO 53-01 Describe pollutants and the composition of air.
- LO 53-02 Describe hydrocarbon and carbon monoxide emissions.
- LO 53-03 Describe NO_x, SO_2, and particulate emissions.
- LO 53-04 Describe how engine design reduces emissions.
- LO 53-05 Describe how the air–fuel ratio can be controlled to affect emissions.
- LO 53-06 Describe precombustion–postcombustion exhaust treatment and the evolution of emission controls.
- LO 53-07 Explain the types and operation of catalytic converters.
- LO 53-08 Explain the types and operation of the positive crankcase ventilation (PCV) system.
- LO 53-09 Explain the operation of the exhaust gas recirculation (EGR) system.
- LO 53-10 Explain the operation of the evaporative emission (EVAP) system.
- LO 53-11 Explain the operation of the heated air intake systems.
- LO 53-12 Inspect and test the PCV system.

Matching

Match the following terms with the correct description or example.

A. Blowby
B. Carbon monoxide
C. Compression-ignition engine
D. Emission
E. EGR system
F. Flame front
G. Hydrocarbon
H. Lambda
I. Particulates

J. PCV valve
K. Photochemical smog
L. PCV system
M. Scavenging effect
N. Stoichiometric ratio
O. Stroke
P. Two-way catalytic converter
Q. Vaporization

_____ 1. The exact ratio between air and fuel at which both are burned completely.

_____ 2. A valve that controls the amount of crankcase ventilation flow that is allowed and varies with changes in manifold pressure.

_____ 3. The rapid burning of the air–fuel mixture that moves outward from the spark plug across the cylinder.

_____ 4. Pressure that leaks past the compression rings during compression and combustion.

_____ 5. A condition caused by moving columns of air, which create a low-pressure area behind them, resulting in a pulling force that is used to pull the remaining burned gases from the combustion chamber.

_____ 6. A system that draws blowby gases from the crankcase into the intake to be burned.

_____ 7. A converter that changes only hydrocarbons and carbon monoxide into harmless elements.

_____ **8.** A system that recirculates a portion of burned gases back into the combustion chamber to displace air and fuel and cool combustion temperatures.

_____ **9.** A molecule made up of hydrogen and carbon. It is considered a harmful vehicle emission.

_____ **10.** An engine that uses the heat of compression to ignite the air–fuel mixture; also known as a diesel engine.

_____ **11.** Small particles of solid matter that are suspended in the air, compromised from small particles of carbon.

_____ **12.** A brown haze that hangs in the sky, typically seen over large cities. It is a major health issue to humans because it affects lung tissue.

_____ **13.** The ability of a liquid to evaporate.

_____ **14.** A poisonous gas released during combustion of rich air–fuel mixtures.

_____ **15.** A gas that is released to the atmosphere; usually refers to a harmful gas.

_____ **16.** The movement of the piston in the engine from top dead center to bottom dead center or vice versa.

_____ **17.** The ratio of air to fuel at which all of the oxygen in the air and all of the fuel are completely burned.

Match the following steps with the correct sequence for inspecting and servicing the PCV system.

A. Step 1   **D.** Step 4

B. Step 2   **E.** Step 5

C. Step 3   **F.** Step 6

_____ **1.** Remove the hose, and check that it is still pliable and not clogged with sludge deposits.

_____ **2.** Reinstall the PCV valve into the valve cover.

_____ **3.** Locate the PCV valve. With the engine idling, remove the PCV valve, and check for the presence of a strong vacuum.

_____ **4.** With the engine idling, block off the breather hose and feel for vacuum building up in the breather hose and crankcase. This indicates that the PCV system can handle the amount of blowby gases. There should also be a slight amount of measurable vacuum in the crankcase at idle, typically measurable at the dipstick tube as 0.5"–2" H_2O. This would indicate that the PCV system is operating and adequate for the blowby produced by the engine.

_____ **5.** Remove the PCV valve, and inspect it for deposits. If there are any issues, replace it with a new one of the same type. If it has a fixed orifice style, make sure the passageway is clean.

_____ **6.** Remove the breather hose from the air cleaner assembly.

Multiple Choice

Read each item carefully, and then select the best response.

_____ **1.** The _____ standard is an automotive self-diagnostic system mandated by the Environmental Protection Agency (EPA) that requires a warning light to alert the driver of an emission system fault.
  A. Clean air
  B. Emission control
  C. On-board diagnostics generation (OBD I)
  D. OBD II

_____ **2.** Oxides of nitrogen are categorized as _____ emission gases.
  A. Nonharmful
  B. Harmful
  C. Debatable
  D. Inert

_____ **3.** _____ is a by-product of complete combustion.
  A. Water
  B. Nitrogen
  C. Argon
  D. Helium

_____ **4.** Since carbon monoxide is absorbed very easily by red blood cells, the _____ has set a permissible exposure limit in the workplace of 35 ppm for an 8-hour shift with a ceiling of 200 ppm for any length of time.
   A. Occupational Safety and Health Administration
   B. National Institute for Occupational Safety and Health
   C. American Conference of Governmental Industrial Hygienists
   D. Society of Automotive Engineers

_____ **5.** Which corrosive compound, emitted into the atmosphere through exhaust, is a major environmental pollutant, coming back to earth in contaminated rainwater, called acid rain?
   A. Oxides of nitrogen
   B. Carbon monoxide
   C. Sulfur dioxide
   D. Hydrocarbons

_____ **6.** What is produced in the atmosphere when unburned hydrocarbons and oxides of nitrogen react chemically with sunlight?
   A. Smog
   B. Acid rain
   C. Sulfur dioxide
   D. Carbon monoxide

_____ **7.** The stoichiometric ratio for compressed natural gas is _____.
   A. 14.7:1
   B. 17.2:1
   C. 6.45:1
   D. 34.3:1

_____ **8.** The _____ is an example of a precombustion emission control system.
   A. PCV system
   B. Secondary air injection system
   C. Heated air intake system
   D. Catalyst system

_____ **9.** _____ catalytic converters contain both a reduction catalyst and an oxidizing catalyst.
   A. Two-way
   B. Three-way
   C. Four-way
   D. One-way

_____ **10.** The following are all types of PCV system, EXCEPT _____.
   A. Fixed orifice
   B. Variable orifice
   C. Separator
   D. Blowby

_____ **11.** The _____ system is designed to ensure hydrocarbons are not released into the atmosphere when fuel in the fuel tank begins to vaporize and build pressure.
   A. Evaporative emission
   B. PCV
   C. EGR
   D. Secondary air injection

_____ **12.** Humans breathe in oxygen and exhale carbon dioxide; trees and plants take in carbon dioxide and give back oxygen.
   A. True
   B. False

_____ **13.** The EPA requires that when a vehicle's emissions deviate from the federal test procedure limit by 3.5 times, the malfunction indicator lamp must turn on and one or more specific diagnostic trouble codes be set in memory.
   A. True
   B. False

_____ **14.** Particulate matter is graded in a size range from 10 nm to 100 μm in diameter; particulates of less than 10 μm are dangerous to humans.
   A. True
   B. False

_____ **15.** Gasoline, diesel, LPG, and natural gas are all hydrocarbon compounds.
   A. True
   B. False

_____ **16.** Exposure to carbon monoxide levels of 400 ppm may be fatal in as little as 30 minutes.
   A. True
   B. False

_____ **17.** Sulfur reduces catalyst efficiency in modern vehicles, and vehicles operating with higher sulfur gasoline have higher emissions than vehicles operating on lower sulfur gasoline.
   A. True
   B. False

_____ **18.** Keeping the air–fuel ratio close to the stoichiometric point produces minimal hydrocarbon and carbon monoxide emissions.
   A. True
   B. False

_____ **19.** The EGR system is a postcombustion emission control system.
   A. True
   B. False

_____ **20.** Any vehicle produced after 1996 monitors catalyst efficiency by using two oxygen sensors, one in front of the catalytic converter and the other in the exhaust stream.
   A. True
   B. False

_____ **21.** The PCV system regulates the flow of blowby gases between the crankcase and the intake manifold.
   A. True
   B. False

_____ **22.** A PCV valve connects the exhaust port, or manifold, and the intake manifold.
   A. True
   B. False

_____ **23.** In the combustion chamber, lower combustion temperatures tend to create more power.
   A. True
   B. False

_____ **24.** The evaporative emission (EVAP) system stores the fuel vapors in the charcoal canister until the engine control module determines that they can be burned in the engine without affecting the drivability of the vehicle.
   A. True
   B. False

_____ **25.** The air diverter valve is used to switch air from upstream to downstream once the engine reaches the specified temperature.
   A. True
   B. False

_____ **26.** _____ systems are designed to limit the pollution caused by the storing and burning of various fuels.
   A. Exhaust
   B. Emission control
   C. Braking
   D. Catalyst

_____ **27.** The _____ completes the combustion process by adding another oxygen molecule to the carbon monoxide to convert it to carbon dioxide.
   A. Breather tube
   B. Feedback system
   C. Vapor line
   D. Catalytic converter

_____ **28.** Victims of _____ poisoning can sometimes look healthy and pink-cheeked because concentrations of carbon monoxide in the bloodstream give the blood a brighter red color than normal.
   A. Sulfur dioxide
   B. Carbon dioxide
   C. Carbon monoxide
   D. Hydrogen

_____ **29.** _____ are claimed to be major contributors to photochemical smog, along with hydrocarbons and sunlight.
   A. Argon emissions
   B. Oxides of sulfur
   C. Carbon monoxide emissions
   D. Oxides of nitrogen

_____ **30.** In spark-ignition engines, _____ are caused by incomplete combustion of rich air–fuel mixtures.
   A. Atoms
   B. Molecules
   C. Particulates
   D. Emissions

_____ **31.** In a combustion chamber where surface temperatures are low, the combustion flame can go out or be _____.
   A. Quenched
   B. Purged
   C. Scavenged
   D. Vacuumed

_____ **32.** Reducing the valve overlap reduces the _____, which is when the exhaust gases moving out of the combustion chamber create a low-pressure area in the exhaust manifold, helping to pull air and fuel into the cylinder.
   A. Purging effect
   B. Scavenging effect
   C. Quenching effect
   D. Activating effect

_____ **33.** The _____, also referred to as lambda, is the ratio of air to fuel at which all of the oxygen in the air and all of the fuel are completely burned.
   A. Octane ratio
   B. Stoichiometric ratio
   C. Median ratio
   D. Vacuum ratio

_____ **34.** The _____ emitted from the internal combustion can be reduced by taking actions both prior to combustion and after combustion.
   A. Noise
   B. Feedback
   C. Pressure
   D. Pollution

_____ **35.** In older vehicles, crankcase vapors were vented directly to the atmosphere through a(n) _____ or road-draft tube.
   A. Blowby tube
   B. Breather tube
   C. Intake tube
   D. Emission tube

_____ **36.** The _____ system was designed by automotive engineers in the 1970s to control the emission of oxides of nitrogen.
   A. Postcombustion
   B. PCV
   C. EGR
   D. Heated air intake

_____ **37.** OBD II made monitoring of the EGR system mandatory; each manufacturer became responsible for ensuring its system runs a self-test called an EGR _____ test to verify that it is operating correctly.

   A. System monitor
   B. Breather tube
   C. Blowby gas
   D. Flame front

_____ **38.** Newer EVAP systems monitor purge operation by reading a _____ pressure sensor.

   A. Catalytic converters
   B. PCV valve
   C. Crankcase emission
   D. Fuel tank

_____ **39.** The _____ intake system collects hot air from around the exhaust manifold and mixes it with outside air entering the air cleaner assembly.

   A. Vapor
   B. Cold air
   C. Heated air
   D. Exhaust

_____ **40.** The arrow in the image is pointing to what part of the PCV system?

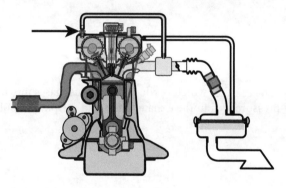

   A. PCV valve
   B. Engine breather hose
   C. Intake manifold
   D. Air cleaner

_____ **41.** The arrow in the image is pointing to what part of the PCV system?

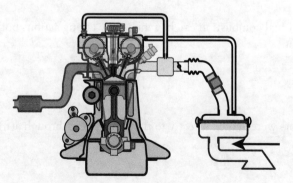

   A. Engine breather hose
   B. Air cleaner
   C. PCV valve
   D. Intake manifold

_____ **42.** The arrow in the image is pointing to what part of the EVAP system?

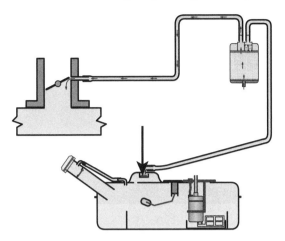

 A. Intake manifold
 B. Vacuum port
 C. Atmospheric vent
 D. Vapor separator valve

_____ **43.** The arrow in the image is pointing to what part of the EVAP system?

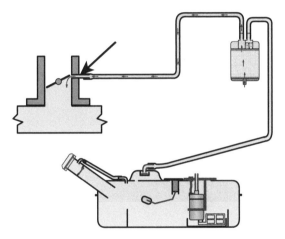

 A. Intake manifold
 B. Vacuum port
 C. Throttle
 D. Charcoal canister

_____ **44.** The arrow in the image is pointing to what part of the heated air intake system?

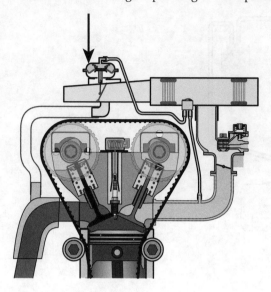

    A. Stove
    B. Stove pipe
    C. Vacuum servo
    D. Bimetallic vacuum switching valve

_____ **45.** The arrow in the image is pointing to what part of the heated air intake system?

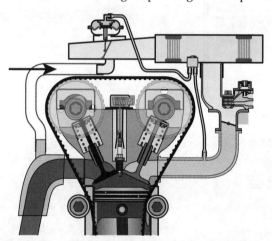

    A. Air intake flow
    B. Stove pipe
    C. Vacuum servo
    D. Bimetallic vacuum switching valve

_____ **46.** Tech A says that the purge valve is part of the EVAP system. Tech B says that OBD II systems must illuminate the malfunctioning indicator lamp when the emissions exceed 1.5 times the federal test procedure. Who is correct?
    A. Tech A
    B. Tech B
    C. Both Tech A and Tech B
    D. Neither Tech A nor Tech B

_____ **47.** Tech A says that carbon monoxide is partially burned fuel. Tech B says that oxides of nitrogen are unburned fuel. Who is correct?
    A. Tech A
    B. Tech B
    C. Both Tech A and Tech B
    D. Neither Tech A nor Tech B

_____ **48.** Tech A says that hydrocarbons are a result of complete combustion. Tech B says that a catalytic converter creates a chemical reaction, changing carbon monoxide and hydrocarbons to water and carbon dioxide. Who is correct?
  A. Tech A
  B. Tech B
  C. Both Tech A and Tech B
  D. Neither Tech A nor Tech B

_____ **49.** Tech A says that allowing hot exhaust gases into the engine through the EGR valve helps to warm up the engine during warm-up. Tech B says that rich fuel mixtures will create low amounts of carbon monoxide. Who is correct?
  A. Tech A
  B. Tech B
  C. Both Tech A and Tech B
  D. Neither Tech A nor Tech B

_____ **50.** Tech A says that burning gasoline in an engine creates water as a by-product of combustion. Tech B says that variable valve timing can be used to reduce oxides of nitrogen, eliminating the need for an EGR valve on some engines. Who is correct?
  A. Tech A
  B. Tech B
  C. Both Tech A and Tech B
  D. Neither Tech A nor Tech B

_____ **51.** Tech A says that the PCV system recirculates exhaust gases into the intake manifold. Tech B says that the PCV system recirculates blowby gases into the intake manifold. Who is correct?
  A. Tech A
  B. Tech B
  C. Both Tech A and Tech B
  D. Neither Tech A nor Tech B

_____ **52.** Tech A says that variable orifice PCV valves allow maximum PCV flow at idle. Tech B says that "lean" means there is not enough air in the air–fuel mixture. Who is correct?
  A. Tech A
  B. Tech B
  C. Both Tech A and Tech B
  D. Neither Tech A nor Tech B

_____ **53.** Tech A says that the EVAP system stores escaping hydrocarbons until the engine can burn them. Tech B says that catalytic converters were designed for regular leaded gas. Who is correct?
  A. Tech A
  B. Tech B
  C. Both Tech A and Tech B
  D. Neither Tech A nor Tech B

_____ **54.** Tech A says that the EGR valve is open fully at idle so that the engine will not die. Tech B says that oxides of nitrogen are created in large amounts when the combustion temperature is above 2500°F (1400°C). Who is correct?
  A. Tech A
  B. Tech B
  C. Both Tech A and Tech B
  D. Neither Tech A nor Tech B

_____ **55.** Two technicians are discussing PCV system failures. Tech A says that the hose to the PCV valve can crack and cause a large vacuum leak. Tech B says that the hose to the PCV valve can become plugged with sludge, restricting PCV flow. Who is correct?
  A. Tech A
  B. Tech B
  C. Both Tech A and Tech B
  D. Neither Tech A nor Tech B

CHAPTER

54 Alternative Fuel Systems

At the start of each chapter, you will find the Learning Objectives from the textbook. These are your objectives as you make your way through the exercises in this workbook and the chapter in your textbook. The following activities have been designed to help you refresh your knowledge of the material in this chapter.

Learning Objectives

After reading this chapter, you will be able to:

- LO 54-01 Identify the types of alternative fuels.
- LO 54-02 Describe the problems alternative fuels address.
- LO 54-03 Describe vehicle emissions and emission standards.
- LO 54-04 Describe battery electric vehicles (BEVs).
- LO 54-05 Describe the types of hybrid and electric vehicles.
- LO 54-06 Describe hybrid drive configurations, enhancements, and operation.
- LO 54-07 Describe hybrid and electric vehicle service precautions, personal protection equipment, and tools.
- LO 54-08 Disable the high-voltage system, service 12-volt battery.

Matching

Match the following terms with the correct description or example.

A. Alternative fuel
B. BEV
C. Continuously variable transmission
D. Displacement-on-demand
E. Electric machine
F. Hybrid electric vehicle (HEV)
G. Idle stop
H. Inverter

I. Parallel hybrid
J. Plug-in hybrid electric vehicle (PHEV)
K. Power splitter
L. Power-on-demand
M. Regenerative braking
N. Retarding effect
O. Series hybrid
P. Series-parallel hybrid

_____ 1. A hybrid electric vehicle that uses the internal combustion engine (ICE) and/or the battery pack for propulsion.

_____ 2. A vehicle powered by battery only.

_____ 3. A feature that shuts down the engine when not needed to save fuel.

_____ 4. A transmission without individual gears or gear ratios.

_____ 5. Another name for a traction motor with regenerative capability.

_____ 6. A feature that turns off the ICE when the vehicle is at a standstill.

_____ 7. A type of braking where the kinetic energy is captured rather than being lost as heat.

_____ 8. A device that converts direct current (DC) to alternating current (AC).

_____ 9. A hybrid vehicle driven simultaneously by an ICE and an electric machine.

_____ 10. A device that receives power from an ICE and electric machine to power a hybrid electric vehicle.

_____ 11. A feature that allows drive motors to act as generators to recharge the traction batteries during deceleration or when braking.

_____ 12. A hybrid electric vehicle in which only one power source, the battery, is used to propel the vehicle for a certain distance, limited by the storage capacity of the battery and the efficiency of the motor.

_____ **13.** A non–petroleum-based motor fuel.

_____ **14.** A vehicle that uses two power sources for propulsion, one of which is electricity.

_____ **15.** A vehicle where a gasoline engine powers a generator to charge the batteries and where an electric motor drives the wheels.

_____ **16.** A feature that allows cylinders to be taken off-line when not needed, such as at vehicle cruise.

Multiple Choice

Read each item carefully, and then select the best response.

_____ **1.** What type of electric motor does not use brushes or commutators?
   A. Permanent magnet
   B. Multiphase
   C. Brushless
   D. Synchronous

_____ **2.** In a _____ electric motor, the stator contains more than one winding, usually three.
   A. Synchronous
   B. Brushless
   C. Multiphase
   D. Permanent magnet

_____ **3.** What type of vehicle uses a combination of electric power and an ICE?
   A. Fuel cell vehicle
   B. Hybrid electric vehicle
   C. BEV
   D. Flexible-fuel vehicle

_____ **4.** In a _____ hybrid drive configuration, the gasoline engine is used only to drive a generator and charge a battery, which produces power to drive an electric motor.
   A. Series
   B. Parallel
   C. Series-parallel
   D. Regenerative

_____ **5.** In a hybrid vehicle, the _____ feature temporarily shuts off the ICE when it is not needed, such as when idling or coasting.
   A. Regenerative braking
   B. Displacement-on-demand
   C. Idle stop
   D. Power-on-demand

_____ **6.** The _____ feature turns off the ICE when the vehicle is at a standstill.
   A. Idle stop
   B. Regenerative braking
   C. Power-on-demand
   D. Displacement-on-demand

_____ **7.** On some vehicles, the _____ is used to change DC from the battery into AC to power the electric motors.
   A. Converter
   B. Inverter
   C. Rectifier
   D. Coupling

_____ **8.** What type of vehicle uses only one power source, the battery, to propel the vehicle for a certain distance before the batteries need charging and the ICE is needed?
   A. Plug-in hybrid electric vehicle
   B. Fuel cell vehicle
   C. Series-parallel hybrid vehicle
   D. Flexible-fuel vehicle

_____ **9.** The high-voltage wiring used in hybrid vehicles is usually _____ in color.
   A. Red
   B. Black
   C. Orange
   D. Green

_____ **10.** In most hybrid vehicles, the _____ auxiliary battery is used to power up the vehicle's computers and run virtually all of the accessories.
   A. 120-volt
   B. 24-volt
   C. 12-volt
   D. 9-volt

_____ **11.** The AC systems used on hybrid and electric vehicles operate at dangerously high voltages, some as high as 500–650 volts.
   A. True
   B. False

_____ **12.** Electric vehicles for road use today use DC motors.
   A. True
   B. False

_____ **13.** U.S. light-duty hybrids are almost exclusively gasoline-electric.
   A. True
   B. False

_____ **14.** A parallel hybrid allows both the engine and the electric motor to drive the vehicle together or either one by itself.
   A. True
   B. False

_____ **15.** Displacement-on-demand deactivates cylinders when operating under cruise or coasting conditions to save fuel.
   A. True
   B. False

_____ **16.** The power-on-demand feature deactivates cylinders when operating under cruise or coasting conditions to save fuel.
   A. True
   B. False

_____ **17.** The motor control unit controls the power divider/splitter so that the drive remains at its most efficient.
   A. True
   B. False

_____ **18.** Plug-in hybrid electric vehicles have a relatively small auxiliary ICE to drive a generator for charging the batteries.
   A. True
   B. False

_____ **19.** Fuel cell vehicles are considered hybrid vehicles because they have two power sources.
   A. True
   B. False

_____ **20.** The Chevy Volt is an example of a fuel cell vehicle.
   A. True
   B. False

_____ **21.** A(n) _____ fuel is essentially anything other than a petroleum-based motor fuel (gasoline or diesel fuel) that is used to propel a motorized vehicle.
   A. Variable
   B. Regenerative
   C. Alternative
   D. Auxiliary

_____ **22.** The electric motor in some powertrains is also used as a(n) _____ to provide partial recharging of batteries when the vehicle is decelerating or braking.
   A. Reflector
   B. Generator
   C. Invertor
   D. Power splitter

_____ **23.** When coupled with sustainable or renewable energy sources, the _____ vehicle can truly serve as an emissions-free vehicle.
   A. Flexible-fuel
   B. Battery electric
   C. Hybrid electric
   D. Liquefied natural gas

_____ **24.** Solar panels, wind generators, and geothermal sources are all examples of _____ energy sources.
   A. Sustainable
   B. Regenerative
   C. High-voltage
   D. Hybrid

_____ **25.** _____ braking occurs when the drive motor(s) act as generators to recharge the traction batteries during deceleration or when braking.
   A. Power
   B. Idle
   C. Displacement
   D. Regenerative

_____ **26.** The continuously _____ transmission does not "shift" from one gear to another in the usual sense; rather, it continuously changes the amounts of speed and torque from the engine and the motor to the wheels.
   A. Idling
   B. Variable
   C. Regenerative
   D. Deactivating

_____ **27.** The _____ effect (vehicle slowing) caused by regenerative braking provides the same deceleration as normal engine braking would.
   A. Displacement
   B. Idle
   C. Regenerative
   D. Retarding

_____ **28.** The main high-voltage battery pack must be _____ before service is performed on or around the high-voltage components.
   A. Connected
   B. Disconnected
   C. Powered on
   D. Removed from the area

_____ **29.** Tech A says that alternative fuels may be liquid or gas (or electric) and contain a variety of latent heat energy for use in the ICE. Tech B says that an alternative fuel vehicle is a vehicle powered by something other than petroleum. Who is correct?
   A. Tech A
   B. Tech B
   C. Both Tech A and Tech B
   D. Neither Tech A nor Tech B

_____ **30.** Tech A says that concept BEVs were introduced more than a century ago. Tech B says that concept BEVs have only been around for approximately the past 10 years. Who is correct?
   A. Tech A
   B. Tech B
   C. Both Tech A and Tech B
   D. Neither Tech A nor Tech B

_____ **31.** Tech A says that high-voltage wires in hybrid and electric vehicles are usually a special color. Tech B says that special high-voltage shoes should be worn when working on high-voltage vehicles. Who is correct?

    A. Tech A

    B. Tech B

    C. Both Tech A and Tech B

    D. Neither Tech A nor Tech B

_____ **32.** Tech A says that in a hybrid vehicle the engine usually drives a generator to provide the electricity, which is stored in batteries that then drive an electric motor. Tech B says that the engine may also drive the wheels directly as well. Who is correct?

    A. Tech A

    B. Tech B

    C. Both Tech A and Tech B

    D. Neither Tech A nor Tech B

_____ **33.** Tech A says that in hybrid vehicles the main high-voltage battery pack must be disconnected before service is performed on or around the high-voltage components. Tech B says that this is not necessary because the system has built-in protection devices. Who is correct?

    A. Tech A

    B. Tech B

    C. Both Tech A and Tech B

    D. Neither Tech A nor Tech B

_____ **34.** Tech A says that on most hybrid vehicles, a 12-volt auxiliary battery is used to power up the vehicle's computers and run most of the accessories. Tech B says that the 12-volt auxiliary battery is typically used to crank the ICE power unit when starting. Who is correct?

    A. Tech A

    B. Tech B

    C. Both Tech A and Tech B

    D. Neither Tech A nor Tech B

_____ **35.** Tech A says that the 12-volt battery in a hybrid vehicle is generally smaller than in a standard vehicle. Tech B says that the battery is more susceptible to parasitic current draw, as they do not have as much current capacity. Who is correct?

    A. Tech A

    B. Tech B

    C. Both Tech A and Tech B

    D. Neither Tech A nor Tech B

_____ **36.** Tech A says that that hybrids have a feature that allows the traction motor's batteries to be charged when the vehicle is braking. Tech B says this is known as regenerative braking. Who is correct?

    A. Tech A

    B. Tech B

    C. Both Tech A and Tech B

    D. Neither Tech A nor Tech B

_____ **37.** Tech A says that when servicing a hybrid electric vehicle, it is OK to leave the power switch on as long as the ICE and electric motor are not operating. Tech B says when performing service on a hybrid electric vehicle, the key fob should be more than 15' away from the vehicle. Who is correct?

    A. Tech A

    B. Tech B

    C. Both Tech A and Tech B

    D. Neither Tech A nor Tech B